Lipids for Functional Foods and Nutraceuticals

Also in the Oily Press Lipid Library:

Volume 15. Lipid Analysis (third edition)
Written by William W. Christie

Volume 14. Confectionery Fats Handbook
Written by Ralph E. Timms

Volume 13. Lipids for Functional Foods and Nutraceuticals
Edited by Frank D. Gunstone

Volume 12. Lipid Glossary 2
Written by Frank D. Gunstone and Bengt G. Herslöf

Volume 11. Lipids in Nutrition and Health: A Reappraisal
Written by Michael I. Gurr

Volume 10. Lipid Oxidation
Written by Edwin N. Frankel

Volume 9. *Trans* Fatty Acids in Human Nutrition
Edited by Jean Louis Sébédio and William W. Christie

Volume 8. Advances in Lipid Methodology – Four
Edited by William W. Christie

Volume 7. Advances in Lipid Methodology – Three
Edited by William W. Christie

Volume 6. Waxes: Chemistry, Molecular Biology And Functions
Edited by Richard J. Hamilton (out of print)

Volume 5. Lipids: Molecular Organization, Physical Functions and Technical Applications
Written by Kåre Larsson

Volume 4. Advances in Lipid Methodology – Two
Edited by William W. Christie

Volume 3. A Lipid Glossary (first edition)
Written by Frank D. Gunstone and Bengt G. Herslöf (out of print)

Volume 2. Advances in Lipid Methodology – One
Edited by William W. Christie

Volume 1. Gas Chromatography and Lipids: A Practical Guide
Written by William W. Christie

Lipids for Functional Foods and Nutraceuticals

Edited by

FRANK D. GUNSTONE
Professor Emeritus
University of St Andrews and
Honorary Research Professor
Scottish Crop Research Institute
Dundee, Scotland

THE OILY PRESS
An imprint of PJ Barnes & Associates
Bridgwater, England

Published in association with
Lipid Technology

Published by **The Oily Press**
an imprint of PJ Barnes & Associates
PO Box 200, Bridgwater TA7 0YZ, England
Tel: +44-1823-698973, Fax: +44-1823-698971
E-mail: editor@pjbarnes.co.uk
Web site: http://www.pjbarnes.co.uk

ISBN 0-9531949-3-0

This book is **Volume 13** in **The Oily Press Lipid Library**

British Library Cataloguing-in-Publication Data
A catalogue record for this book is available from the British Library

Front cover: the graphic shows the Japanese government (FOSHU) approval symbol for use on 'food specified for health use' (see Chapter 1 for more details).

Typeset in 10½/12pt Times New Roman by
Ann Buchan (Typesetters), Shepperton, Middlesex, UK

Preface

Functional foods and nutraceuticals have become "hot topics" with an increasing number of books, college courses, and meetings devoted to this area of interest. In the field of nutrition there is a new emphasis on shifting from avoiding certain "harmful" foods to the more positive view of eating those foods and ingredients – including lipids – that promote good health. This may be seen as a contribution to holistic medicine. Healthy foods are to be preferred to pills and potions.

The idea that diet can contribute to good health is attractive on many counts. It is attractive to health-conscious individuals and especially to the growing number of those living beyond their appointed "three score years and ten" who want, not merely to stay alive, but to enjoy health and strength in their advancing years. It is attractive to governments concerned about increasing health costs and it is attractive to food producers anxious to develop new products with added value.

Foods can be regarded as functional if they can be satisfactorily demonstrated to contain bioactive molecules that reduce the risk of disease or act positively to promote good health. They are foods that are consumed normally, preferably in a form and at levels that are not likely to deviate greatly from the norm. Many producers of functional foods will wish to make a health claim and to do this they will have to be aware of recent research which supports their claims. They will find much to help them in the current volume.

Lipids are essential components in the diet and nutritionists should be more active in arguing the positive aspects of these molecules. They are required in their own right and also for the important minor components such as carotenoids, tocols, and sterols which accompany the oils and fats. The first chapter in the book is an overview of the development of functional foods and this is followed by chapters devoted to materials which are either lipids or are compounds that are fat-soluble and co-occur with oils and fats. These include: carotenoids, tocopherols, phytosterols, omega-3 (n-3) oils, diacylglycerols, structured lipids, sesame oil, rice bran oil, and oils or preparations containing palmitoleic acid, γ-linolenic acid, stearidonic acid, or conjugated linoleic acid. These topics will be of interest to those who produce lipids and those who seek to incorporate them into appropriate foods, and this book is directed primarily to these two groups.

The authors come from Japan, China, North America, Australia, Malaysia, and several European countries. I am grateful to them for accepting my invitation to contribute and for agreeing to be bullied so that contributions of quality were prepared according to the timetable worked out by the editor and the publisher. We, the writers, are grateful to the publisher, Peter J. Barnes, and copy editor, Beverley White, for their help and understanding and for converting our edited manuscripts to a pleasing volume of text. Finally we offer it to our readers in the hope that they will find it useful.

Frank Gunstone
September 2002

Contents

List of Contributors

Marek Adamczak, Assistant Professor, University of Warmia and Mazury in Olsztyn, Institute of Food Biotechnology, J. Heweliusz St. 1, 10-718 Olsztyn, Poland

Ortwin Bode, Director of Research, Clover Corporation Ltd, PO Box 192, Sydney, NSW 2229, Australia

Uwe Bornscheuer, Professor and Head of the Department of Technical Chemistry and Biotechnology, Institute for Chemistry and Biochemistry, University of Greifswald, Soldmannstrasse 16, D-17487 Greifswald, Germany

Yaakob B. Che Man, Professor and Deputy Dean, Faculty of Food Science and Biotechnology, Universiti Putra Malaysia, 43400 UPM Serdang, Selangor, Malaysia

Hamish Drummond, Chairman, Clover Corporation Ltd, PO Box 192, Sydney, NSW 2229, Australia

Claire Fernie, Research Chemist, Scottish Crop Research Institute, Invergowrie, Dundee DD2 5DA, UK

Frank D. Gunstone, Honorary Research Professor, Scottish Crop Research Institute, Invergowrie, Dundee DD2 5DA, UK

Clifford Hall III, Department of Cereal and Food Sciences, North Dakota State University, Fargo, ND 58105, USA

Anu Hopia, Vice President, R&D, Raisio Benecol Ltd, PO Box 101 (Raisionkaari 60), FIN-21201 Raisio, Finland

Heikki Kallio, Professor (Food Chemistry), Department of Biochemistry and Food Chemistry, University of Turku, FIN-20014 Turku, Finland

Duo Li, Professor of Food Science & Nutrition, Department of Food Science, Hangzhou University of Commerce, Hangzhou, China

Noboru Matsuo, Senior Scientist, Biological Science Laboratories, Kao Corporation, 2606 Akabane, Ichikai-machi, Haga-Gun, Tochigi 321-3497, Japan

Pia Salo, Medical Adviser, Raisio Benecol Ltd, PO Box 101 (Raisionkaari 60), FIN-21201 Raisio, Finland

Andrew J. Sinclair, Professor of Food Science, Department of Food Science, RMIT University, GPO Box 2476V, Melbourne, VIC 3001, Australia

Mohamed M. Soumanou, Assistant Professor, Polytechnic College University, Research Laboratory for Applied Chemistry and Biology (LARECBA), Abomey-Calavi University, 01 BP 2009, Cotonou, Benin

William L. Stone, Director Pediatric Research, Department of Pediatrics, James H. Quillen College of Medicine, East Tennessee State University, Johnson City, TN 37614-0578, USA

Chin-Ping Tan, Lecturer, Department of Food Technology, Faculty of Food Science and Biotechnology, Universiti Putra Malaysia, 43400 UPM Serdang, Selangor, Malaysia

Hiroyuki Watanabe, Senior Scientist, Biological Science Laboratories, Kao Corporation, 2606 Akabane, Ichikai-machi, Haga-Gun, Tochigi 321-3497, Japan

Ingmar Wester, Vice President, Principal Scientist, Sterol Technology and Regulatory, Raisio Benecol Ltd, PO Box 101 (Raisionkaari 60), FIN-21201 Raisio, Finland

Baoru Yang, Senior Research Chemist (Food Chemistry), Department of Biochemistry and Food Chemistry, University of Turku, FIN-20014 Turku, Finland

John Young, Information Director, Leatherhead Food RA, Randalls Road, Leatherhead, Surrey KT22 7RY, UK

Glossary

AA	arachidonic acid (20:4*n*-6)
ACE	vitamins A, C and E
ACO	acyl-coenzyme A oxidase
ACP	acyl carrier protein
ACS	acyl-coenzyme A synthase
ADHD	attention-deficit/hyperactivity disorder
AHA	American Heart Association
ALA	α-linolenic acid (18:3*n*-3)
ANOVA	analysis of variance
ANZFA	Australia New Zealand Food Authority (now replaced by Food Science Australia New Zealand, FSANZ)
ASA	acetylsalicylic acid (aspirin)
ATP	adenosine triphosphate
a_w	water activity
BMI	body mass index (kg/m^2)
bpm	(heart) beats per minute
CALA	conjugated linolenic acid (18:3)
cAMP	cyclic adenosine monophosphate
cDNA	copy deoxyribonucleic acid
CHD	coronary heart disease
CLA	conjugated linoleic acid (18:2)
CoA	coenzyme A
COMA	Committee on Medical Aspects of Food and Nutrition Policy (UK) (now replaced by the Scientific Advisory Committee on Nutrition, SACN)
COX	cyclooxygenase
CT	computed tomography
CYP	cytochrome P-450 enzyme

DAG	diacylglycerol(s)
DGLA	dihomo-γ-linolenic acid (20:3*n*-6)
DHA	docosahexaenoic acid (22:6*n*-3)
DNA	deoxyribonucleic acid
DPA	docosapentaenoic acid (22:5*n*-3)
EFAD	essential fatty acid deficiency
EPA	eicosapentaenoic acid (20:5*n*-3)
EU	European Union
FDA	Food and Drug Administration (USA)
FFA	free fatty acids
FOSHU	food(s) for specified health use (Japan)
GC	gas chromatography
GLA	γ-linolenic acid (18:3*n*-6)
GRAS	generally recognized as safe (USA)
GTPase	guanosine triphosphatase
HDL	high-density lipoprotein(s)
HEPE	hydroxyeicosapentaenoic acid
HIV	human immunodeficiency virus
4-HNE	4-hydroxynonenol
HPEPE	hydroperoxyeicosapentaenoic acid
HPLC	high-performance liquid chromatography
HTST	high-temperature, short-time (processing)
HU	Hounsfield units (a measure of density)
IU	international units
IUB	International Union of Biochemistry (now the International Union of Biochemistry and Molecular Biology, IUBMB)
IUPAC	International Union of Pure and Applied Chemistry
LA	linoleic acid (18:2*n*-6)
LDL	low-density lipoprotein(s)
LPL	lipoprotein lipase
LT	leukotriene(s)
LX	lipoxin(s)

MAG	monoacylglycerol(s)
MDA	malondialdehyde
MLM	triacylglycerol with a medium-chain fatty acid at the *sn*-1 and *sn*-3 positions, and a long-chain fatty acid at the *sn*-2 position
mRNA	messenger ribonucleic acid
MS	mass spectrometry
MTBE	methyl *tert*-butyl ether
MUFA	monounsaturated fatty acid(s)
NADH	nicotinamide adenine dinucleotide, reduced form
NADPH	nicotinamide adenine dinucleotide phosphate, reduced form
NCEP	National Cholesterol Education Program (an American Heart Association initiative)
NLEA	Nutrition Labeling and Education Act (USA)
NMR	nuclear magnetic resonance (spectroscopy)
OPO	1,3-dioleoyl-2-palmitoyl-*sn*-glycerol
PAI-1	plasminogen activator inhibitor 1
PG	prostaglandin(s)
PGE_1	prostaglandin E_1
PPAR-α	peroxisome proliferator-activated receptor α
PPL	porcine pancreatic lipase
PPO	polyphenol oxidase
PS	phytosterol(s)
PUFA	polyunsaturated fatty acid(s)
RE	retinol equivalent(s)
RLP	remnant-like lipoprotein particle(s)
rpm	revolutions per minute
RT-PCR	reverse transcriptase polymerase chain reaction
SC-CO_2	supercritical carbon dioxide
SD	standard deviation
SEM	standard error of the mean
SFE	supercritical fluid extraction

SREBP-1	sterol regulatory element binding protein 1
sTAG	structured triacylglycerol(s)
TAG	triacylglycerol(s)
TBARs	thiobarbituric acid-reactive substances
TNF-α	tumour necrosis factor α
α-TTP	α-tocopherol transfer protein
TX	thromboxane
UCP-2	uncoupling protein 2
VLDL	very-low-density lipoprotein(s)
WHO	World Health Organization

Note on fatty acid nomenclature:

Throughout this book, an abbreviated nomenclature is used to designate the structures of fatty acids. For example, γ-linolenic acid (GLA), the commonly used term for all-*cis*-6,9,12-octadecatrienoic acid, can also be designated as 18:3*n*-6. This indicates a fatty acid with 18 carbon atoms and 3 double bonds, with the first double bond being found on the 6th carbon from the methyl group. It is understood, in the absence of other indicators, that all the double bonds of all fatty acids have the *cis* configuration and are methylene-interrupted; i.e. the unsaturated centres are separated from each other by one CH_2 group:

$$CH_3(CH_2)_4\overset{cis}{CH=CH}CH_2\overset{cis}{CH=CH}CH_2\overset{cis}{CH=CH}(CH_2)_4COOH$$

CHAPTER 1

Introduction

JOHN YOUNG

Leatherhead Food Research Association (LFRA), Leatherhead, UK

A. Preamble

It has been recognized for centuries that diet and health are inextricably linked, but only comparatively recently have the relationships between various dietary components and disease prevention begun to be more fully understood and appreciated.

Awareness of this issue in the West has recently been raised by considerable media attention surrounding the practice of incorporating physiologically active materials into foods and drinks in Japan, leading to the development of a new category of foodstuffs commonly referred to as either functional foods or nutraceuticals.

The Japanese vision surrounding functional foods is that it should be possible to manufacture and market a variety of everyday foodstuffs containing physiologically active ingredients that will deliver health benefits to the population at large, thereby helping to control government health-care expenditure.

It follows that foods can have an expanded role in human health that goes well beyond providing an adequate source of nutrition. This concept has served to blur the distinction between pharmaceuticals and foods, bringing new companies into the "food" area and making food companies more conscious of the health-promoting properties of foodstuffs.

It is worth emphasizing that functional foods/nutraceuticals are part of an increasingly complex jigsaw that recognizes that many foodstuffs, both processed and "natural", have key roles to play in enhancing and maintaining health. As a consequence, recognizing the importance of foods that enhance well-being, of which functional foods are a subset, should form an integral component of any future nutrition strategy.

B. The concept of functional food

The word "functional" implies that functional foods are designed for, or suited to, a particular operation or use. Four interrelated factors have given momentum to the concept of functional foods:

- A growing body of scientific evidence that supports associations between specific food components and specific physiological responses associated with health status and disease control.
- Increasing consumer education and consciousness of the role of diet in promoting good health.
- Advances in food technologies, particularly those relating to biotechnology, that potentially provide the food industry with increasingly sophisticated methods to control and manipulate the physical and chemical composition of foodstuffs.
- A growing appreciation of the value-added opportunities that functional foods offer to food manufacturers, pharmaceutical companies, and food ingredient suppliers.

However, functional foods are not the exclusive preserve of modern food technology. The concept implies the recognition and promotion of the positive health properties of a food product, traditional or novel, and thus could include a variety of conventional foods (e.g. fermented milk drinks and yoghurts) as well as compositionally enhanced foods.

C. Definition of a functional food

A widely accepted definition of a functional food can be found in the consensus document produced by the Concerted Action on Functional Food Science in Europe, which was coordinated by the International Life Sciences Institute (Diplock *et al.*, 1999). This states that foods can be regarded as "functional" if they can be satisfactorily demonstrated to have a beneficial effect on one or more target functions in the body, beyond adequate nutritional effects, that leads to an improved state of health and well-being and/or to a reduction in disease risk. Functional foods must remain foods and they must achieve their effects in amounts that would be consumed normally in the diet. They are not pills or supplements.

Japan is the only country to date that has a legal definition of functional foods. The Japanese Ministry of Health's Functional Food Study Committee defines functional foods as "foods for specified health use", commonly referred to as FOSHU. The FOSHU concept is based on knowledge concerning the relationships between particular foods or food components and certain expected health benefits. FOSHU have been licensed to bear a label claiming that a person who consumes them may expect to obtain the specified health benefit.

Thus it is the potential to associate a particular health claim with a food that makes it "functional".

D. The history of functional foods

Throughout Europe and the rest of the developed world there has been a significant increase in interest in health and healthy eating. The food industry has responded to the call for healthy eating in a number of ways. One major response was to remove or reduce the content of ingredients perceived as unhealthy, such as fat, sugar and salt. This development has become known as the "low and light" movement. A second response was fuelled when consumer demand for additive-free foods drove manufacturers in certain countries, notably the UK, to remove (in some cases unadvisedly) certain ingredients perceived as "unnatural" (e.g. preservatives) – which itself led to an increasing demand for more "natural" alternatives. The third, current, response is the recognition and promotion of foods (functional foods) which have the potential to improve general health and to reduce the risk of developing life-threatening diseases, of which heart disease is arguably the most significant/important.

The concept of functional foods is attributed to the Japanese, having been spurred on by the Japanese government's 1984 move to consider the limitations and cost-effectiveness of curative medicine's contribution to a healthier society. In parallel, in Europe there has been a smaller, but nevertheless highly significant, number of food product launches carrying "new style" health claims (e.g. cholesterol reduction, immune system enhancement), though without Japan's legislative framework and with less well-informed consumers. While there has also been notable functional food activity in North America, it has not been on the same scale as in Europe and Japan due to a greater focus on using supplements as the delivery vehicle.

The Japanese Ministry of Education, Science and Culture began investigating the idea that eating and drinking certain foods could significantly improve the working of human body functions. In cooperation with the Ministry of Agriculture, Forestry and Fisheries and the Ministry of Health and Welfare, it was assigned to develop the concept of functional foods. A December 1989 report from the Ministry of Health and Welfare, showing that the average fibre content of the Japanese diet had declined over the previous 35 years from 22.7 g/day in 1951 to 17.37 g/day in 1985, caused an upsurge in demand for functional foods in Japan. A number of dietary fibre drinks were already on the market when the report was published. The report brought credence to the health claims being made by manufacturers, and led to strong growth of the fibre drinks category. The Ministry of Health and Welfare also expressed concerns about calcium deficiency, resulting in much publicity and the launch of a plethora of calcium-fortified foods.

In 1989, the Ministry of Health and Welfare released a mid-term progress report that offered a general interpretation of the term functional food, which drew a significant demarcation line between this category and health foods. In September 1991, the Japanese government established its system of licensing FOSHU. The Japanese authorities currently recognize 10 categories of health-enhancing functional food components: dietary fibres; oligosaccharides; sugar alcohols; polyunsaturated fatty acids; peptides and proteins; glycosides, isoprenoids and vitamins; choline; alcohols and phenols; lactic acid bacteria; and minerals. An additional category covers miscellaneous components that cannot be classified in the other categories listed.

E. Factors influencing demand for functional foods

1. Emerging consumer trends and attitudes relating to diet and health

Of a number of factors fuelling the growing interest in functional foods, it is consumers' growing appreciation and understanding of the positive role of diet in health maintenance and well-being that is arguably the key driving force. Other factors that warrant mention are the expectation of a longer, more active life, and in the UK a growing interest in self-medication due to increasing prescription costs and a lively interest in complementary medicine. While currently of only minor influence, interest in the use of self-diagnostics in illness prevention may become increasingly important in the future.

A study conducted by the UK's Leatherhead Food Research Association (LFRA) towards the end of 1998 investigated the levels of interest in and attitudes relating to healthy/functional foods in France, Germany and the UK (Hilliam and Boyle, 1998). Supporting the concept of functional foods was the finding that UK and German respondents considered diet to be the most important contributor to their health, followed by exercise and genetic make-up. Interestingly, the French preferred exercise as the key contributor to health; however, of diet, exercise and genetic make-up, diet was perceived as the most important overall contributor to health (Figure 1).

These findings are of considerable significance as they suggest that the consumption of functional foods is probably seen by many consumers as the most viable and practical route towards maintaining health. However, for those engaged in marketing and product development, it is also important to realize that there are differences in attitudes between countries.

Also of interest from the LFRA study was the implication that consumer interest in health claims is likely to be high when associated with conditions that generate a high degree of concern, such as heart disease, and where consumers feel that they have a fair degree of influence. Although there appears to be some overlap in the top five potential health concerns for French, German and UK consumers (Table 1), there are sufficient differences to suggest that a pan-

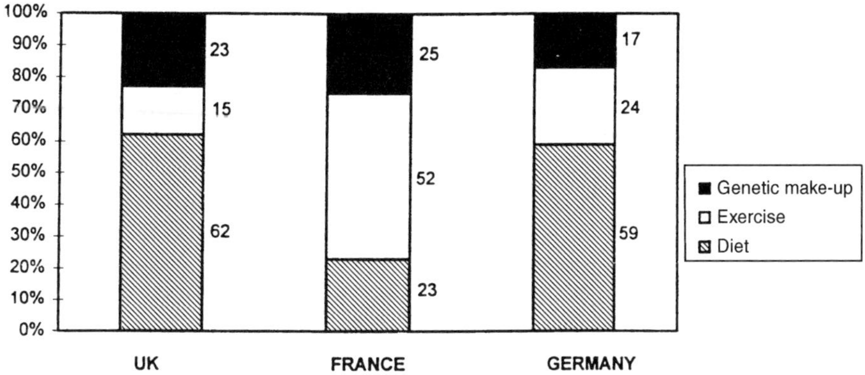

Figure 1. Factors perceived to be most important to health. Data taken from LFRA (1998).

European functional foods marketing strategy needs to take account of these national differences. For example, while UK consumers place great emphasis on claims related to heart disease (e.g. cholesterol reduction), German consumers appear more interested in claims relating to general disease resistance.

2. *Socio-economic trends*

Central to the development of the functional food concept in Japan was the government's view that there might be limits to the cost-effectiveness of curative medicines in creating a healthier society. Assuming that the same underlying philosophy applies in other developed economies (and there appears to be no reason why it should not), it becomes important to consider the impact of various social trends on demand for functional foods. Social trends of particular significance are changes in population age profiles, life expectancy, the incidence of various medical conditions, and levels of health support.

Table 1. Top five health concerns by country, 1998[a]

	UK	France	Germany
1.	Breast cancer	Lack of energy	Breast cancer
2.	Stress	Stress	Lung cancer
3.	Lack of energy	Breast cancer	Bowel/colon cancer
4.	Heart disease	Migraine	Memory decline
5.	Osteoporosis	Coughs/colds/flu	Heart disease

[a]Data taken from LFRA (1998).

For the foreseeable future there will be considerable shifts in demographics and lifestyle, of which the most significant will the large increase in the number of elderly people. For example, in the USA the percentage of adults over the age of 65 is expected to nearly double within 30 years, thereby placing an additional burden on the already strained health care system.

Western Europe is also an ageing society, reflected in the fact that over the period 1999–2020 the number of people aged 65 or over is forecast to increase by around 35%.

Clearly, an ageing population can be expected to place a greater burden on health and social services in years to come. Measures that can be taken to reduce or prevent chronic illnesses, including dietary strategies, appear attractive not only from the perspective of the individual but also to governments struggling to keep health-care costs under control.

Many of the additional burdens being placed on health services are the result of significant increases in life expectancy achieved over the past 40 years. These have come about largely through advances in medical science rather than through changes to the diet. Overall improvements in life expectancy are forecast to continue for the foreseeable future. Bearing in mind that, with few exceptions, the risk of cancer increases with age, and that the population in the West is ageing, it is possible to forecast with some certainty that the health services will be faced with more cases to treat in years to come. Epidemiological data suggest that an avoidance of overweight and an increase in the consumption of fresh fruit and vegetables is the best route to reducing the incidence of cancer. But there are those who consider a more practical approach to be the incorporation of cancer-inhibiting constituents from plants (phytochemicals) into everyday foods.

3. Policy initiatives

It is now universally accepted that the health of a population is influenced by diet. Because of the potential benefits to people's quality of life, to the economy, and to their popularity rating, governments and associated representative bodies continue to place greater emphasis on diet and health. Indicative of this is the growing number and scope of government nutritional policy recommendations, as put forward by public health agencies and government-funded initiatives. For example, in the USA alone, there were 11 such initiatives between 1988 and 1995. Of these, the most significant was the introduction of the Nutrition Labeling and Education Act (NLEA) in 1990, which allowed regulations to be promulgated authorizing the use of specific health-related claims on food products, where they can be supported by valid and substantial scientific evidence. Claims currently approved in the USA include those relating to: calcium and osteoporosis; sodium and hypertension;

dietary lipids and cancer; low-fat foods and heart disease; fibre-containing foods and heart disease; fibre-containing foods and cancer; fruits and vegetables and cancer; folate and neural tube defects; sugar alcohols and dental caries; soluble fibres from psyllium, oat bran or oat meal and reduced risk of heart disease; soy and heart disease; and plant stanol/sterol esters and heart disease.

Although these are "generic" health claims, and are not exclusive to particular functional foods, the granting of permission to use claims supported by scientific evidence is important. The food industry considers that the NLEA has some serious drawbacks. For example, the wording of the prescribed claims is long and complex, and there is no provision for owner-specific claims. However, through its introduction the USA can be considered to be ahead of Europe, and indeed many other regions of the world, in developments toward a wider recognition of health-related and disease-related claims on everyday foods and drinks.

Having watched with interest developments in the USA, many agencies around the world are currently considering the introduction of similar measures, albeit mainly through voluntary codes of practice rather than through changes in food regulation. The enormity of the task of modifying European Union (EU) food law to accommodate similar health claims has favoured codes of practice as the most pragmatic approach to date.

4. *Pressure groups and the media*

Consumer interest groups, particularly those in the UK, were extremely effective in raising doubts and concerns about functional foods in the mid-1990s, both at government level and in the media. A reflection of this was a series of negative headlines in the broadsheet and tabloid newspapers. Although not as vociferous or as frequent, similar headlines appeared following the introduction of functional foods into a number of other countries, both within and outside Europe.

However, as the media have become more informed about functional foods and their associated issues, I am of the opinion that the style and tone of reporting have become more balanced. Indeed, the women's press in particular has shown considerable interest in the topic, with a clear emphasis on the potential benefits of functional foods.

This highlights the need of the food industry to consider the views and likely stance of consumer interest groups and the media as part of its functional food strategy.

F. The market for functional foods

The functional foods market is undoubtedly developing strongly in most of the countries and regions of the world where there is a mature processed foods market. Estimates of the size of the market vary considerably due to differing

Table 2. Market for functional foods in selected countries/regions, 1999[a]

	Market value (US$billions)	
Region	**Strict definition**	**Broad definition**
Europe	1.79	4–8
USA	1.80	15
Japan	2.13	14
Australia	0.05	NA
Total	5.77	≥33

[a]Data taken from LFRA (2000). NA, Data not available.

definitions of what is a functional food. Using the strict definition that functional foods should make specific health claims of some kind on the packaging and/or in the advertising, LFRA estimated that the market in Europe, the USA, Japan and Australia had a combined value of over US$5.7 billion in 1999 (Hilliam and Young, 2000). Taking a broader definition of the market to include the wide range of "healthy" products not necessarily making claims, but often perceived as functional foods, the market value may be estimated as high as US$30 billion (Table 2).

Markets have developed very differently in different regions. In Europe, products aimed at improving gut health, particularly those containing probiotics, dominate the market. In the USA, there is much wider interest in anti-cancer products and the use of botanicals, while gut health products remain relatively undeveloped. The greater use of dietary supplements in the USA has also influenced market development, with increasing numbers of products being launched that straddle the line between dietary supplements and food products.

The Japanese market is entirely different from those of Europe and North America, reflecting the more established nature of the market, and the greater links between food and health already made by Japanese consumers. The stronger presence of pharmaceutical companies in the Japanese functional food market may also be a contributory factor. The size of the market for foods approved to carry claims (FOSHU) is clearly documented, and is currently just over US$2 billion. However, there are also many products on the market that have not yet applied for FOSHU approval or where the process is under way. The existence of a wide range of functional products across all sectors of the food and drinks market makes it difficult to quantify the total size of the functional foods market in its broad sense.

The functional foods industry is almost as difficult to define as the market. While some sectors are very fragmented, with small local brands from specialist suppliers leading playing roles, other sectors are dominated by major multinationals, such as Nestlé and Unilever, competing for market share. However, as the market develops there are clear signs that it is consolidating in favour of the major multinationals, particularly as they reposition existing brands on a health-promoting platform. Two examples are Nestlé's promotion of a number of wholegrain cereal brands (e.g. 'Cheerios' and 'Shredded Wheat') as beneficial to heart health, and Danone's promotion of its 'Bio' yoghurt brand as beneficial to gut health. At the same time Danone renamed the product 'Bio Activia', to ensure differentiation from other 'bio' product lines.

Many new functional food brands have appeared in recent years, with mixed success. Some notable successes are Danone's 'Actimel' probiotic yoghurts and milk drinks, selected products from Nestlé's 'LC1' range, and Unilever's 'pro-activ' cholesterol-lowering margarine. Notable failures include the 'Aviva' range from Novartis and 'Ribena Juice & Fibre' from SmithKline Beecham, both of which indicate that a strong brand name and a pharmaceutical heritage are not sufficient guarantees for success.

With the large amount of investment required to enter the market, various approaches are being taken. As well as companies, particularly dairy and cereal companies, developing and launching functional varieties within their existing markets, joint ventures between food and pharmaceutical companies and food and ingredients companies to enter and/or develop particular areas are also being set up. There has also been considerable acquisition activity, particularly in the US market, enabling companies to enter new areas. An interesting recent joint venture was that between the food ingredients supplier Protein Technologies International and the UK high street retailer Marks & Spencer, leading to the launch of the '&more' range of cholesterol-lowering foods and beverages.

G. Dairy products

While probiotic dairy products dominate the European functional foods market, this is not the case in the USA, where the market remains comparatively unexploited. In Europe, the market is worth around US$1.3 billion a year, and is growing in most countries, mainly in the area of probiotic drinks rather than spoonable products. With the chilled dairy products market much less developed in the USA than in Europe, it is not surprising that the US market for probiotic yoghurts and other probiotic dairy lines is relatively small. In the USA, interest in probiotics has so far largely been confined to the dietary supplements market, although the recent withdrawal of a supplement version of Nestlé's LC1 brand indicates that there is still some way to go to gain US consumer acceptance of the probiotic concept.

However, it is important to note that the lack of health claims extolling the virtues of probiotic cultures in yoghurts is the key reason why this category is comparatively underdeveloped in the USA compared with Europe. Ironically, most of the major chilled cup yoghurt products in the USA contain probiotic cultures, but this is not mentioned in most marketing campaigns. Of the US$2.1 billion worth of probiotic-containing yoghurts sold in the USA every year, it is estimated that only 6% of the products actually state that they contain probiotic cultures.

It is difficult to quantify the Japanese functional dairy sector, where live cultures (probiotics) are used in almost all fermented milk drinks and desserts, often in conjunction with numerous other ingredients, including prebiotics and vitamins and minerals. The market for 'Yakult'-style fermented milk drinks alone is worth over US$800 million/year, and the market for FOSHU products making gut health claims (likely to be predominantly dairy products) is about US$1.75 billion.

With its greater market development, the functional dairy products sector tends to be relatively concentrated compared with other functional food categories. It is dominated by the leading dairy companies and multinationals, particularly Danone and Nestlé, which have a significant presence, particularly in Europe. However, one emerging functional food category that is probably even more dominated by a relatively small number of large companies is that of functional margarines and spreads.

H. Functional margarines and spreads

While a number of margarines and spreads have been introduced that claim or infer benefits to gut health or bone health, the major focus of activity in this sector has been in phytosterol-enriched heart-benefit products. The first of these were launched in Finland in 1995, but the sector did not really take off until 1999, when a number of products were launched in the USA, Europe and Australia. While heart-benefit spreads have been introduced based on *n*-3 (also known as omega-3 or ω-3) fatty acids, these have not enjoyed anything like the commercial success of their phytosterol-enriched counterparts.

Despite a high profile and major expenditure on consumer awareness and education programmes, sales of the new cholesterol-lowering spreads remain limited in terms of the market as a whole, at less than US$300 million worldwide at the end of 2000 (LFRA, 2001a). The largest market is in Europe, and accounts for about US$200 million of the total. The US market, where the products have faced particular difficulties, is worth US$60 million. The products have also fared relatively well in Australia, where they are worth US$23 million. The UK now has the largest market in Europe, ahead of the much longer-established Finnish market and that in the Benelux countries (Belgium, the Netherlands and Luxembourg), which started developing at a similar time to that in the UK.

There are currently only three players in the sector. Unilever was first on the market in the USA, with its phytosterol margarine sold under the name of 'Take Control' in the USA and as 'Flora/Becel/Fruit d'Or/pro.activ' in Europe and

Australia, plus the newer and much less developed markets of New Zealand and Brazil. 'Benecol', launched by Raisio in Finland in 1995, but now licensed by McNeil Consumer Products (a subsidiary of Johnson & Johnson) outside Scandinavia, was launched in the USA just after Take Control, but was first onto the UK and Continental European markets. There is a third contender, Goodman Fielder's 'Meadow Lea Logicol' in Australia and New Zealand, where vegetable oil-derived plant stanol and sterol esters were formally approved as novel food ingredients in edible oil spreads by the Australia New Zealand Food Authority (ANZFA; now renamed Food Standards Australia New Zealand) in June 2001.

The market for heart-benefit oils and fats products is reviewed in more detail by region and country below.

1. UK

While the UK yellow fats market is in overall decline, interest in the sector has been maintained in recent years by two sectors: dairy spreads, and so-called health margarines or spreads – primarily reduced-fat and polyunsaturated lines. More recently, in 2000, there was some evidence that consumers were returning to butter and to dairy spreads with a buttery taste. There was also interest in the new cholesterol-lowering spreads sector, pioneered by McNeil's Benecol in 1999, and boosted further with the launch of Unilever/Van den Bergh's Flora pro.activ in the summer of 2000. In total, the cholesterol-lowering spreads market grew from under £15 million at the end of 1999 to about £60 million at the beginning of 2001, boosted by strong promotional support programmes. This is equivalent to a volume of over 7000 tonnes.

Overall, sales in the yellow fats market rose just 1% in value in 2000, though this followed a period of continuous decline, which took the market from £923 million in 1996 to £818 million in 1999. Within the overall market, butter and margarine both increased their shares in 2000, while that of spreads fell (Table 3).

The difficulties of defining functional food sectors are nowhere more apparent than in the UK spreads market, where product repositioning has created a wildly fluctuating market in recent years, as the brand leader Flora has started to use on-pack claims.

In 1997, it started to use the claim "Can help to lower cholesterol as part of a low-fat diet", which made it a functional food and instantly gave the functional spreads market a value of over £90 million a year. In early 1998, this claim was dropped, but reappeared later that year as "As part of a healthy diet can help to lower cholesterol and to maintain a healthy heart". This claim had changed again by the beginning of 2000. The "healthy heart" reference was dropped, just leaving the cholesterol-lowering claim. In June 2000, it was totally repackaged prior to the launch of Flora pro.activ cholesterol-lowering

Table 3. UK yellow fats market by value, 1999–2000[a]

	1999		2000	
Product	**Value (£millions)**	**Market share (%)**	**Value (£millions)**	**Market share (%)**
Butter	338	41	350	42
Margarine	69	9	80	10
Spreads	411	50	396	48
Total	818	100	826	100

[a]Data taken from LFRA (2001a).

spreads in the UK, emphasizing a "with sunflower goodness" message and using a new 'Unique Flora Balance' logo on the 'Original', 'Light', 'Buttery', and 'Low Salt' variants.

Until 1999, relatively few totally new functional spread lines had been launched, and those that had appeared had been unsuccessful. They were mainly *n*-3-fortified lines, led by 'Pact' manufactured by MD Foods; there was also an own-brand version from supermarket chain Tesco. Pact was originally launched in 1995 in a heart-shaped pack, with claims relating to maintaining a healthy heart. Following complaints to the Advertising Standards Authority various changes were made, including a complete repackaging in 1997, when it was marketed under a general health platform with no specific claims. Sales were always very limited. The *n*-3-enriched spreads sector never even reached a value of £5 million, and the products were eventually withdrawn.

In 1999, a new contender appeared on the UK functional spreads market, however – Benecol, a cholesterol-lowering spread containing plant stanol esters [see Chapter 7]. It had already been a success for Raisio in its home market of Finland, but was subsequently licensed to McNeil Consumer Products (a subsidiary of Johnson & Johnson), and the UK became its first international market, just before the Netherlands and the USA. It featured the claim "Helps actually lower cholesterol as part of a healthy diet". Strong promotion and a high-profile advertising campaign featuring a major TV celebrity resulted in considerable interest in the marketplace. Despite the high price of the product (£2.49 for a 250 g pack) and the limited target market (those with a high blood cholesterol problem), the brand achieved considerable success. Sales were over £13 million by the end of 1999, equivalent to nearly 2.7% of the margarines and spreads market, and 1.7% of the total yellow fats market. By the end of the first quarter of 2000, Benecol claimed a 3.8% value share in the margarines and spreads market, rising to over 5% by the beginning of 2001 – equivalent to a value of nearly £24 million. With its premium pricing, of course, its volume share is considerably lower.

The Benecol range now includes an olive oil variant, launched in autumn 2000, as well as the Original and Light (40% fat) versions in its spreads portfolio. The manufacturer has also extended the brand into other areas, including cream cheese spreads (initially launched with the fat spreads, but since reformulated and relaunched), as well as yoghurts and snack bars – opening up a unique market sector that McNeil describes as “cholesterol management”.

Meanwhile, a new product had been waiting since 1999 for clearance from the EU Commission for a European launch. This was Unilever’s Flora pro.activ spread, made with plant sterols, and already available in the USA under the Take Control brand name. It was finally launched in the UK in August 2000, with strong promotional support. Priced 20% lower than its rival Benecol, Flora pro.activ was claiming market leadership within 6 months of its launch, outselling Benecol by a ratio of more than 2:1. UK sales of Flora pro.activ are currently running at around £30 million/year.

2. *Germany*

Functional spreads did not appear on the German market until September 2000, when Unilever subsidiary Union launched the pro.activ cholesterol-lowering phytosterol half-fat margarine under the becel brand. The brand is estimated to have gained a 6% share of the margarine market.

3. *France*

An early contender in the heart-health sector in France was a new-style margarine with *n*-3 fatty acids, launched onto the market in 1997 under the existing ‘Cema Primavere’ brand. Phytosterol-enriched margarines did not appear in France until September 2000, when Unilever subsidiary Astra Fralib launched the pro.activ cholesterol-lowering phytosterol spread under its Fruit d’Or umbrella brand, at a premium price of €3.04.

4. *Benelux*

The only other European countries, apart from the UK and Finland, to have seen cholesterol-lowering spreads prior to 2000 are those of the Benelux region, where Benecol had its first Continental European launches in June 1999 just after its UK debut. The launch in the Netherlands was less successful than in the UK, taking just 1.6% of margarines and spreads sales by the end of 1999, and 1.8% during its first full year on the market – equivalent to sales of about €5 million. In Belgium, Benecol has taken a higher share, albeit of a smaller market, with about 4% by value of a total spreads and margarines market worth €124 million, giving sales in the region of €5 million.

The much-heralded arrival of Unilever’s Becel pro.activ in these markets in

2000, and the continued promotion of Benecol, have undoubtedly caused further market development. Becel pro.activ sales overtook Benecol's in mid-2001, and now take about two-thirds of the market. The total market value in 2001 was approximately €30 million.

5. *Finland*

Finland has the highest market penetration worldwide for cholesterol-reducing spreads, reflecting the fact that it is the home of Benecol. Benecol was instantly successful, gaining sales of over €50 million within a year of its launch. It sold at about seven times the price of standard margarines. Sales are currently in the region of €130 million, representing a volume of about 1000 tonnes.

6. *USA*

The US functional spreads market took major steps forward in 1999, with the appearance of phytosterol-based spreads in the US$1.7 billion margarine and spreads market. Two products, both featuring health claims about cholesterol reduction, entered the market in May 1999, and have been in close competition ever since.

Approval from the US Food and Drug Administration (FDA) for Take Control (manufactured by Lipton, a subsidiary of Unilever) came first. It contains a natural soybean extract (containing plant sterols), and carries the claim: "helps promote healthy cholesterol levels". It marked Lipton's first entry into the US functional foods market, although the company started studying the potential of soybean extracts as ingredients for improving the properties of foods in the late 1980s.

Benecol also received FDA approval for its cholesterol-lowering stanol ester ingredient, a hydrogenated sterol sourced from wood pulp. Approval was originally delayed when the FDA ruled that it could not be labelled as a "dietary supplement" and would have to undergo the same safety procedures as regular food products.

Both products were launched on the US market with premium pricing. Benecol, in particular, was relatively expensive, at US$5.99 for a 250 g pack – US$1.50 more than Take Control, and more than double the price of non-functional brands in the categories where it competes. Benecol faced problems almost immediately. Manufacturer McNeil Consumer Products cancelled its TV advertising campaign in late 1999, because it had not succeeded in sufficiently boosting sales. At the same time, the company reported that it was "changing focus and looking at new ways to reach consumers", recognizing the need to drive up the perceived value of the product further – possibly by gaining recommendations for the range from physicians.

By the middle of October 1999, after less than 6 months on the market, sales of phytosterol margarines were worth about US$22 million/year. This was less than 1.5% of the US$1.7 billion total margarines and spreads market in the USA, and less than 0.7% of the US$3.3 billion total yellow fats market. Within the phytosterol sector, Benecol took two-thirds of the market, with Take Control accounting for the remainder. Sales were very limited, however, with Benecol taking less than 1% of the market by value, and an even smaller share by volume.

By the end of August 2000, sales in the sector had nearly trebled to just over US$60 million/year, but this was still equivalent to just 3.6% of the total margarines and spreads market by value, and nearly 1.9% of the total yellow fats sector. Benecol and Take Control each took a similar market value.

Both products were originally launched with labels stating that they helped "promote healthy cholesterol levels". In September 2000, the FDA issued an interim final ruling allowing more specific health claims linking foods containing plant stanol esters or other sterols or with a reduced risk of coronary heart disease. This was the first time that the FDA had allowed a direct health claim of this type to be used prior to issuing a final ruling, reflecting the importance of coronary heart disease to public health as well as the strength of the scientific evidence for the effectiveness of plant sterols. It was hoped that this ruling would boost the sales of products containing plant sterols, but this does not appear to have been the case.

7. Japan

There is a strong emphasis on healthy products across the entire oils and fats market in Japan. There appears to be a high level of consumer awareness of the benefits of particular ingredients, notably docosahexaenoic acid (DHA), that have been used in large numbers of products since the mid-1990s. As well as having heart health benefits, DHA is also claimed to be useful in aiding memory, concentration and learning, enhancing the nervous system, and preventing cancer – but most DHA-fortified products are simply promoted as being good sources of DHA.

Over the past few years, emphasis has tended to move towards products that can lower blood cholesterol levels, with recent launches including margarines and oils with phytosterols, oils with oleic acid and, more recently, red koji. Kao's 'Econa' cooking and frying oils containing diacylglycerols and phytosterols have FOSHU approval to use blood cholesterol-lowering claims [Chapter 5]. There appears to be more emphasis on developing heart-benefit liquid oils in Japan, rather than the margarines and spreads that are being promoted strongly elsewhere around the world.

8. Australia and New Zealand

The first phytosterol spreads were launched in Australia in 1999, led by

Unilever's global launch of Flora pro.activ, which had already appeared in the USA under the Take Control brand name. Although it was priced four times higher than standard spread brands (at Aus$3.99 for 250 g), it was an instant success, taking a 7% value share of the spreads market within its first 7 weeks on the market. It was followed in October 1999 by a launch from Unilever's main rival in the yellow fats market in the region, Goodman Fielder, which introduced Logicol under the Meadow Lea umbrella brand. Launches of both brands followed in New Zealand within a few months.

The new products appeared to have a positive effect on the market in 2000, with the margarine sector growing 7% by value following several years of decline. Overall, the phytosterol margarine market more than doubled in 2000, reaching a value of about Aus$40 million, equivalent to about 14% of the margarine category, and over 9% of the yellow fat spreads sector as a whole.

The two brands currently on the market are both sub-brands of the Australian margarine category leaders: Flora and Meadow Lea. Meadow Lea claims leadership of the market as a whole, with a brand share of just under 18%. In the phytosterol margarine sector, however, Flora's pro.activ was the first on the market and is the clear leader, with over 70% of sector sales – equivalent to about 10% of the total margarine market.

In June 2001, plant sterol esters were formally approved for use in spreads by ANZFA. However, the products must contain an advisory statement that they are unsuitable for infants, children and pregnant women, and that people using cholesterol-reducing medications should seek medical advice before using the spreads.

A more recent development in Australia has been the appearance of margarines containing fish oils, led by the March 2001 launch of 'Seachange Margarine' by Caines Foods. The margarine uses fish oil as a source of DHA and eicosapentaenoic acid (EPA), two *n*-3 fatty acids believed to have roles in optimal brain function and heart health.

I. Bakery and cereal products

Activity in the functional bakery and cereals products market has been very fragmented to date, and has tended to concentrate on two areas: fortification with vitamins and minerals, and the benefits of high-fibre and oat-containing products.

The US market for breakfast cereals is particularly developed, reflecting the FDA's approval of claims relating to high fibre content and reduced blood cholesterol levels and improved heart health. This has boosted the bran and fibre cereals market, which is now worth about US$1.5 billion/year, although not all products necessarily make specific claims. Quaker, the company which gained FDA approval for the health claims in the mid-1990s, certainly does;

Quaker also puts health claims on its hot and cold cereals sold in the UK. In Europe, Nestlé has repackaged its UK wholegrain cereals, including Shreddies, Shredded Wheat, and Cheerios, with claims about heart health.

More recent initiatives have focused on the calcium content of children's cereals, emphasizing the role of calcium in building strong teeth and bones. This movement has been led by Kellogg and has been applied to its children's ranges in the USA and a number of European countries.

Market activity in functional bakery products has been even more fragmented. Biscuits are the key line in Europe, particularly in France and Spain, where there are existing markets for dietetic biscuits. Use of prebiotics and probiotics, while not common, appears to be increasing, particularly in Germany. Fortification with *n*-3 fatty acids is also evident across a number of markets, having found applications in breads, bread mixes, and pasta. Soy-based ingredients are also growing in importance due to their reported contribution to heart health and women's health.

J. Soft drinks

While sports drinks and energy drinks dominate the functional soft drinks sector in most countries, there are a growing number of other functional lines coming onto the market. Sales of these other drinks in Japan, the USA, Australia, and key European countries totalled nearly 1.87 billion litres in 2000, equivalent to a market value of approximately US$4 billion, or 26% of total functional soft drinks sales (LFRA, 2001b). This sector of the market is relatively small and undeveloped in most countries, however, and is highly fragmented, with relatively high levels of product activity not yet reflected in terms of market size.

The Japanese market is the most established by far. It consists largely of health and functional drinks in small bottles, although more recently lines are starting to appear in standard soft drinks packaging. Many of the brands on the market target gut health through the use of prebiotics and dietary fibre. Bone health is also a major focus, using various calcium sources in combination with promoters of calcium absorption such as calcium citrate malate.

Germany is the most developed functional drinks market in Europe, due to the success of drinks fortified with vitamins A, C and E, otherwise known as "ACE drinks". Sales of ACE lines, including newer products such as those fortified with vitamins B, C and E and DHA, are worth in excess of €100 million/year.

The US functional beverages market has seen considerable activity, with the introduction of many nutrient-fortified and mineral-enhanced lines, plus products with herbal and botanical ingredients. The total market is worth about US$750 million/year, with the relatively mature calcium-fortified juices market accounting for about US$450 million of the total.

K. Functional food ingredients

A number of key ingredients or classes of ingredients underpin current functional food development; these are summarized below.

1. Probiotics

Probiotics are live microorganisms that improve the health status of the host, and exert their effects primarily in the intestinal tract. A broad range of applications of probiotic cultures has emerged, and the application of probiotic cultures is restricted only by the ability of the cultures to survive in the different food types. Probiotic bacteria in the genera *Lactobacillus* and *Bifidobacterium* are most commonly used. The most important and well-documented health benefits of these genera are their effects on lactose digestion, intestinal microbial balance and dysfunction, immune modulation, hypersensitivity, carcinogenesis, and blood cholesterol levels.

2. Prebiotics

Prebiotics are naturally occurring non-digestible food ingredients, such as oligosaccharides, that can selectively stimulate the growth and/or activity of naturally present or introduced beneficial bacterial species in the colon. In addition, there is strong scientific evidence that they can aid stool bulking and relieve constipation. There is also promising preliminary scientific evidence that they can improve the bioavailability of minerals such as calcium and magnesium, prevent intestinal disorders and infections, reduce blood triacylglycerol and cholesterol levels, reduce the risk of developing colon cancer, and modulate the immune system. Prebiotics used commercially include inulin, fructo-oligosaccharides, galacto-oligosaccharides, and soy oligosaccharides.

3. Polyunsaturated fatty acids (PUFA)

A number of PUFA are essential micronutrients or have been ascribed particular health benefits, including oleic acid, linoleic acid, γ-linolenic acid (GLA), EPA and DHA [see Chapters 8, 9 and 10]. There is strong and well-documented evidence that *n*-3 (commonly referred to as omega-3) long-chain PUFA such as EPA and DHA can lower the risk of coronary heart disease. There is also growing evidence that they can have beneficial effects related to inflammatory diseases, early development, neurological diseases and cognitive performance.

4. Phytochemicals

There are hundreds, possibly thousands, of chemical constituents in plants that may have health-promoting properties. Increasingly, food and drinks companies are looking to exploit the association between phytochemicals and reduced risks of developing cancer or cardiovascular disease. Phytochemicals that are currently attracting the most attention include the antioxidant vitamins C, E and β-carotene; polyphenols; sterols; carotenoids (particularly lycopene and lutein); and the flavonoids (particularly phytoestrogens) [see Chapters 2, 3 and 4].

5. Vitamins and minerals

For certain vitamins and minerals there are indications that an intake beyond the recommended daily intake required to prevent deficiencies may be beneficial for the prevention of certain diseases. For instance, it has been established that increased intake of folic acid during pregnancy can reduce the incidence of neural tube defects in the newborn. There is also evidence to suggest that the risk of developing cardiovascular disease may be decreased by increasing vitamin E and folic acid intakes, but this remains to be firmly established. Calcium-enriched products are also believed to have a role to play in bone health. These are just a few of the growing number of links being made between vitamins and minerals and disease risk reduction.

6. Fibres

Fibres are a diverse group of soluble and insoluble substances of chemical and morphological complexity that have different physiological activities. Key benefits currently ascribed to insoluble fibres (e.g. wheat bran and maize bran) are reduced risk of developing bowel cancer and heart disease. Important sources of soluble fibre are gums from grains and a wide range of polysaccharides. On the basis of the naturally rich soluble fibre content of some grains, such as oats, a number of breakfast cereals claim to lower blood cholesterol levels.

L. Factors critical to the success of functional foods

The success of all food and drinks products is influenced by such factors as advertising, product quality (particularly taste), pricing, and market positioning. Manufacturers of functional foods face additional challenges, notably the substantiation of health claims, an often unclear regulatory environment, and communicating potentially complex concepts in language that consumers understand and trust.

1. Safety

The foremost requirement of all food manufacturers is to ensure that their products are safe, and manufacturers of functional foods are no different from the rest. Furthermore, it must be recognized that the incorporation of novel raw materials that have a physiological action, such as lowering blood cholesterol levels or modifying the gut flora, will require a more formal safety evaluation both by the company and by government regulatory agencies.

It is also important to recognize that, in future, nutritional assessments of novel foods must be carried out against a backdrop of rapidly advancing knowledge about the role of diet in the causation and prevention of many diseases. Consequently, it is recommended that an integral part of any company's functional food strategy should be the preparation of a comprehensive scientific dossier covering both health claims and nutritional safety.

2. Proof of efficacy

There is now universal agreement that sound scientific research underpinning health claims is critical to the future success of functional foods. The Code of Practice on Health Claims on Foods resulting from the UK Joint Health Claims Initiative established in June 1997 will serve to clarify the type of evidence required to support functional food claims.

The Code, which has the support of enforcement and regulatory authorities, industry bodies, and consumer and public interest organizations, lays down the following guidelines on the source and nature of the scientific evidence:

- The health claim must be based on a systematic review of all the scientific evidence relating to the validity of the health claim, including published scientific literature.
- The conclusion of the review should be based on the totality of the evidence, not just data that support the health claim; on human studies or evidence, not just biochemical, cellular or animal studies; and on the most methodologically sound studies available.
- The conclusion of the review should be based on experimental studies in humans, but observational studies may be sufficient in some circumstances. Additional information could be expected to come, as appropriate, from: human studies, including epidemiological and clinical studies and chemical extrapolation from food usage in disease states; animal studies; biochemical and cellular studies; and any other relevant information sources.
- Research should assess the effects of foods on the health status of human subjects. That is, the outcome measure in clinical or other human studies should be an improvement in some indicator of well-being or the lessening of some disease.

- Recognizing that gathering full clinical evidence on foods can often be difficult, lengthy or expensive, it is acceptable under the Code to provide evidence of the effects of the food on a bio-marker – that is, some intermediate physiological state or process known to have a strong correlation with some relevant indicator of well-being. For example, a reduction in blood cholesterol level is closely associated with a reduced risk of developing heart disease. It would therefore be acceptable for a food proven to lower blood cholesterol levels to carry an indirect claim that it helps to maintain a healthy heart.

3. Consumer education

For functional foods to realize their full potential, consumers will need to be educated about their various health claims and how they should be used within the context of a normal dietary pattern. Lack of knowledge, confusion and misunderstanding within the media, law enforcement agencies, the medical profession, and consumer interest groups emphasize the need for more broadly based communication strategies. While many governments and associated agencies are giving increasing emphasis to improving consumers' understanding of the role of food and other lifestyle factors on general health and well-being, the ultimate responsibility will continue to rest with the manufacturer and/or retailer. Governments should assist by creating an environment that ensures consumer safety as a first priority, as well as lending credence to the role of certain dietary constituents in reducing disease risk.

As the market for functional foods has developed, companies have become only too aware of the need to support products with point-of-sale literature, particularly when the product is presenting a new health claim or using new ingredients.

To handle all the aspects of the communication process across all appropriate channels and target groups could be considered too large and complex a task for all but the largest food companies. Therefore, a number of companies have taken the strategic decision to employ public relations agencies.

4. Market positioning

As most manufacturers are keen to position functional foods as normal foodstuffs carrying a health claim, it follows that they should be physically positioned in stores alongside traditional products, where possible. However, this may prove problematic for products that appear as new concepts to the consumer. One such example when it was launched in the UK was the fermented milk drink Yakult. Because it was neither a yoghurt, nor a traditional milk drink, it required dual siting, both alongside the drinking milks and in the yoghurt cabinet.

5. Price

To charge too high a premium for products whose intended role is to provide a common health benefit could lay the industry open to criticism. This is particularly true now that growing attention is being given to the comparatively poor diet of the lower socio-economic groups. Clearly, any company that adopts a pricing structure that puts functional foods out of the reach of such groups could be criticized for creating a class of "elitist" foods.

However, it is important to note that certain markets appear more tolerant of paying high premiums for certain functional foods. For example, in Finland Benecol spread sells extremely well, even though it costs six times as much as a normal margarine.

6. Health claims strategy

Experience to date suggests that the more strident the health claim made, the greater the onus on the manufacturer to be able to support the claim by way of a strong scientific dossier, ideally including data from human studies.

An example of a strident health claim strategy supported by strong in-house research is the cholesterol-lowering claim for Flora pro.activ spread. At the other end of the spectrum, companies may take a more prudent approach based on generic scientific knowledge and/or government dietary guidelines. For example, Dean Farms in the UK highlighted the findings of an important UK government report (COMA, 1994), which recommended doubling intakes of *n*-3 fatty acids without increasing intakes of *n*-6 fatty acids, when marketing its 'Columbus' eggs enriched in *n*-3 fatty acids.

M. Conclusions

"Functional foods" is a term used by industry and the media to describe a new category of foodstuffs which, while still in its infancy in the West, is arousing much debate and controversy. There are those who consider that the introduction and wider availability of functional foods will offer significant healthcare benefits over the long term, while others consider it a cynical attempt by the industry to add value to its business by exploiting consumer weakness for health claims on food products.

However, a bright future for functional foods is indicated by the high priority that industry, academia and governments have placed on strengthening the knowledge base that supports their introduction and development. Although many questions remain, functional foods appear to have the potential to make a significant contribution towards controlling healthcare expenditure and expanding healthy or complaint-free life expectancy. At the very least they have a valuable role to play in advancing consumer knowledge about the importance of diet in maintaining health and reducing disease risk.

Companies planning to do business in functional foods should not underestimate the additional complexities and challenges they face compared with standard food products. Indeed, the high failure rate for functional food launches to date probably indicates a failure to recognize this.

It is worth stressing that the oils and fats industry has a significant role to play in the future growth of the global functional foods market. However, crucial to the sustained and profitable involvement of the oils and fats industry will be growing consumer acceptance of the principle that there are "good" fats. One of the major challenges that the oils and fats industry faces is to convince consumers that oils and fats have a positive role to play in health maintenance.

References

COMA (Committee on Medical Aspects of Food and Nutrition Policy) (1994) *Nutritional Aspects of Cardiovascular Disease (Report on Health and Social Subjects No. 46),* Her Majesty's Stationery Office.

Diplock, AT, Aggett, PJ, Ashwell, M, Bornet, F, Fern, EB and Roberfroid, MB (1999) Scientific concepts of functional foods in Europe. Consensus document. *Br. J. Nutr.*, 81, S1–S27.

LFRA (1998) *Functional Foods and the European Consumer. Implications for European Product and Market Development*, LFRA, Leatherhead.

LFRA (2000) *Functional Food Markets, Innovation and Prospects. A Global Analysis*, LFRA, Leatherhead.

LFRA (2001a) *The Market for Heart Benefit Foods. A Global Analysis*, LFRA, Leatherhead.

LFRA (2001b) *Functional Soft Drinks. A Global Analysis*, LFRA, Leatherhead.

CHAPTER 2

Carotenoids

YAAKOB B. CHE MAN AND CHIN-PING TAN

Universiti Putra Malaysia, Selangor, Malaysia

A. Introduction

Of the various classes of pigments in nature, the carotenoids are among the most important (Pfander, 1992). Carotenoids are yellow, orange and red pigments present in many common fruits and vegetables (Astorg, 1997). Animals are incapable of biosynthesizing carotenoids, though many are coloured by carotenoids from their diet (Gordon and Bauernfeind, 1982). These coloured features include feathers of birds, the skin of goldfish, and the flesh of salmon.

Carotenoids, structurally similar to vitamin A, are a group of more than 600 natural compounds. However, only about 50 have provitamin A activity (Olson, 1989; Palace *et al.*, 1999). More than 60 different carotenoids have been identified in vegetable products consumed by humans (Khachik *et al.*, 1992; Bauernfeind, 1972), and approximately 20 are present in quantifiable amounts in human serum and tissues. Hydrocarbon carotenoids (α-carotene, β-carotene and lycopene), monohydroxy xanthophylls (e.g. β-cryptoxanthin), and dihydroxy xanthophylls (e.g. lutein and zeaxanthin) are among the main carotenoids present in human plasma and tissues (Krinsky *et al.*, 1990). They have received considerable attention because of their various functions in human health. Most nutrition research studies have focused on the six carotenoids found in the highest concentrations in blood: β-carotene, lycopene, α-carotene, lutein, zeaxanthin and β-cryptoxanthin (Basu *et al.*, 2001).

Carotenoids have been reported to play a role as antioxidants in the lipid phase by trapping free radicals or by quenching singlet oxygen (Edge *et al.*, 1997). However, the best-known biological function of carotenoids is their long-established role as provitamin A. α-Carotene, β-carotene and β-cryptoxanthin all have provitamin A activity, whereas lycopene, lutein and zeaxanthin have little or no such activity (Block, 1994; Stahl and Sies, 1996). In addition to serving as a source of vitamin A, dietary carotenoids are considered to play a role in the prevention of common chronic diseases such as cardiovascular disease, age-related macular degeneration, and cancers (Cooper *et al.*, 1999a,

1999b). The protective effects of carotenoids depend on their chemical structures, which can differ in the length of the polyene chromophore, the nature of the end groups, and the various substituents that they contain (Britton, 1995).

Functional foods are foods and/or food components that provide health benefits beyond those that can be attributed to basic nutritional functions (Hasler, 1998; Roberfroid, 1999; Diplock *et al.*, 1999). They have been shown to play a role in preventing or delaying the onset of certain diseases, especially chronic diseases. Therefore, it is apparent that carotenoids will play an important role in functional foods. The functional food market is growing at a rate of 15–20% per year, and the industry was recently claimed to be worth US$33 billion worldwide (Hilliam, 2000). The sale of carotenoids was earlier estimated to be approximately US$500 million, and the market is increasing (Johnson and Schroeder, 1996).

The majority of carotenoids used in foods are chemically synthesized, but some are also available from natural extracts. In addition, β-carotene and astaxanthin are produced commercially by microbial fermentation, but their market share is not very significant (Sandmann *et al.*, 1999).

B. Structure, nomenclature and chemistry

Carotenoids are a class of hydrocarbons (carotenes) and their oxygenated derivatives (oxycarotenoids or xanthophylls) (Weedon and Moss, 1995). They contain eight isoprenoid units joined in such a manner that the arrangement of isoprenoid units is reversed at the centre of the molecule: the two central methyl groups are in a 1,6-position, while the remaining non-terminal methyl groups are in a 1,5-position (Pfander, 1992). The chemical structures of the six common dietary carotenoids are given in Figure 1.

The carotenoids represent a large group of isoprenoid structures with many different structural characteristics and biological activities. All are derived from the same basic C_{40} isoprenoid skeleton by modifications such as hydrogenation, dehydrogenation, cyclization, oxidation, or any combination of these processes (Britton *et al.*, 1995). Typical C_{40} carotenoids carry cyclic β-ionone end groups that can be substituted by oxo, hydroxy, and epoxy groups at different positions (Sandmann, 2001). Carotenoids of green, photosynthetically active plant tissue are classified as primary carotenoids, while those of red fruits and flowers have been termed secondary carotenoids. Primary carotenoids include β-carotene, violaxanthin, and neoxanthin. Secondary carotenoids include α-carotene, β-cryptoxanthin, zeaxanthin, and antheraxanthin (Lichtenhaler, 1987).

Rules for the nomenclature of carotenoids (semi-systematic names) were approved by the International Union of Pure and Applied Chemistry (IUPAC) in 1974 (Weedon and Moss, 1995). However, because they are familiar and

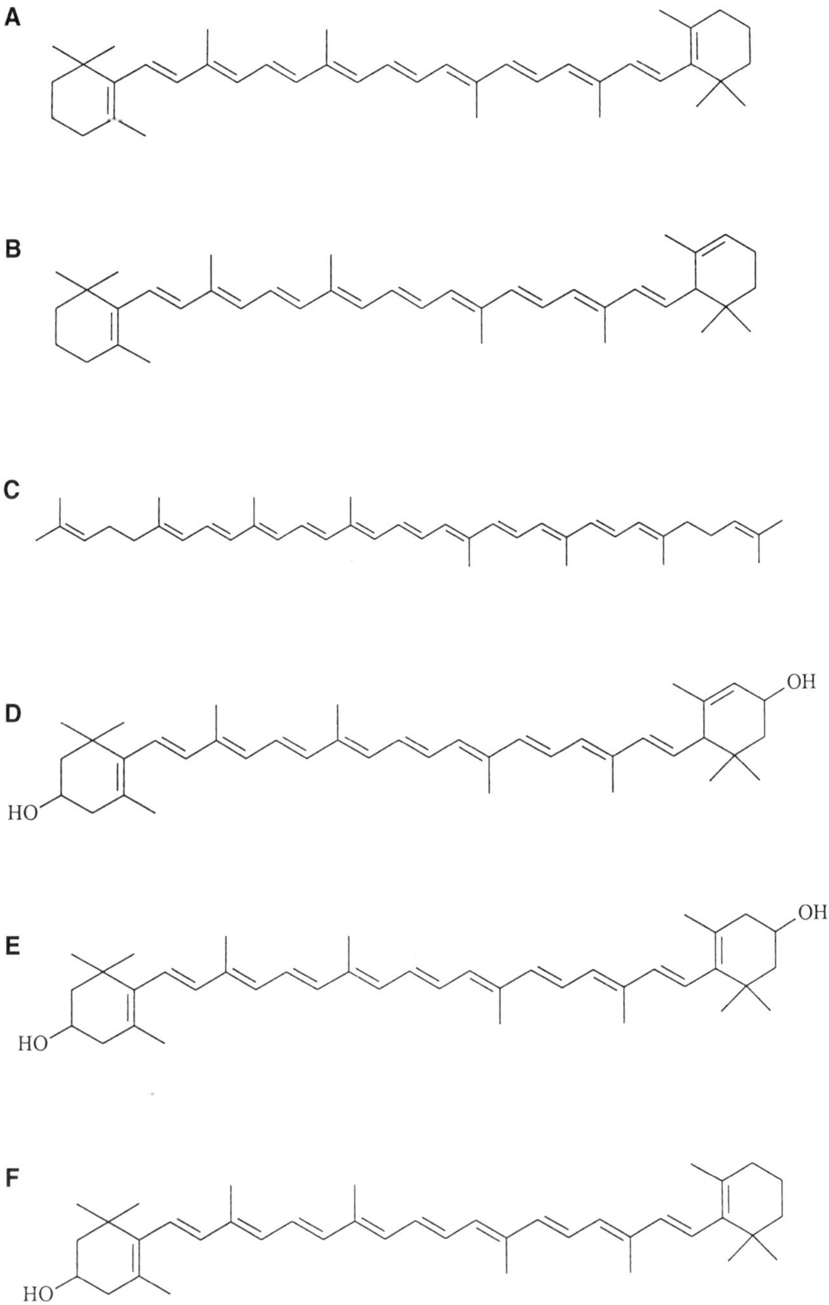

Figure 1. Chemical structure of some common carotenoids. Abbreviations: A, β-carotene; B, α-carotene; C, lycopene; D, lutein; E, zeaxanthin; F, β-cryptoxanthin.

Table 1. Comparison of trivial and semi-systematic names for carotenoids in higher plants[a]

Trivial name	Semi-systematic name
α-Carotene	β,ε-Carotene
α-Carotene	β,β-Carotene
β-Cryptoxanthin	β,β-Caroten-3-ol
Lutein	β,ε-Caroten-3,3′-diol
Zeaxanthin	β,β-Caroten-3,3′-diol
Violaxanthin	5,6,5′,6′-Diepoxy-5,6,5′,6′-tetrahydro-β,β-caroten-3,3′-diol
Neoxanthin	5′,6′-Epoxy-6,7-didehydro-5,6,5′,6′-tetrahydro-β,β-caroten-3,5,3′-triol

[a]Data taken from Goodwin (1992).

short, trivial names are still popular among the scientific community. Table 1 shows a comparison of semi-systematic and trivial names for some major carotenoids in the plant kingdom.

Some carotenoids are lipophilic compounds and can be stored biologically in a lipophilic environment. These are generally insoluble in water unless strongly polar groups are present, as in the case of the dicarboxylic acid norbixin and the carotenoid sulphates. Carotenoids are sensitive to alkali and very sensitive to air and light, especially at high temperatures (Deshpande *et al.*, 1995). Pfander (1992) recommended that all operations involving carotenoids should be carried out in an inert atmosphere (nitrogen or vacuum), at low temperature (room temperatures up to 20°C), in darkness or diffuse light, under acid-free conditions, and using peroxide-free solvents. Purified carotenoids should be stored as solids under nitrogen in a refrigerated condition. A detailed description of the isolation and analysis of carotenoids is provided by Schiedt and Liaaen-Jensen (1995). In general, the isolation method is mainly determined by the nature of the biological source, the ease of solvent extraction, and the properties (e.g. colour, stability and solubility) and quantities of the carotenoids.

C. Sources of carotenoids

1. Natural sources

As already mentioned, carotenoids are among the most important and widely distributed pigments in the plant kingdom, although their presence in green leaves is often masked by chlorophyll. Although certain photosynthetic and non-photosynthetic bacteria, fungi and algae are also able to produce carotenoids, this chapter focuses mainly on carotenoids that are produced by higher plants, which are the important food sources of these materials.

The total production of carotenoids in nature has been estimated at 10^8 tonnes/

year. The four major carotenoids are fucoxanthin (found in marine algae) and lutein, violaxanthin and neoxanthin (found in green leaves) (Haugan *et al.*, 1992). The content of some common carotenoids in various fruits and vegetables is shown in Table 2. More recently, O'Neill *et al.* (2001) compiled a carotenoid database for more than 60 fruit and vegetable products from a variety of sources. The information was compiled using available data from Europe, but on occasions when values were not available, US values are included (Mangels *et al.*, 1993). For additional information, a thorough study of natural sources of carotenoids from plants and oils has been published by Ong and Tee (1992). This report is based on more than 100 references, and lists carotenoids from four major categories of natural sources: leaves and other vegetables, fruits, roots and tubers, and vegetable oils. Chug-Ahuja *et al.* (1993) also compiled a carotenoid database for 2458 fruits, vegetables, and multi-component foods containing fruits and vegetables, giving values for the five most common carotenoids (α-carotene, β-carotene, β-cryptoxanthin, lycopene and lutein).

In plants, carotenoids are held in the chloroplasts, where they are found as complexes formed through non-covalent binding with proteins. Within the chloroplasts, carotenoids have important functional roles as accessory light-harvesting pigments and as photo-protective agents.

Fruits and vegetables differ greatly in their carotenoid content and profile. For example, red fruits and vegetables are the most common sources of lycopene. Tomatoes and tomato-based products account for more than 85% of dietary lycopene (Rao and Agarwal, 1998). The lycopene content of tomatoes differs with variety and increases with ripeness, the colour serving as a good indicator of the lycopene content (Nir *et al.*, 1993; Scott and Hart, 1995). Until the onset of ripening, a tomato fruit contains no lycopene, but during the subsequent period of colour development it can form more than 1 mg carotenoids per day (Thompson, 1993). Watermelon, pink guava, pink grapefruit, papaya and apricots also contain abundant amounts of lycopene (Ong and Tee, 1992; Mangels *et al.*, 1993; Scott and Hart, 1995). A group of researchers from the Phytonutrients Laboratory (Beltsville Human Nutrition Research Center, Agricultural Research Service, Beltsville, Maryland, USA) recently found that lycopene levels in some varieties of watermelon are equal to or greater than those in raw tomatoes (Arnold, 2002). They also showed that the bioavailability of lycopene from watermelon was higher than from raw tomatoes. Autumn olive (*Elaegnus umbellata*) also has a high level of lycopene, which is approximately 10 times higher than that of raw tomato. Carrots contain high levels of both α-carotene and β-carotene, and cantaloupe and dried apricots both contain high levels of β-carotene (Table 2).

Lutein is the dihydroxy form of α-carotene, and is distributed among a variety of vegetables. Zeaxanthin is the dihydroxy form of β-carotene, and is present in corn (maize) and various vegetables. The highest amounts of lutein and zeaxanthin are found in egg yolk and in corn (Sommerburg *et al.*, 1998).

Food type	Carotenoid content (mol%)						
	α-Carotene[b]	β-Carotene[b]	Lutein	Lycopenes[c]	Cryptoxanthins[d]	Zeaxanthin	Neoxanthins and violaxanthin[e]
Apple (red)	5	17	19	13	23	1	22
Broccoli	0	27	22	0	49	0	3
Brussels sprouts	0	11	27	0	39	2	20
Butternut squash	5	0	37	0	34	0	24
Cantaloupe	0	87	1	3	0	0	9
Carrots	43	55	2	0	0	0	0
Celery	13	0	32	1	40	2	12
Corn (maize)	0	0	60	0	5	25	9
Cucumber	0	4	38	0	38	4	16
Dried apricots	0	87	1	0	9	0	2
Egg yolk	0	0	54	0	4	35	8
Green beans	1	5	22	0	42	3	27
Green grapes	0	7	25	0	52	7	10
Green kidney beans	0	0	0	0	28	0	72
Green lettuce	16	0	15	0	36	0	33
Green pepper	0	12	36	0	20	3	29
Honeydew melon	0	48	17	0	0	18	18
Kiwi fruit	0	8	54	0	0	0	38
Mango	0	20	2	6	4	16	52
Nectarine	0	48	6	0	23	6	18
Orange	8	11	7	11	12	15	36
Orange juice	3	8	15	0	25	20	28
Orange pepper	8	21	8	0	22	37	4
Peas	0	5	41	0	21	0	33
Peach	10	50	5	0	8	8	20
Pumpkin	0	21	49	0	0	0	30
Red grapes	1	6	33	0	29	4	27
Red pepper	24	3	7	8	2	0	56
Red seedless grapes	3	16	43	5	4	10	23
Spinach	0	16	47	4	19	0	14
Tomato	0	12	6	82	0	0	0
Tomato juice	12	16	11	57	2	2	0
Yellow pepper	1	0	12	0	1	0	86
Yellow squash	28	9	44	0	0	0	19
Zucchini squash	0	5	47	0	24	5	19

[a]Data taken from Sommerburg *et al.* (1998). [b]All possible combinations of *trans* and *cis* isomers. [c]Lycopene and related isomers. [d]Includes α-cryptoxanthin, β-cryptoxanthins, and their isomers. [e]Includes neoxanthin, violaxanthin, and their related isomer lutein 5,6-epoxide.

Lutein is also the major carotenoid in kiwi fruit, red seedless grapes, zucchini squash and pumpkin, while zeaxanthin is the major carotenoid in orange pepper. β-Cryptoxanthin is also distributed in foodstuffs such as broccoli and green grapes, and is one of the major carotenoids detectable in human blood.

Recently, Hornero-Méndez and Mínguez-Mosquera (2000) highlighted an alternative carotenoid source in *Rosa mosqueta* hips. Six major carotenoids were identified. The average composition was estimated as 497.6 mg/kg β-carotene, 391.9 mg/kg lycopene, 703.7 mg/kg rubixanthin, 289.2 mg/kg gazaniaxanthin, 183.5 mg/kg β-cryptoxanthin, 266.6 mg/kg zeaxanthin, and 67.1 mg/kg minor carotenoids (dry-weight basis).

Red palm oil is also a convenient and rich source of natural carotenoids (Choo, 1996). It is obtained from the fruits of the palm tree, *Elaeis guineensis*, of which a large number of subspecies are known. Palm oil extracted from a commercially planted species (*E. guineensis* var. Tenera) in Malaysia was reported to have a carotenoid content of about 500–700 ppm (Goh *et al.*, 1985). Choo (1994) reported that palm oil is the richest natural plant source of carotenoids, with α-carotene and β-carotene being the major ones present. Tables 3 and 4 show the carotenoid contents of different oil palm species and palm oil fractions.

Table 3. Carotenoid contents of various oil palm species[a]

Species	**Carotenoid content (ppm)[b]**
Elaeis oleifera	4347
E. oleifera × *E. guineensis* Dura	1846
E. oleifera × *E. guineensis* Pisifera	1289
E. oleifera × *E. guineensis* Dura × *E. guineensis* Pisifera	864
E. guineensis Pisifera	380
E. guineensis Dura	948
E. guineensis Tenera	610

[a]Contents given for fresh fruits; data taken from Choo *et al.* (1989). [b]Total carotenoids were estimated using absorption spectrophotometry at 446 nm.

Table 4. Carotenoid contents of selected palm oil fractions[a]

Fraction	**Carotenoid content (ppm)[b]**
Crude palm oil	630–700
Crude palm olein	680–760
Crude palm stearin	380–540
Second pressed oil	1800–2400
Residual oil from fibre	4000–6000

[a]Data taken from Choo *et al.* (1989). [b]Total carotenoids were estimated using absorption spectrophotometry at 446 nm.

2. *Synthesis of carotenoids*

Carotenoids can be produced by chemical synthesis or by extraction from natural materials present in fruits and vegetables. In 1954, β-carotene was first manufactured on an industrial scale, and since then the commercial synthesis of carotenoids has been developed continuously (Britton *et al.*, 1995). Carotenoids are mainly produced because of the demand for their colouring properties. In general, these "nature-identical" carotenoids are regarded as toxicologically safe. Moreover, it is becoming increasing evident that synthetic carotenoids may have important functional roles in the human body. The major producers of synthetic carotenoids use two fundamentally different strategies: synthesis from vitamin A, or by preparing a specific trimethylcyclohexanone end group synthon for each carotenoid. An essential part of the industrial process is the transformation of the pure, crystalline, lipophilic, synthetic carotenoids into formulations suitable for industrial applications. For example, a microcrystalline dispersion of the carotenoid in an edible fat is used in margarine manufacture, but a microdispersion in a hydrophilic protective colloid is used for aqueous preparations such as fruit juices.

The supply of carotenoids by chemical synthesis and from natural resources is restricted to only a few carotenoids. Moreover, some dietary carotenoids are not present in abundant amounts. For example, zeaxanthin is a rare carotenoid which, together with lutein, forms an essential component of the macular pigment in the eye. Scientific knowledge on the biosynthesis of carotenoids is well established. As stated by Sandmann (2001), the advantage of modifying and engineering the carotenoid biosynthetic pathways is to provide sources for the isolation of desired carotenoids or to generate food plants with increased carotenoid content. The general pathway for the biosynthesis of carotenoids (via the sequence: phytoene formation, dehydrogenation, cyclization, and introduction of oxygen functions) and the mechanisms and stereochemical course of some of the basic reactions have been elucidated. However, little is known about the biochemistry of the final structural modifications that give rise to the hundreds of diverse natural carotenoids. The industrial production of natural carotenoids through microbial biotechnology is already established mainly through the exploitation of some microalgae (particularly *Dunaliella* species) which can synthesize large amounts of carotenoids. With continuous advances in the field of biotechnology it is expected that in future there will be some genetically engineered, high-producing strains of bacteria or yeasts capable of producing a variety of natural carotenoids.

Progress in the cloning of carotenogenic genes opens up the possibility of modifying and engineering the carotenoid biosynthetic pathways in microorganisms and in plants. The potential for carotenoid production in heterologous microorganisms has been shown using several different molecular biology approaches. Advances in the cloning of carotenogenic genes include the engineering of crop plants to manipulate carotenoid biosynthesis. To obtain high-yield carotenoid production, the focus should be on increasing the precursor supply, maintaining the balance of interacting metabolic pathways, and on the use of organisms or targeting of tissues that are capable of incorporating lipophilic molecules.

Today, microbial synthesis offers a promising method for the production of carotenoids (Johnson and Schroeder, 1996). Recently, Misawa and Shimada (1998) successfully synthesized carotenoids from the non-carotenogenic yeast *Candida utilis*. *C. utilis* has systematically been genetically modified as a production host for lycopene, β-carotene and astaxanthin. The potential of *Escherichia coli* as a heterologous carotenoid production system has been reviewed recently (Johnson and Schroeder, 1996). The large-scale extraction and purification of carotenoids from *E. coli* pose a number of specific problems that still have to be resolved (Schiedt and Liaaen-Jensen, 1995). Further research is necessary to optimize the growth parameters in fermenters (Sandmann *et al.*, 1999). Although the production of carotenoids in suitable microorganisms is a versatile approach, the major problem for reaching high yields is the storage capacity of the hosts (Sandmann, 2001). The commercial success of heterologous carotenoid production depends on solving these key problems.

Investigations of the production of red pigments by *Rhodotorula glutinis* on raw substrates of agro-industrial origin represent the first approach to the utilization of these raw materials for biotechnological purposes (Buzzini, 2001). A multivariate statistical approach was employed for the optimization of conditions for carotenoid production by *R. glutinis* DBVPG 3853 from a substrate containing concentrated rectified grape must as the sole carbohydrate source (Buzzini, 2000). Several experimental parameters (pH, and concentrations of carbohydrate, yeast autolysate and salt) were tested at two levels. Carotenogenesis was most sensitive to initial pH and to yeast autolysate concentration. This study determined an optimum condition (pH=5.78, [yeast autolysate]=4.67 g/l) that maximized both the concentration of total carotenoids and that of β-carotene (6.9 mg/l and 1.1 mg/l of culture fluid, respectively) after 120 h of fermentation. Buzzini (2001) suggested that the red yeasts belonging to the genus *Rhodotorula* may have industrial relevance as carotenoid producers. He also suggested that oligosaccharides and dextrins of corn syrup could be used profitably for pigment production by co-cultures of *R. glutinis* DBVPG 3853 and *Debaryomyces castellii* DBVPG 3503. The fed-batch co-cultures gave a total carotenoid production of 8.2 mg/l.

Several attempts have been made to genetically manipulate carotenoid biosynthesis in crop plants, thereby increasing their nutritional value, but care has to be taken not to disturb other isoprenoid pathways (Sandmann, 2001). In 2000, Römer *et al.* modified the carotenoid pathway in tomato to produce different carotenoid products. Ye *et al.* (2000) engineered a carotenogenic pathway into rice endosperm tissue normally devoid of carotenoids. Recently, a bacterial phytoene synthase gene was overexpressed in a seed-specific manner and the protein product targeted to the plastid in rapeseed (*Brassica napus*) (Shewmaker *et al.*, 1999). The resultant embryos from these transgenic plants were visibly orange, and the mature seed showed up to a 50 fold increase in carotenoid content. In the transgenic seeds, concentrations of carotenoids (mainly α-carotene and β-carotene) of >1 mg/g fresh weight accumulated, yielding an oil containing 2 mg/g carotenoids. Although the genetic manipulation of rapeseed to increase the carotenoid content to high levels was a success, a major obstacle to modifying carotenoid metabolism more precisely is the limited knowledge of how the expression of endogenous carotenogenic genes is regulated in higher plants (Sandmann, 2001).

D. Technological advances in carotenoid extraction

Carotenoids are lipophilic substances generally extracted using nonpolar solvents. The choice of solvent depends on the biological material, its pretreatment, the carotenoid composition, and whether the objective is partial or complete extraction (Schiedt and Liaaen-Jensen, 1995). Water-immiscible solvents such as petroleum or ethyl ether are usually used for extraction from dried tissues. For fresh materials, acetone or ethanol is used; these are both extracting and dehydrating solvents. Polar solvents such as acetone, methanol or ethanol are also good for extracting xanthophylls, but not carotenes. Moh *et al.* (1999) showed that hexane is a better solvent than isopropanol or acetone to extract carotenoids (mainly β-carotene) from crude palm oil using a water-jacketed glass chromatographic column. Although nonpolar solvents are recommended, De Ritter and Purcell (1981) postulated that, for fresh plant samples with a low moisture content, complete extraction can be achieved using a mixture of slightly polar plus nonpolar solvents. In general, the solvent should be pure and free from oxygen, acids, halogens and peroxides to avoid degradation of carotenoid compounds.

Currently, industrial extraction consists of pressing the material to milled or pelleted form, followed by extraction with hexane (Delgado-Vargas *et al.*, 2000). The modern solvent-based process usually consists of the extraction of oleaginous material by successive countercurrent washes with hexane (Rosenthal *et al.*, 1996). The extracted meal is then carried on a sealed conveyor for solvent recovery in enclosed vessels using jacket or sparged steam. Hexane is eliminated from the oil in rising-film evaporators and, finally, by vacuum distillation (Serrato, 1981). However, there are two problems with this process: half the pigment is lost during the extraction process, and there are safety hazards. Alternative solvents have also been employed. For example, ethanol has been used to extract carotenoids from chilli guajillo puya flour (Santamaría *et al.*, 2000). During extraction some carotenoids may undergo isomerization, as in the preparation of marigold oleoresin, when there is a slight increase in the content of *cis*-lutein ester due to the hot processing steps involved during concentration and solvent removal (Antony and Shankaranarayana, 2001).

Various methods have been developed for the recovery and concentration of carotenoids from red palm oil (Choo, 1994). Choo *et al.* (1989) recovered carotenoids from crude palm oil and from the volatile methyl ester fraction obtained from palm oil by various methods, including vacuum distillation, molecular distillation, and adsorption chromatography (using C_{18} reverse-phase and carbon columns). Baharin *et al.* (1998) also concentrated palm carotene from crude palm oil in a single-stage chromatographic process using a synthetic porous polymer. The carotene was concentrated to an approximately 10^5 ppm solution, which is about 160 times the original concentration in crude palm oil. Carotene recovery was between 40% and 65% depending on the chromatographic conditions. They considered that this carotene recovery method might be suitable for use as an edible palm oil pretreatment process, due to its efficient mass recovery of a valuable bioresource. Baharin *et al.* (2001) also studied the effects on the carotene extraction system of the quality of the crude palm oil, the carotene composition, and carotenoid stability during storage. More recently, Latip *et al.* (2000, 2001) described different types of synthetic adsorbents and adsorption and solvent extraction processes for extracting carotene from crude palm oil.

Enzymatic extraction or aqueous extraction has been proposed as an alternative to the use of organic extraction solvents. For example, cell-wall-degrading enzymes can be used to extract oil by solubilizing structural cell-wall components of oilseeds (Rosenthal *et al.*, 1996). Today, the food industry uses enzymatic methods to obtain a variety of products. The main advantages of such methods are specificity and the use of moderate temperature and pH conditions; consequently the final product is virtually unchanged in molecular structure (Delgado-Vargas *et al.*, 2000). However, reports of the application of enzyme technology to pigment extraction are scarce. Santamaría *et al.* (2000) studied a two-stage, enzyme-mediated process for the selective extraction of capsaicinoids and carotenoids from chilli guajillo puya flour. The first stage released up to 60% of the capsaicinoids, and the second stage permitted the recovery of 83% of the carotenoids.

Aqueous extraction has been used since 1950 because of its safety and cost-saving features (Delgado-Vargas *et al.*, 2000). The effects of enzyme treatments on the aqueous extraction of marigold (*Tagetes erecta*) meal at different dehydration levels (yielding meals with a final moisture content of 5%, 10%, 15% or 20%, dry-weight basis) to enhance carotenoid extraction were evaluated by Delgado-Vargas and Paredes-López (1997a). Carotenoid extraction was more efficient as a result of the enzyme treatments, and the total amount of carotenoids extracted from enzyme-treated samples was statistically ($P<0.05$)

higher for the marigold meal with 5% moisture content than for the meals with higher moisture contents. Delgado-Vargas and Paredes-López (1997b) also evaluated the effects of five different commercial enzymes on the extraction of carotenoids from fresh marigold flowers. The efficiency of pigment extraction was greater for all the enzyme-treated samples than for untreated samples, and the highest carotenoid yields were obtained using the enzyme 'Econase-cep'. This enzyme may also have applications in enhancing carotenoid extraction at the industrial level.

Supercritical fluid extraction (SFE) technology is an ideal technique for the production of high-purity pigment products, and has been widely studied. There are no solvent residues in the end products and no degradation of bioactive compounds during processing (Dunford, 2001). SFE is a valuable procedure for the production of lipid-based food products such as carotenoids. It has several advantages: it is rapid; solvent strength can be controlled; and the solvents (gases) are friendly to the environment and have low toxicity (Hawthorne, 1990; Chester *et al.*, 1992).

Favati *et al.* (1988) extracted lutein and carotene from leaf protein concentrates using CO_2 as the SFE solvent, and showed that the SFE conditions can be manipulated to allow selective extraction. Spanos *et al.* (1993) extracted β-carotene from sweet potato by SFE and found the technique to be 3–5 times more efficient than the traditional method. Barth *et al.* (1995) compared SFE with traditional extraction processes for carrot carotenoids. The traditional method extracted more α-carotene than SFE, but the total recovery of vitamin A equivalents was higher using SFE and the extraction time was reduced from 6 h to 1 h. More recently, Careri *et al.* (2001) used SFE to extract carotenoids from *Spirulina pacific* algae, seeking the optimal SFE conditions to maximize carotenoid recoveries, and comparing the quantitative results obtained using SFE with those obtained using liquid–liquid extraction methods. Again, SFE proved to be more effective than the traditional solvent extraction method. Gnayfeed *et al.* (2001) extracted ground paprika with supercritical CO_2 and subcritical propane at different pressure and temperature conditions, to estimate the effects of changing the SFE parameters on the yield and composition of carotenoids, tocopherols and capsaicinoids. Under mild conditions, subcritical propane was superior to supercritical CO_2 for the extraction of carotenoids and tocopherols, but less efficient for the extraction of capsaicinoids.

E. Functional properties of carotenoids

1. Colour

The importance of carotenoids for colouration in living organisms has already been mentioned. For example, apart from their universal occurrence in the photosynthetic apparatus of phototrophic bacteria, carotenoids are also often responsible for yellow, orange or red colours in non-phototrophic bacteria, moulds and yeasts. The importance of the carotenoids in these microorganisms lies mainly in protection against damage by light and oxygen.

Various natural pigments, including carotenoids, are widely used as additives to provide an attractive colour to manufactured foods (Nir *et al.*, 1993; Antony and Shankaranarayana, 2001; Baker, 2001). Natural extracts have been used for this purpose for centuries, especially annatto, saffron, tomato, paprika and red palm oil. In the food industry, carotenoids are prepared in oily or aqueous media, including emulsions, colloidal suspensions, and complexes with proteins (Kläui, 1979). These preparations find widespread applications in margarine, butter, fruit juices and beverages, canned soups, dairy products, desserts and mixes, preserves and syrups, sugar and flour confectionery, salad dressings, meat, pasta, and egg products. Carotenoids are also widely used as additives in animal feeds to impart the desired colour to animal tissues or to derived products, or to provide adequate vitamin supplies.

2. *Antioxidant activity*

A considerable amount of work has been carried out to understand the quenching of singlet oxygen by carotenoids and how this reaction protects against singlet oxygen-mediated photo-oxidation reactions. The antioxidant potential of carotenoids was first described by Monaghan and Schmitt (1932), and 36 years later, Foote and Denny (1968) demonstrated that β-carotene could inhibit photosensitized oxidation and was an efficient quencher of singlet oxygen. Farmillo and Wilkinson (1973) showed that electron exchange energy transfer quenching is the principal mechanism of carotenoid photoprotection against singlet oxygen. Chemical quenching also occurs, resulting in destruction of the carotenoids (Liebler, 1993). In general, carotenoids with nine or more conjugated double bonds can absorb the energy from singlet oxygen. The energy-rich carotenoid then releases the absorbed energy in the form of heat and reverts to its normal energy level (Deshpande *et al.*, 1996).

Di Mascio *et al.* (1989, 1992) showed that lycopene is the most efficient singlet oxygen quencher of the biological carotenoids, including β-carotene, and this observation has been confirmed by other studies (Conn *et al.*, 1991). Lycopene has attracted attention due to its biological and physicochemical properties, especially related to its effects as a natural antioxidant. Lycopene has no provitamin A activity, but exhibits a physical quenching rate constant with singlet oxygen almost twice as high as that of β-carotene (Shi and Le Maguer, 2000). However, lycopene is not necessarily a better antioxidant, because lycopene is itself more susceptible to oxidation than β-carotene and may therefore be used up more quickly.

Carotenoids can also react with free radicals and thus act as chain-breaking antioxidants, although they do not have the specific characteristic structural features normally associated with this class of antioxidants. The extensive system of conjugated double bonds makes these compounds very susceptible to attack by peroxyl radicals. Although there is increasing evidence that carotenoids are very effective quenchers of peroxyl radicals, the mechanism of their antioxidant action has not yet been defined (Polozza and Krinsky, 1992).

In 1984, Burton and Ingold proposed that β-carotene is not a conventional antioxidant, and that at high oxygen partial pressures (>150 mm Hg, 20% O_2) and high carotenoid concentrations it may even act as a pro-oxidant. In 100% oxygen, a β-carotene concentration of 5×10^{-4} M reduced the oxidation rate of methyl linoleate more effectively than either higher (5×10^{-3} M) or lower (5×10^{-5} M) concentrations. Clearly, the concentration of carotenoids, oxygen, and the nature of the environment are all important in assessing the antioxidant or pro-oxidant properties of carotenoids. Carotenoids can autoxidize when the oxygen tension increases, and are thus most effective as antioxidants at the low oxygen tensions typical of physiological levels found in tissues. Many variables contribute to the overall singlet oxygen quenching ability of carotenoids, including the solvent and the structural characteristics of the carotenoid, such as the number of conjugated double bonds and the nature of the end groups.

While showing that carotenoids are extremely good quenchers of singlet oxygen *in vitro*, the studies so far discussed are not necessarily directly relevant to biological environments, and little work has been carried out to test how effectively carotenoids protect cells against singlet oxygen damage. Tinkler *et al.* (1994) showed that four water-solubilized carotenoids bound to human lymphoid cells offered protection from rose bengal or *meso*-tetra(4-sulfonatophenyl)porphine sensitized singlet oxygen. Their results showed that greatest protection was afforded by lycopene, followed by astaxanthin, β-carotene, and canthaxanthin.

The relative antioxidant activities of carotenoids in multilamellar liposomes, as assayed by inhibition of the formation of thiobarbituric acid-reactive substances (TBARs), were: lycopene>α-carotene>β-cryptoxanthin >zeaxanthin≈ β-carotene>lutein (Stahl *et al.*, 1998). Mixtures of carotenoids were more effective than the single compounds, and this synergistic effect was most pronounced when lycopene or lutein was present. The superior antioxidant activity of mixtures of carotenoids may be related to the specific positioning of different carotenoids within cell membranes.

In many edible oil systems, carotenoids have been shown to prevent or delay lipid oxidation (Kiritsakis and Dugan, 1985; Fakourelis *et al.*, 1987; Lee and Min, 1988; Jung and Min, 1991). Henry *et al.* (1998) studied the thermal and oxidative degradation of carotenoids in an oil model system to determine their relative stabilities. The rates of degradation were as follows: lycopene>all-*trans*-β-carotene≈9-*cis*-β-carotene>lutein, suggesting that lycopene is the most susceptible to degradation and lutein has the greatest stability in the model system tested. However, contradictory results have been obtained from studies of the effects of carotenoids on the oxidation of oils. β-Carotene has been shown to act as a pro-oxidant during lipid oxidation, both in the light (Terao *et al.*, 1980; Faria and Mukai, 1983; Warner and Frankel, 1987) and in the dark (Suzuki *et al*, 1989). Similarly, Haila *et al.* (1996) reported that lutein acts as a pro-oxidant in triacylglycerols, both in the dark and the light. In a comprehensive study of the effects of β-carotene and lutein on bulk purified triacylglycerols, Subagio and Morita (2001) showed that β-carotene suppressed the oxidation of triacylglycerols when added at a low concentration, but increased the rate of lipid oxidation when added at higher concentrations. On the other hand, lutein accelerated triacylglycerol oxidation even at low concentrations: the more lutein added, the faster the triacylglycerol oxidation rate.

These contradictory results, showing evidence of both antioxidant and pro-oxidant activity, may suggest that the antioxidant behaviour of carotenoids is closely related to their own oxidation. Studies of the kinetics of carotenoid oxidative degradation under different conditions, as well as of the structures of the products formed, have allowed the mechanism of the oxidation process to be clarified (Yanishilieva *et al.*, 1998).

When the effects of lycopene, lutein, annatto and γ-tocopherol on triacylglycerol oxidation were each examined by measuring their peroxide values, the results showed that lutein and lycopene were pro-oxidants, whereas the natural food colour annatto and γ-tocopherol both effectively inhibited hydroperoxide formation (Haila *et al.*, 1996). The pro-oxidant effects of lutein and lycopene were inhibited, and loss of the yellow carotenoid colour was retarded, when γ-tocopherol was added together with the carotenoids. Moreover, a combination of lutein and γ-tocopherol was more efficient than γ-tocopherol alone in inhibiting hydroperoxide formation, indicating that lutein and γ-tocopherol can act synergistically to prevent lipid oxidation. Palozza and Krinsky (1991) showed that β-carotene and α-tocopherol act synergistically as radical scavengers in living systems such as rat liver microsomes. However, the consumption of α-tocopherol increased in the presence of β-carotene, indicating that the α-tocopherol protects the β-carotene.

These results suggest that the potential pro-oxidant effects of carotenoids should always be considered when carotenoids are proposed for use as colourants in lipid-containing foods. Carotenoids have special and remarkable properties, and their functions and actions vary significantly in different systems. Britton (1995) considered it essential that studies of the antioxidant properties of carotenoids be based on chemical principles and on the relationships between structure, properties and function.

3. *Disease prevention*

Carotenoids, some of which have provitamin A properties, have a range of diverse biological functions and actions, especially in relation to human health.

For example, carotenoids are known to be crucial for normal vision and have been associated with reducing the risk of developing several degenerative diseases including various cancers. There have been many studies on the effects of carotenoids on various diseases, and excellent reports have been prepared by Palace *et al.* (1999), Cooper *et al.* (1999a, 1999b), Rao and Agarwal (1998), and Nishino *et al.* (2000).

Le Marchand *et al.* (1993) studied the effects of dietary intakes of several carotenoids on lung cancer patients. After adjusting for smoking and other risk factors, they reported that the lowest risk of lung cancer occurred in those with the highest intakes of β-carotene, α-carotene and lutein. In their study of the effects of palm carotene on chemical carcinogenesis, Okuzumi *et al.* (1992) showed that palm carotene suppressed mouse epidermal ornithine decarboxylase activity induced by glycocholic acid. Another study showed that men with the highest blood levels of carotenoids had 36% fewer heart attacks and deaths than those with the lowest levels of carotenoids (Morris *et al.*, 1994).

Epidemiological evidence has suggested that dietary carotenoids may inhibit certain types of cancer. This may be due to their provitamin A activity and/or because carotenoids have the ability to act as antioxidants (Peto *et al.*, 1981; Stahl and Sies, 1996). The antioxidant activity of carotenoids is conferred by the hydrophobic chain of polyene units that can quench singlet oxygen, neutralize thiyl radicals, and combine with and stabilize peroxyl structures (Palace *et al.*, 1999).

Epidemiological investigations have also shown that cancer risk is inversely related to the consumption of green and yellow vegetables and fruits. Recently, Nishino *et al.* (2000) studied the cancer prevention activities of several natural carotenoids in foods (α-carotene, lutein, lycopene, zeaxanthin and β-cryptoxanthin) in comparison with β-carotene, and found that some of them showed more potent activity than β-carotene. For example, α-carotene showed higher activity than β-carotene in suppressing tumorigenesis in the skin, lung, liver and colon of mice. In general, they concluded that not only β-carotene, but also other carotenoids, in our daily foods have anticarcinogenic activity and that mixtures of these carotenoids may be useful for cancer prevention. The level of dietary lycopene shows an inverse association with prostate cancer risk (Stanley, 2002). A recent human dietary intervention trial provides good evidence that lycopene protects against prostate cancer by protecting DNA against free radical mediated damage (Chen *et al.*, 2001).

Many studies have shown that an increased intake of fruits and vegetables rich in vitamin A is associated with a reduced risk of developing cancer. Peto *et al.* (1981) hypothesized that the association between vitamin A and reduced disease risk is due to the portion of the vitamin A derived from β-carotene. However, two major intervention trials subsequently showed that β-carotene supplementation in fact increased lung cancer in smokers (Alpha-Tocopherol, Beta-Carotene Cancer Prevention Study Group, 1994; Mayne *et al.*, 1994).

After these findings were announced, several β-carotene intervention trials were halted. However, β-carotene did protect against lung cancer in non-smokers. Since then, several studies have been initiated to try to understand these results. Böhm *et al.* (1997) reported that vitamin C could repair carotenoid radical cations in methanol, and suggested that, in biological systems, carotenoid radical cations may be reoriented within the cell membranes and become accessible to vitamin C in the aqueous phase. As smoking can lower blood vitamin C levels by up to 40% (Edge *et al.*, 1997), these findings may provide a good reason why β-carotene protected against lung cancer in non-smokers but not in smokers.

Omenn *et al.* (1996) suggested that a direct approach to addressing the question of whether or not carotenoids are responsible for the observed association between high fruit and vegetable intake and reduced cancer risk would be to conduct further intervention trials, in which average dietary levels of β-carotene or other dietary carotenoids or a placebo are provided to individuals who normally have low-quintile dietary carotenoid intakes. Copper *et al.* (1999a) suggested that a more indirect route to characterizing the role of dietary carotenoids would be to identify and quantify other phytochemicals and their effects on disease risk. This approach would yield a better understanding of the many other factors associated with high fruit and vegetable intake and, by inference, the portion of any change in risk that is attributable to carotenoids.

F. Applications and stability of carotenoids in food products

The dietary use of carotenoids involves their application in nutrient or dietary supplements, as functional food ingredients, and as natural food colourants and antioxidants, as well as in cosmetic formulations, for pharmaceutical purposes, and in animal feeds. The commercial demand for carotenoids is mainly met by chemical synthesis and, in part, by extraction from natural sources. A number of studies on the thermal stability of carotenoids in processed fruits and vegetables have found that the hydrocarbon carotenoids such as lycopene, α-carotene and β-carotene are relatively heat stable (Khachik *et al.*, 1992). However, there are many conflicting studies on the stability of carotenoids during food processing, and it is difficult to reconcile the differences in their findings. A common limitation is a failure to distinguish between chemical degradation and geometrical isomerization of the parent molecules (Nguyen and Schwartz, 1999).

The effects of traditional processing practices on the retention of α-carotene, β-carotene, and total provitamin A carotenoids were studied using amaranth, cowpea, peanut, pumpkin and sweet potato leaves (Mosha *et al.*, 1997). The traditional practice of sun drying and storage in ventilated containers resulted in a significant ($P<0.05$) decrease in the concentration of total carotenoids, β-carotene, and α-carotene for all the vegetables. Conventional blanching and cooking resulted in a significant ($P<0.05$) increase in the concentration of carotenoids in the cowpea, peanut and pumpkin leaves; while in amaranth and sweet potato greens, thermal processing resulted in a significant ($P<0.05$) decrease in carotenoid concentrations. Cooking also enhanced the quantity of carotenoids extracted. Thus, blanching and cooking could be considered as advantageous processes for increasing the amount of provitamin A carotenoids in vegetables. In another study, Yadav and Sehgal (1997) investigated the effects of selected processing and storage methods on the concentration of β-carotene in bathua and fenugreek leaves. Methods included storage of leaves with or without polythene bags for 24 h or 48 h in a refrigerator at 5°C, storage at 30°C in polythene bags, drying (sun or oven), blanching (for 5, 10 or 15 min), open-pan cooking, and pressure cooking. The results showed that storing leaves in a refrigerator, drying in an oven, blanching for 5 min, and cooking in a pressure cooker resulted in better retention of β-carotene.

The effects of dietary supplementation and processing treatments on the stability of cholesterol and paprika carotenoids in egg powders were studied by Lai *et al.* (1996). Cholesterol oxidation and loss of carotenoids in eggs dried with a direct gas-fired spray dryer were greater ($P<0.05$) than in eggs dried using an indirect (electric) heating system. Dietary supplementation of hens with α-tocopherol acetate (200 mg/kg feed) significantly increased ($P<0.01$) the oxidative stability of cholesterol and carotenoids in eggs dried with the direct heating system.

Accumulating epidemiological evidence continues to show that lycopene (found at high levels in tomatoes, grapefruits and watermelons) is associated with a reduced risk of developing certain chronic diseases and cancers (Nguyen and Schwartz, 1998, 1999). The effect of thermal processing on the stability of lycopene in tomato products has not yet been rigorously addressed. Nguyen and Schwartz (1998) indicated that, in contrast to β-carotene, lycopene is relatively resistant to heat-induced geometrical conversion during typical processing of tomatoes and related products. They also observed that the presence of fat, the change in percentage of solids, and the severity of heat treatment were not contributing factors in the formation of lycopene isomers in tomato products – except at extreme conditions not regularly employed in the food industry or during food preparation. Thus, while lycopene is stable in the tomato matrix, sample handling techniques should be carefully evaluated to minimize the formation of lycopene *cis* isomers in organic solutions. Rao and Agarwal (1998) recently suggested that heating tomatoes might improve the availability of lycopene, and thus improve its chances of altering oxidative stress.

Shi and Le Maguer (2000) recently showed that the main causes of tomato lycopene degradation during processing are isomerization and oxidation. Thermal processing generally causes some loss of lycopene in tomato-based foods. In general, dehydrated and powdered tomatoes have poor lycopene stability unless carefully processed and promptly placed in a hermetically sealed and inert atmosphere for storage. Frozen foods exhibit excellent lycopene stability throughout their normal shelf life.

In 1995, Chen *et al.* studied the effects of various processing methods on changes in the colour and vitamin A content of carrot juice. Results showed that canning (121°C, 30 min) resulted in the greatest destruction of carotenoids, followed by high-temperature, short-time (HTST) heating at 120°C for 30 s, acidification plus HTST heating at 110°C for 30 s, and acidification plus HTST heating at 105°C for 25 s. The vitamin A content decreased with increasing temperature and heating time. The main breakdown products formed were 13-*cis*-β-carotene, followed by 13-*cis*-lutein and 15-*cis*-β-carotene. In the canning process (121°C for 30 min), concentrations of 9-*cis*-β-carotene and 13,15-di-*cis*-β-carotene also increased, mainly due to conversion of 13-*cis*-β-carotene.

The effects of various cooking methods on carotenoid levels in raw and cooked (microwaved, boiled, steamed or stewed) green vegetables (broccoli, spinach, and green beans) and tomatoes have been extensively studied (Khachik *et al.*, 1992). The epoxy carotenoids were somewhat sensitive to prolonged heating for 1 h, but lutein and hydrocarbon carotenoids such as neurosporene, α-carotene, β-carotene, lycopene, ζ-carotene, phytofluene and phytoene survived the heat treatments. Similar results were obtained in an HPLC (high-performance liquid chromatography) study of the carotenoid composition of the 18 vegetables (raw and cooked) most frequently consumed in Spain. Carotenoid saponification resulted in some losses, mainly of xanthophylls (Granado *et al.*, 1992). Boiling did not alter the carotenoid profile of the samples, and the total amounts of carotenoids were higher after boiling than in raw samples.

A recent study of the effects of storage temperature on pigment degradation in green beans (*Phaseolus vulgaris* cultivars 'Perona' and 'Boby') showed that carotenoid pigments suffered moderate changes during cold storage (Monreal *et al.*, 1999). A second study analysed the effects of microwave energy on carotenoid, chlorophyll and anthocyanin levels in papaya, strawberry and kiwi purées (de Ancos *et al.*, 1999). Various microwave powers and treatment times were used to inactivate the oxidoreductases, peroxidase, and polyphenol oxidase in the purées, and changes in the pigment levels were evaluated by HPLC. Microwave treatments produced small modifications in the quantitative and qualitative composition of carotenoids in papaya, and of anthocyanins in strawberry.

Cano and Marin (1992) studied differences in pigment profiles between fresh (unripe and ripe), frozen and canned kiwi fruit slices, using thin-layer chromatography (TLC), HPLC, UV–visible spectroscopy, and chemical tests. Pigments present in fresh and frozen kiwi fruit slices were xanthophylls (9′-*cis*-neoxanthin, *trans*-violaxanthin, *cis*-violaxanthin, auroxanthin, lutein epoxide, and lutein), chlorophylls and their derivatives, and one hydrocarbon carotenoid (β-carotene). There was no significant difference between the carotenoid profile of kiwi stored for 6 months at –18°C and that of the fresh fruit, except that additional antheraxanthin was detected in the frozen fruit. Thermal treatment of the canned kiwi fruit resulted in some degradation of xanthophylls such as neoxanthin and violaxanthin.

The effects of the drying process on fresh peppers were studied by Mínguez-Mosquera and Hornero-Méndez (1994) and Mínguez-Mosquera *et al.* (1994). In general, when drying peppers in the presence of light there is a net synthesis of carotenoids, which increases the concentration from that initially found in the fresh pepper. However, once biosynthesis is complete, light has a strong degradative effect on carotenoids and its presence is undesirable. Therefore, combining a first drying step in the light with a second one in darkness should result in the highest possible carotenoid concentration in the peppers without subsequent degradation.

The carotenoid concentration of paprika also depends on the variety of pepper used. Fresh ripe fruits, dried fruits, and ground paprika from the 'Agridulce' pepper variety consistently have a higher carotenoid content than those from the 'Bola' variety (Mínguez-Mosquera and Hornero-Méndez, 1994). Variations in the carotenoid stabilities of different industrially processed paprika samples are mainly due to intrinsic differences between the pepper varieties used, rather than to processing factors.

Non-thermal food processing has gained importance in recent years as a potential technology to replace or complement traditional thermal processes. The effects of high-pressure processing on the carotenoid composition of purées of two persimmon fruit cultivars ('Rojo Brillante' and 'Sharon') were recently evaluated by de Ancos *et al.* (2000). The total carotenoid content was higher in untreated Rojo Brillante purée (22.11 mg/g) than in untreated Sharon purée (15.22 mg/g), and purées of both untreated cultivars showed similar carotenoid patterns after saponification. After high-pressure treatments (50 MPa or 300 MPa for 15 min at 25°C for Rojo Brillante; 50 MPa or 400 MPa for 15 min at 25°C for Sharon), the total amount of extractable carotenoids increased by 9–27%, and the content of total vitamin A retinol equivalents (RE) increased by 75–87 RE/100 g. However, the high-pressure treatment used in this study showed non-uniform behaviour, with not all parts of the purée samples receiving the pre-determined pressure treatment.

Berset and Marty (1992) studied the degradation of all-*trans*-β-carotene during extrusion cooking. When purified all-*trans*-β-carotene was incorporated into cornstarch, extensive degradation occurred during extrusion cooking. Losses were much lower in commercial preparations of β-carotene, in which the colourant was coated in a matrix. The presence of starch and water, as well as the incorporation of oxygen, may play important roles in the oxidative degradation of the pigment.

The carotenoid contents of foods also vary as a function of the stage of ripening and the environmental growing conditions. Márkus *et al.* (1999) investigated changes in the quality attributes of red peppers (*Capsicum annum* L. var. Km-622) as a function of ripening and some technological factors. They found that the contents of carotenoids and bioantioxidants (vitamin C and tocopherols) were influenced by the weather conditions during the production season: a rainy and cool season yielded fruits with more β-carotene, but a lower content of red xanthophyll diesters, than a relatively dry and warm season. Harvesting the peppers when unripe partially retarded carotenoid biosynthesis. Centrifugation of the peppers prior to drying caused a marked loss of vitamin C; consequently, carotenoid stability was impaired during storage of the ground red pepper.

G. Conclusions and future research

The society of abundance which now characterizes most of the industrialized world is facing new challenges from exponentially growing health care costs, increased life expectancy, improved scientific knowledge, the development of new technologies, and major changes in lifestyle (Roberfroid, 2000). Increasing health awareness and corresponding changes in lifestyle are the main factors influencing the development of new functional foods. Today, consumer demand for healthy products provides an opportunity to develop carotenoid-rich foods as new functional foods. During the past decade, carotenoids have emerged as potential protective agents against many chronic diseases. As a result, the demand for carotenoids is increasing and new methods are being developed for their production. The development of large-scale carotenoid extraction, purification and concentration procedures that result in minimal loss of activity is highly desirable for the food industry. With advances in the field of biotechnology, various improvements have been carried out to produce carotenoids for use in various industries. These include improvements in traditional chemical synthesis, and mass-culture production of natural carotenoids from algae, yeast, fungi, and recombinant DNA systems. New chemical regulators of carotenoid biosynthesis and isolated or immobilized enzyme systems are also being developed for carotenoid production.

The science behind the effects of dietary carotenoids on human health is complex, and there are unlikely to be simple and straightforward answers to the outstanding questions in the field today. In 1995, Woteki showed that the overall dietary intake of carotenoids has increased since the 1970s, but that this increase is not consistent across all age groups. In general, groups at high risk of having an inadequate dietary intake of carotenoids are the poor, tobacco users, non-supplement users, and the elderly. Moreover, the efficiency of carotenoid absorption from the diet is strongly dependent on the type of food, the way the food is processed, and the amount of accompanying dietary lipids (Sandmann, 2001). Therefore, optimized nutrition is key to maximizing the physiological functions of carotenoids for each individual, and to ensuring both maximum well-being and health with a minimum risk of disease throughout the lifespan.

At the same time, advances in food ingredient technologies have stimulated an explosion of innovative nutrition products by food manufacturers (Hardy, 2000). Continued development in food processing technologies will prove invaluable for improving nutrient stability and the organoleptic quality of foods. The development of non-thermal food processing methods is important, preventing the degradation of nutrients such as carotenoids in foods.

Although food manufacturers have the freedom to innovate, they must also avoid marketing products with questionable effects. One common problem is that the biologically active ingredients in functional foods are often present in insufficient doses to give the desired beneficial effects. The raw materials are often costly, and this means that only small amounts are generally used in functional foods (Ernst, 2001). Food manufacturers must clearly demonstrate that new carotenoid-rich food products are safe for their intended use – and also that any health claims are substantiated by well-documented scientific evidence.

Especially confusing are the apparent increases in cardiovascular and cancer-related mortality that have been found in β-carotene supplementation trials. More work is required to reconcile the different results from intervention trials, studies of biological mechanisms, and epidemiological observations (Palace *et al.*, 1999). Furthermore, interactions among different carotenoid compounds and between carotenoids and other micronutrients will be of continued interest, and future research on carotenoids should incorporate this important area. Research on carotenoids is currently being carried out in various scientific fields (e.g. food science and technology, nutrition, biotechnology, medical science), and the sharing of information on carotenoids between these disciplines will help to promote the progress of knowledge about these important functional ingredients.

References

Alpha-Tocopherol, Beta-Carotene Cancer Prevention Study Group (1994) The effect of vitamin E and β-carotene on the incidence of lung cancer and other cancers in male smokers. *New Engl. J. Med.*, 330, 1029–1035.

Antony, JIX and Shankaranarayana, ML (2001) Lutein. A natural colourant and a phytonutrient for eye health protection. *The World of Food Ingredients*, April/May, 64–67.

Arnold, J (2002) Watermelon packs a powerful lycopene punch. *Ag. Res. Mag.*, 50(6).

Astorg, P (1997) Food carotenoids and cancer prevention: an overview of current research. *Trends Food Sci. Technol.*, 8, 406–413.

Baharin, BS, Rahman, KA, Karim, MIA, Oyaizu, T, Tanaka, K, Tanaka, Y and Takagi, S (1998) Separation of palm carotene from crude palm oil by adsorption chromatography with a synthetic polymer adsorbent. *J. Am. Oil Chem. Soc.*, 75, 399–404.

Baharin, BS, Latip, RA, Che Man, YB and Rahman, RA (2001) The effect of carotene extraction system on crude palm oil quality, carotene composition, and carotene stability during storage. *J. Am. Oil Chem. Soc.*, 78, 851–855.

Baker, RTM (2001) Canthaxanthin in aquafeed applications: is there any risk? *Trends Food Sci. Technol.*, 12, 240–243.

Barth, MM, Zhou, C, Kute, KM and Rosenthal, GA (1995) Determination of optimum conditions for supercritical fluid extraction of carotenoids from carrot (*Daucus carota* L.) tissue. *J. Agric. Food Chem.*, 43, 2876–2878.

Basu, HN, Del Vecchio, AJ, Flider, F and Orthoefer, FT (2001) Nutritional and potential disease prevention properties of carotenoids. *J. Am. Oil Chem. Soc.*, 78, 665–675.

Bauernfeind, JC (1972) Carotenoid vitamin A precursors and analogs in foods and feeds. *J. Agric. Food Chem.*, 20, 456–473.

Berset, C and Marty, C (1992) Formation of nonvolatile compounds by thermal degradation of β-carotene: protection by antioxidants. *Methods Enzymol.*, 213, 129–142.

Block, G (1994) Nutrient sources of provitamin A carotenoids in the American diet. *Am. J. Epidemiol.*, 139, 290–293.

Böhm, F, Edge, R, Land, EJ, McGarvey, DJ and Truscott, TG (1997) Carotenoids enhance vitamin E antioxidant efficiency. *J. Am. Chem. Soc.*, 119, 621–622.

Britton, G (1995) Structure and properties of carotenoids in relation to function. *FASEB J.*, 9, 1551–1558.

Britton, G, Liaaen-Jensen, S and Pfander, H (1995) Carotenoids today and challenges for the future. In: *Carotenoids Vol. 1A: Isolation and Analysis* (G Britton, S Liaaen-Jensen and H Pfander, eds), Birkhäuser Verlag, Basel, pp.13–26.

Burton, GW and Ingold, KU (1984) β-carotene: An unusual type of lipid antioxidant. *Science*, 224, 569–573.

Buzzini, P (2000) An optimization study of carotenoid production by Rhodotorula glutinis DBVPG 3853 from substrates containing concentrated rectified grape must as the sole carbohydrate source. *J. Ind. Microbiol. Biotechnol.*, 24, 41–45.

Buzzini, P (2001) Batch and fed-batch carotenoid production by Rhodotorula glutinis – Debaryomyces castellii co-cultures in corn syrup. *J. Appl. Microbiol.*, 90, 843–847.

Cano, MP and Marin, MA (1992) Pigment composition and color of frozen and canned kiwi fruit slices. *J. Agric. Food Chem.*, 40, 2141–2146.

Careri, M, Furlattini, L, Mangia, A, Musc, M, Anklam, E, Theobald, A and von Holst, C (2001) Supercritical fluid extraction for liquid chromatographic determination of carotenoids in Spirulina Pacific algae: a chemometric approach. *J. Chromatogr. A*, 912, 61–71.

Chen, BH, Peng, HY and Chen, HE (1995) Changes of carotenoids, color, and vitamin A contents during processing of carrot juice. *J. Agric. Food Chem.*, 43, 1912–1918.

Chen, L, Stacewicz-Sapuntzakis, M, Duncan, C, Sharifi, R, Ghosh, L, van Breemen, R, Ashton, D and Bowen, PE (2001) Oxidative DNA damage in prostate cancer patient consuming tomato sauce-based entrees as a whole-food intervention. *J. Natl Cancer Inst.*, 93, 1872–1879.

Chester, TL, Pinkston, JD and Raynie, DE (1992) Supercritical fluid chromatography and extraction. *Anal. Chem.*, 64, 153R–170R.

Choo, YM (1994) Palm oil carotenoids. *Food Nutr. Bull.*, 15, 130–137.

Choo, YM (1996) Antioxidants in red palm oil. *Malaysian Oil Sci. Technol.*, 5, 15–16.

Choo, YM, Yap, SC, Ong, ASH, Ong, CK, Ooi, CK and Goh, SH (1989) Extraction of carotenoids from palm oil. In: *Proceedings of the AOCS World Conference of Edible Fats and Oils: Basic Principles and Modern Practices* (DR Erikson, ed.), AOCS Press, Champaign, pp. 436–440.

Chug-Ahuja, JK, Holden, JM, Forman, MR, Mangles, AR, Beecher, GR and Lanza, E (1993) The development and application of a carotenoid database for fruits, vegetables, and selected multicomponent foods. *J. Am. Diet. Assoc.*, 93, 318–323.

Conn, PF, Schalch, W and Truscott, TG (1991) The singlet oxygen and carotenoid interaction. *J. Photochem. Photobiol. B*, 41, 189–200.

Cooper, DA, Eldridge, AL and Peters, JC (1999a) Dietary carotenoids and lung cancer: a review of recent research. *Nutr. Rev.*, 57, 133–145.

Cooper, DA, Eldridge, AL and Peters, JC (1999b) Dietary carotenoids and certain cancers, heart disease, and age related macular degeneration: a review of recent research. *Nutr. Rev.*, 57, 201–214.

de Ancos, B, Cano, MP, Hernandez, A and Monreal, M (1999) Effects of microwave heating on pigment composition and colour of fruit purees. *J. Sci. Food Agric.*, 79, 663–670.

de Ancos, B, Gonzalez, E and Cano, MP (2000) Effect of high pressure treatment on the carotenoid composition and the radical scavenging activity of persimmon fruit purees. *J. Agric. Food Chem.*, 48, 3542–3548.

De Ritter, E and Purcell, AE (1981) Carotenoid analytical method. In: *Carotenoids as Colorants and Vitamin A Precursors* (JC Bauernfeind, ed.), Academic Press, New York, pp. 815–882.

Delgado-Vargas, F and Paredes-López, O (1997a) Enzymatic treatment to enhance carotenoid content in dehydrated marigold flower meal. *Plant Foods Hum. Nutr.*, 50, 163–169.

Delgado-Vargas, F and Paredes-López, O (1997b) Effects of enzymatic treatments on carotenoid extraction from marigold flowers (*Tagetes erecta*). *Food Chem.*, 58, 255–258.

Delgado-Vargas, F, Jiménez, AR and Paredes-López, O (2000) Natural pigments: carotenoids, anthocyanins, and betalains – characteristics, biosynthesis, processing, and stability. *Crit. Rev. Food Sci. Nutr.*, 40, 173–289.

Deshpande, SS, Deshpande, US and Salunkhe, DK (1995) Nutritional and health aspects of food antioxidants. In: *Food Antoxidants*, (DL Madhavi, SS Deshpande and DK Salunkhe, eds), Marcel Dekker Inc, New York, pp. 361–470.

Di Mascio, P, Kaiser, S and Sies, H (1989) Lycopene as the most efficient biological carotenoid singlet oxygen quencher. *Arch. Biochem. Biophys.*, 274, 532–538.

Di Mascio, P, Sunquist, AR, Devasagayam, TPA and Sies, H (1992) Assay of lycopene and other carotenoids as singlet oxygen quenchers. *Methods Enzymol.*, 213, 429–438.

Diplock, AT, Aggett, PJ, Ashwell, M, Bornet, F, Fern, EB and Roberfroid, MB (1999) Scientific concepts of functional foods in Europe: consensus document. *Br. J. Nutr.*, 81 (Suppl. 1), S1–S27.

Dunford, NT (2001) Health benefits and processing of lipid-based nutritionals. *Food Technol.*, 55(11), 38–44.

Edge, R, McGravey, DJ and Truscott, TG (1997) The carotenoids as antioxidants – a review. *J. Photochem. Photobiol. B*, 41, 189–200.

Ernst, E (2001) Functional foods, neutraceuticals, designer foods: innocent fad or counterproductive marketing ploy? *Eur. J. Clin. Pharmacol.*, 57, 353–355.

Fakourelis, N, Lee, EC and Min, DB (1987) Effect of chlorophyll and β-carotene on the oxidation stability of olive oil. *J. Food Sci.*, 52, 234–235.

Faria, JAF and Mukai, MK (1983) Use of gas chromatographic reactor to study lipid photooxidation. *J. Am. Oil Chem. Soc.*, 60, 77–81.

Farmillo, A and Wilkinson, F (1973) On the mechanism of quenching of singlet oxygen in solution. *J. Photochem. Photobiol.*, 18, 447–450.

Favati, F, King, JW, Friedrick, JP and Eskins, K (1988) Supercritical CO_2 extraction of carotene and lutein from leaf protein concentrates. *J. Food Sci.*, 53, 1532–1536.

Foote, CS and Denny, RW (1968) Chemistry of singlet oxygen. VIII. Quenching of β-carotene. *J. Am. Chem. Soc.*, 90, 6233–6235.

Gnayfeed, MH, Daood, HG, Illíes, V and Biacs, PA (2001) Supercritical CO_2 and subcritical propane extract of pungent paprika and quantification of carotenoids, tocopherols and capsaicinoids. *J. Agric. Food Chem.*, 49, 2761–2766.

Goh, SH, Choo, YM and Ong, ASH (1985) Minor constituents of palm oil. *J. Am. Oil Chem. Soc.*, 62, 237–240.

Goodwin, TW (1992) Biosynthesis of carotenoids: an overview. *Meth. Enzymol.*, 214, 2330–2340.
Gordon, HT and Bauernfeind, JC (1982) Carotenoids as food colorants. *Crit. Rev. Food Sci. Nutr.*, 18, 59–97.
Granado, F, Olmedilla, B, Blanco, I and Rojas-Hidalgo, E (1992) Carotenoid composition in raw and cooked Spanish vegetables. *J. Agric. Food Chem.*, 40, 2135–2140.
Haila, KM, Lievonen, SM and Heinonen, MI (1996) Effects of lutein, lycopene, annatto and γ-tocopherol on autoxidation of triglycerides. *J. Agric. Food Chem.*, 44, 2096–2100.
Hardy, G (2000) Nutraceuticals and functional foods: introduction and meaning. *Nutrition*, 16, 688–697.
Hasler, CM (1998) Functional Foods: their role in disease prevention and health promotion. *Food Technol.*, 52(11), 63–70.
Haugan, JA, Akermann, T and Liaaen-Jensen, S (1992) Isolation of fucoxanthin and peridinin. *Methods Enzymol.*, 213, 231–245.
Hawthorne, SB (1990) Analytical-scale supercritical fluid extraction. *Anal. Chem.*, 62, 633A–642A.
Henry, LK, Catignani, GL and Schwartz, SJ (1998) Oxidative degradation kinetics of lycopene, lutein, and 9-*cis* and all-*trans*-β-carotene. *J. Am. Oil Chem. Soc.*, 75, 823–829.
Hilliam, M (2000) Functional foods: how big is the market? *World of Food Ingredients*, 12, 50–53.
Hornero-Méndez, D and Mínguez-Mosquera, MI (2000) Carotenoid pigments in Rosa mosqueta hips, an alternative carotenoid source for foods. *J. Agric. Food Chem.*, 48, 825–828.
Johnson, EA and Schroeder, WA (1996) Microbial carotenoids. *Adv. Biochem. Eng. Biotechnol.*, 53, 119–178.
Jung, MY and Min, DB (1991) Effect of quenching mechanism of carotenoids in the photosensitized oxidation of soybean oil. *J. Am. Oil Chem. Soc.*, 68, 653–658.
Khachik, F, Goli, MB, Beecher, GR, Holden, J, Lusby, WR, Tenorio, MD and Barrera, MR (1992) Effects of food preparation on qualitative and quantitative distribution of major carotenoid constituents of tomatoes and several green vegetables. *J. Agric. Food Chem.*, 40, 390–398.
Kläui, H (1979) Carotenoids and their applications. In: *Natural Colours for Food and Other Uses* (JN Counsell and JA Dunastable, eds), Elsevier, London, pp.91–122.
Kiritsakis, A and Dugan, LR (1985) Studies in photooxidation of olive oil. *J. Am. Oil Chem. Soc.*, 62, 892–896.
Krinsky, NI, Russett, MD, Handelman, GJ and Snodderly, DM (1990) Structural and geometrical isomers of carotenoids in human plasma. *J. Nutr.*, 120, 1654–1662.
Lai, SM, Gray, JI, Partridge, JA and Flegal, CJ (1996) Stability of cholesterol and paprika carotenoids in egg powders as influenced by dietary and processing treatments. *J. Sci. Food Agric.*, 72, 171–178.
Latip, RA, Baharin, BS, Che Man, YB and Rahman, RA (2000) Evaluation of different types of synthetic adsorbents for carotene extraction from crude palm oil. *J. Am. Oil Chem. Soc.*, 77, 1277–1281.
Latip, RA, Baharin, BS, Che Man, YB and Rahman, RA (2001) Effect of adsorption and solvent extraction process on the percentage of carotene extraction from crude palm oil. *J. Am. Oil Chem. Soc.*, 78, 83–87.
Lee, EC and Min, DB (1988) Quenching mechanism of β-carotene on the chlorophyll sensitized photooxidation of soybean oil. *J. Food Sci.*, 53, 1894–1895.
Le Marchand, L, Hankin, JH, Kolonel, LN, Beecher, GR, Wilkens, LR and Zhao, LP (1993) Intake of specific carotenoids and lung cancer risk. *Cancer Epidemiol.*, 2, 183–187.
Lichtenhaler, HK (1987) Chlorophylls and carotenoids: pigments of photosynthetic biomembranes. *Methods Enzymol.*, 148, 350–382.

Liebler, DC (1993) Antioxidant reactions of carotenoids. *Ann. New York Acad. Sci.*, 691, 20–31.

Mangels, AR, Holden JM, Beecher, GR, Forman, MR and Lanza, E (1993) Carotenoid content of fruits and vegetables: an evaluation of analytical data. *J. Am. Diet. Assoc.*, 93, 284–296.

Márkus, F, Daood, HG, Kapitany, J and Biacs, PA (1999) Change in the carotenoid and antioxidant content of spice red pepper (paprika) as a function of ripening and some technological factors. *J. Agric. Food Chem.*, 47, 100–107.

Mayne, ST, Janerich, DT, Greenwald, P, Chorost, S, Tucci, C, Zaman, MB, Melamed, MR, Kiely, M and McKneally, MF (1994) Dietary β-carotene and lung cancer risk in U.S. smokers. *J. Natl Cancer Inst.*, 86, 33–38.

Mínguez-Mosquera, MI and Hornero-Méndez, D (1994) Comparative study of the effect of paprika processing on the carotenoids in peppers (*Capsicum annuum*) of Bola and Agridulce varieties. *J. Agric. Food Chem.*, 42, 1555–1560.

Mínguez-Mosquera, MI, Jarén-Galán, M and Garrido-Fernández, J (1994) Competition between the process of biosynthesis and degradation of carotenoids during the drying of peppers. *J. Agric. Food Chem.*, 42, 645–648.

Misawa, N and Shimada, H (1998) Metabolic engineering for the production of carotenoid in non-carotenogenic bacteria and yeasts. *J. Biotechnol.*, 59, 169–181.

Moh, MH, Che Man, YB, Badlishah, BS, Jinap, S, Saad, MS and Abdullah, WJW (1999) Quantitative analysis of palm carotene using Fourier transform infrared and near infrared spectroscopy. *J. Am. Oil Chem. Soc.*, 76, 249–254.

Monaghan, BR and Schmitt, FO (1932) The effects of carotene and of vitamin A on the oxidation of linoleic acid. *J. Biol. Chem.*, 96, 387–395.

Monreal, M, de Ancos, B and Cano, MP (1999) Influence of critical storage temperatures on degradative pathways of pigments in green beans (Phaseolus vulgaris cvs. Perona and Boby). *J. Agric. Food Chem.*, 47, 19–24.

Morris, DL, Kritchevsky, SB and Davis, CE (1994) Serum carotenoids and coronary heart disease. The lipid research clinics coronary primary prevention trial and follow-up study. *J. Am. Med. Assoc.*, 272, 1439–1441.

Mosha, TC, Pace, RD, Adeyeye, S, Laswai, HS and Mtebe, K (1997) effect of traditional processing practices on the content of total carotenoid, β-carotene, α-carotene and vitamin A activity of selected Tanzanian vegetables. *Plant Foods Hum. Nutr.*, 50, 189–201.

Nguyen, ML and Schwartz, SJ (1998) Lycopene stability during food processing. *Exp. Biol. Med.*, 218, 101–105.

Nguyen, ML and Schwartz, SJ (1999) Lycopene: chemical and biological properties. *Food Technol.*, 53, 38–45.

Nir, Z, Hartal, D and Raveh, Y (1993) Lycopene from tomatoes. A new commerical natural carotenoid. *Int. Food Ingredients*, 3, 45–51.

Nishino, H, Tokuda, H, Murakoshi, M, Satomi, Y, Masuda, M, Onozuka, M, Yamaguchi, S, Takayasu, J, Tsuruta, J, Okuda, M, Khachik, F, Narisawa, T, Takasuka, N and Yano, M (2000) Cancer prevention by natural carotenoids. *BioFactors*, 13, 89–94.

Okuzumi, J, Nishino, H, Murakoshi, M, Yamane, T, Kitao, Y, Inagake, M, Ohya, K, Yoshida, M and Takahashi, T (1992) Palm carotene inhibits tumor-promoting activity of bile acids and intestinal carcinogenesis. *Oncology*, 49, 492–497.

Olson, JA (1989) Provitamin A function of carotenoids: the conversion of β-carotene into vitamin A. *J. Nutr.*, 119, 105–108.

Omenn, GS, Goodman, GE, Thornquist, MD, Malmes, J, Cullen, MR, Glass, A, Keogh, JP, Meyskens, FL, Valanis, B, William, JH, Barnhart, S and Hammar, S (1996) Effects of a combination of beta carotene and vitamin A on lung cancer and cardiovascular disease. *New Engl. J. Med.*, 334, 1150–1155.

O'Neill, ME, Carroll, Y, Corridan, B, Olmedilla, B, Granado, F, Blanco, I, Van den Berg, H, Hininger, I, Rousell, AM, Chopra, M, Southon, S and Thurnham, DI (2001) A European carotenoid database to assess carotenoid intakes and its use in a five-country comparative study. *Br. J. Nutr.*, 85, 499–507.

Ong, ASH and Tee, ES (1992) Natural sources of carotenoids from plants and oils. *Methods Enzymol*, 213, 142–167.

Palace, VP, Khaper, N, Qin, Q and Singal, PK (1999). Antioxidant potentials of vitamin A and carotenoids and their relevance to heart disease. *Free Radic. Biol. Med.*, 26, 746–761.

Palozza, P and Krinsky, NI (1991) The inhibition of radical-initiated peroxidation of microsomal lipids by both α-tocopherol and β-carotene. *Free Radic. Biol. Med.*, 11, 407–414.

Palozza, P and Krinsky, NI (1992) Antioxidant effects of carotenoids *in vivo* and *in vitro*: an overview. *Methods Enzymol.* 213, 403–420.

Peto, R, Doll, R, Buckley, JD and Sporn, MB (1981) Can dietary β-carotene materially reduce human cancer rates?. *Nature*, 290, 201–208.

Pfander, H (1992) Carotenoids: an overview. *Methods Enzymol.*, 213, 3–22.

Rao, AV and Agarwal, S (1998) Bioavailability and in vivo antioxidant properties of lycopene from tomato products and their possible role in the prevention of cancer. *Nutr. Cancer*, 31, 199–203.

Roberfroid, MB (1999) What is beneficial for health? The concept of functional food. *Food Chem. Toxicol.*, 37, 1039–1041.

Roberfroid, MB (2000) A European consensus of scientific concepts of functional foods. *Nutrition*, 16, 689–691.

Römer, S, Fraser, PD, Kiano, WJ, Shipton, CA, Misawa, N, Schuch, W and Bramley, PM (2000) Elevation of the provitamin A content of transgenic tomato plants. *Nature Biotechnol.*, 18, 666–669.

Rosenthal, A, Pyle, DL and Niranjan, K (1996) Aqueous and enzymatic processes for edible oil extraction. *Enzym. Microb. Technol.*, 19, 402–420.

Sandmann, G (2001) Genetic manipulation of carotenoid biosynthesis: strategies, problems and achievements. *Trends Plant Sci.*, 6, 14–17.

Sandmann, G, Albrecht, M, Schnurr, G, Knörzer, O and Böger, P (1999) The biotechnological potential and design of novel carotenoids by gene combination in *Escherichia coli. Trends Biotechnol.*, 17, 233–237.

Santamaría, RI, Reyes-Duarte, MD, Bárzana, E, Fernando, D, Gama, FM, Mota, M and López-Munguía, A (2000) Selective enzyme-mediated extraction of capsaicinoids and carotenoids from chili guajillo puya (*Capsicum annum* L.) using ethanol as solvent. *J. Agric. Food Chem.*, 48, 3063–3067.

Schiedt, K and Liaaen-Jensen, S (1995) Isolation and analysis. In: *Carotenoids. Vol. 1A: Isolation and Analysis* (G Britton, S Liaaen-Jensen and H Pfander, eds), Birkhäuser Verlag, Basel, pp. 81–108.

Scott, KJ and Hart, DJ (1995) Development and evaluation of an HPLC method for the analysis of carotenoids in foods and the measurement of caroteniod content of vegetables and fruits commonly consumed in the UK. *Food Chem.*, 54, 101–111.

Serrato, AG (1981) Extraction of oil from soybeans. *J. Am. Oil Chem. Soc.*, 58, 157–159.

Shewmaker, CK, Sheehy, JA, Daley, M, Colburn, S and Ke, DY (1999) Seed-specific overexpression of phytoene synthase: increase in carotenoids and other metabolic effects. *Plant J.*, 20, 401–412.

Shi, J and Le Maguer, M (2000) Lycopene in tomatoes: chemical and physical properties affected by food processing. *Crit. Rev. Biotechnol.*, 20, 293–334.

Sommerburg, O, Keunen, JEE, Bird, AC and van Kuijk, JGM (1998) Fruits and vegetables that are sources for lutein and zeaxanthin: the macular pigment in human eyes. *Br. J. Ophthalmol.*, 82, 907–910.

Spanos, GA, Chen, H and Schwartz, SJ (1993) Supercritical CO_2 extraction of β-carotene from sweet potatoes. *J. Food Sci.*, 58, 817–820.

Stahl, W and Sies, H (1996) Lycopene – A biologically important carotenoid for humans. *Arch. Biochem. Biophys.*, 336, 1–9.

Stahl, W, Junghan, A, de Boer, B, Driomina, ES, Briviba, K and Sies, H (1998) Carotenoid mixtures protect multilamellar liposomes against oxidative damage: synergistic effects of lycopene and lutein. *FEBS Lett.*, 427, 305–308.

Stanley, J (2002) Tomatoes, lycopene and prostate cancer. *Lipid Technol.*, 14 (4), 85–87.

Subagio, A and Morita, N (2001) Instability of carotenoids is a reason for their promotion of lipid oxidation. *Food Res. Int.*, 34, 183–188.

Suzuki, T, Usuki, R and Kaneda, T (1989) The role of carotenoids in the oxidative deterioration of edible oils. *J. Jpn. Oil Chem. Soc.*, 38, 486–491.

Terao, J, Yamauchi, R, Murakami, H and Matsushita, S (1980) Inhibitory effects of tocopherols and β-carotene on singlet initiated photooxidation of methyl linoleate and soybean oil. *J. Food Process. Preserv.*, 4, 79–93.

Thompson, GA, Jr (1993) Responses of lipid metabolism to developmental changes and environmental perturbation. In: *Lipid Metabolism in Plants* (TS Moore, Jr, ed.), CRC press, Boca Raton, pp. 591–619.

Tinkler, JH, Böhm, F, Schalch, W and Truscott, TG (1994) Dietary carotenoids protect human cells from damage. *J. Photochem. Photobiol. B*, 41, 189–200.

Warner, K and Frankel, EN (1987) Effect of β-carotene on light stability of soybean oil. *J. Am. Oil Chem. Soc.*, 64, 213–218.

Weedon, BCL and Moss, GP (1995) Structure and Nomenclature. In: *Carotenoids Vol. 1A: Isolation and Analysis* (G Britton, S Liaaen-Jensen and H Pfander, eds), Birkhäuser Verlag, Basel, pp.27–70.

Woteki, CE (1995) Application of antioxidants in physiologically functional foods. Consumption, intake pattern, and exposure. *Crit. Rev. Food Sci. Nutr.*, 35, 143–147.

Yadav, SK and Sehgal, S (1997) Effect of home processing and storage on ascorbic acid and β-carotene content of bathua (*Chenopodium album*) and fenugreek (*Trigonella foenum graecum*) leaves. *Plant Foods Hum. Nutr.*, 50, 239–247.

Yanishilieva, NV, Aitzermuller, K and Raneva, VG (1998) β-Carotene and lipid oxidation. *Lipid (Fett)*, 100, 442–462.

Ye, X, Al-Babili, S, Klöti, A, Zhang, J, Lucca, P, Beyer, P and Potrykus, I (2000) Engineering the provitamin A (β-carotene) biosynthetic pathway into (carotenoid-free) rice endosperm. *Science*, 287, 303–305.

CHAPTER 3

Tocopherols, tocotrienols and vitamin E

WILLIAM L. STONE[1] and ANDREAS PAPAS[2]

[1]*East Tennessee State University, Johnson City, Tennessee, USA;* [2]*Eastman Chemical Co, Kingsport, Tennessee, USA*

A. Vitamin E is more than α-tocopherol

The general biochemistry of vitamin E has been extensively reviewed (Azzi and Stocker, 2000; Brigelius Flohe and Traber, 1999; Stone and Papas, 1997). This review will primarily focus on the biochemical and health aspects of vitamin E in food. It is important, however, to contrast the chemical forms of vitamin E found in plants with that found in mammals and in dietary supplements. Vitamin E is often equated with α-tocopherol, which is the primary lipid-soluble antioxidant in human plasma, and the most potent form in protecting the fetuses of pregnant female rats from death and resorption. We will explore the hypothesis that vitamin E supplements containing a chemical distribution of tocopherols and tocotrienols similar to that found in plant oils may be superior to supplements containing only α-tocopherol or α-tocopheryl acetate. Finally, this review will summarize the very recent application of nutritional genomics to vitamin E synthesis in plants.

The term vitamin E is best used as a generic description of all tocopherol and tocotrienol derivatives exhibiting the biological activity of α-tocopherol. Vitamin E, therefore, refers to at least eight compounds: four tocopherols and four tocotrienols (Figure 1). Tocopherols have a saturated phytyl chain attached to the 2-position of the chromanol ring, whereas tocotrienols have a farnesyl group attached at this position. The farnesyl side chain has three unsaturated double bonds. The α-, β-, γ- and δ-tocopherols and the respective tocotrienols differ in the number and positions of methyl group residues on the chromanol ring. [Two additional tocol compounds, desmethyl tocotrienol and didesmethyl tocotrienol, appear to be unique to rice bran oil, and are discussed in further detail in Chapter 4 of this book.] A clear explanation of the nomenclature of tocopherols and related compounds can be found at web site: http://www.chem.qmw.ac.uk/iupac/misc/toc.html

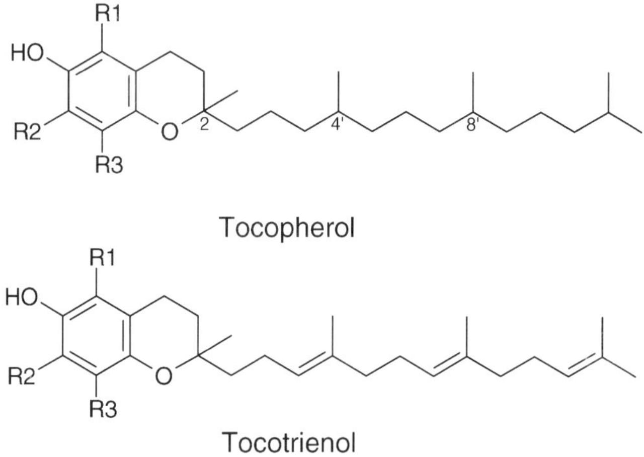

Figure 1. The chemical structures of tocopherols and tocotrienols. For α-tocopherol and α-tocotrienol, R1=R2=R3=CH_3; for β-tocopherol and β-tocotrienol, R1=R3=CH_3, R2=H; for γ-tocopherol and γ-tocotrienol, R1=H, R2=R3=CH_3; for δ-tocopherol and δ-tocotrienol, R1=R2=H, R3=CH_3.

B. Vitamin E in food is different from vitamin E in most supplements

Surprisingly, the chemical forms of vitamin E in foods are usually quite different from the chemical form found in supplements obtained in health food stores. In Western diets, the primary form of vitamin E is γ-tocopherol, which is present in a typical diet at levels about 2 fold higher than α-tocopherol (Bieri and Evarts, 1974). Tables 1–4 list the vitamin E contents of major food sources (Papas, 1999). This information is taken from *The Vitamin E Factor* by Dr Andreas Papas, and the reader is referred to this excellent book for a more extensive discussion of the vitamin E contents of various foods. In general, plant oils (Table 1) and nuts (Table 2) are excellent sources of γ-tocopherol. Moreover, all forms of naturally occurring vitamin E in food have a particular stereochemistry in which the methyl groups in the 2, 4′ and 8′ positions are all in the *R* configuration. In contrast, vitamin E supplements most often contain only α-tocopheryl acetate – usually only synthetic α-tocopheryl acetate, in which there is an equimolar mixture of all possible racemates (all-*rac*-α-tocopheryl acetate). In α-tocopheryl acetate, the configuration at the 2, 4′ and 8′ positions can be *R* or *S*. Only one-eighth of all-*rac*-α-tocopheryl acetate is the naturally occurring *R*,*R*,*R*-α-tocopheryl form. Work from a number of laboratories has shown that *R*,*R*,*R*-α-tocopheryl acetate has about twice the bioavailability of all-*rac*-α-tocopheryl acetate in both infants (Stone *et al.*, 2002a) and adults

Table 1. The vitamin E content of oils and fats[a]

	Tocopherols (mg/100 g)					Tocotrienols (mg/100 g)					Grand total	IU per
Oil/fat	**α**	**β**	**γ**	**δ**	**Total**	**α**	**β**	**γ**	**δ**	**Total**	**(mg/100 g)**	**100 g[b]**
Soybean	10		59	26	**96**					**0**	**96**	**24**
Corn (maize)	11	5	60	2	**78**					**0**	**78**	**29**
Canola	17		35	1	**53**					**0**	**53**	**30**
Sunflower	49		5	1	**55**					**0**	**55**	**73**
Peanut	13		22	2	**37**					**0**	**37**	**23**
Cottonseed	39		39		**78**					**0**	**78**	**64**
Safflower	39		17	24	**80**					**0**	**80**	**61**
Palm	26		32	7	**65**	14	3	29	7	**53**	**118**	**49**
Coconut	0.5			0.6	**1**	0.5		2	0.6	**3**	**4**	**1**
Olive	20	1	1		**22**					**0**	**22**	**30**
Evening primrose	16		42	7	**65**					**0**	**65**	**30**
Wheat germ	121	65	24	25	**235**	2	17			**19**	**254**	**233**
Rice	12	4	5		**21**	18	2	57		**77**	**98**	**30**
Barley	35	5	5		**45**	67	12	12		**91**	**136**	**87**
Oat	18	2	5	5	**30**	18		3		**21**	**51**	**38**
Butter	2	0	0	0	**2**	0	0	0	0	**0**	**2**	**3**
Lard	1.2				**1.2**	0.7				**0.7**	**1.9**	**2**
Margarine	7	0	51	3	**62**	0	0	0	0	**0**	**62**	**18**

[a]Data taken from Papas (1999). Blank entries indicate that the compounds were not detectable. [b]Note that in many samples the total concentration of international units (IU) of α-tocopherol equivalents is less than half of the total of all tocopherols and tocotrienols.

(Burton *et al.*, 1998; Ferslew *et al.*, 1993).

Most clinical studies have been conducted with all-*rac*-α-tocopheryl acetate rather than a mixture of tocopherols and tocotrienols as found in food. As detailed below, this could raise problems since an ever-increasing body of evidence suggests that the different tocopherols have distinct chemical and biological properties. Moreover, the acetate ester of α-tocopherol is inactive as an antioxidant since the phenolic hydroxyl group on position 6 of the chroman ring is essential for this function. The acetate group does, however, protect vitamin E from loss of activity during storage. In adults, α-tocopheryl acetate is rapidly hydrolysed by intestinal esterase into the active α-tocopherol form. It is not clear if premature infants or newborns have sufficient intestinal esterases to effectively convert α-tocopheryl acetate to the free and bioactive form (Stone, 1998).

The addition of α-tocopheryl acetate to food would not protect food constituents from oxidative damage, as would the naturally occurring forms of vitamin E.

Table 2. The Vitamin E content of nuts, seeds, legumes and grains[a]

	Tocopherols (mg/100 g)					Tocotrienols (mg/100 g)					Grand total	IU per
Food	**α**	**β**	**γ**	**δ**	**Total**	**α**	**β**	**γ**	**δ**	**Total**	**(mg/100 g)**	**100 g[b]**
Almonds	27	0	1		**28**	1				**1**	**29**	**41**
Walnuts	1		17	2	**20**					**0**	**20**	**3**
Peanuts	11		8		**19**					**0**	**19**	**18**
Pistachios	3	0	30	1	**34**	1		4		**5**	**39**	**10**
Pecans	1		20		**21**					**0**	**21**	**4**
Cashews	0		4	0	**4**					**0**	**4**	**1**
Wheat	1	1			**2**	0	3			**3**	**5**	**2**
Wheat germ	17	8			**25**	1				**1**	**26**	**32**
Barley	1				**1**	1	1			**2**	**3**	**1**
Oats	1				**1**	1				**1**	**2**	**2**
Rice, brown	1				**1**	1		1		**2**	**3**	**2**
Corn (maize)	1		4		**5**	0.2	0.4	0.5		**1**	**6**	**2**
Soybeans	2	0.5	13	5	**21**					**0**	**21**	**7**
Soybean meal	0.3				**0.3**					**0**	**0.3**	**0**
Lima beans			7	1	**8**					**0**	**8**	**1**
Lentils			4		**4**					**0**	**4**	**1**
Chick peas	1		7		**8**					**0**	**8**	**3**
Northern beans			4		**4**					**0**	**4**	**1**

[a]Data taken from Papas (1999). Blank entries indicate that the compounds were not detectable. [b]Note that in many samples the total concentration of international units (IU) of α-tocopherol equivalents is less than half of the total of all tocopherols and tocotrienols.

This is an important consideration, since many of the oxidative by-products formed from cooking oils could have deleterious health consequences.

C. Vitamin E and lipid peroxidation

Vitamin E is the primary lipid-soluble antioxidant found in foods and in human blood and tissues. It is well known that vitamin E inhibits the process of lipid peroxidation (Figure 2) in oils and in biological lipid–protein complexes such as biological membranes or circulating lipoproteins. The hydroxyl radical (•OH), generated by the Fenton reaction, can be a source of initiating free radicals for the process of lipid peroxidation. During lipid peroxidation, highly reactive lipid hydroperoxides, peroxyl radicals, and reactive aldehydes, such as malondialdehyde (MDA) and 4-hydroxynonenol (4-HNE), are generated. The peroxyl radicals support chain reactions that can rapidly damage oils, biological membranes or lipoproteins containing polyunsaturated fatty acids (PUFA). The higher the PUFA content of an oil or a biological membrane, the greater its susceptibility to lipid peroxidation. Fish oil, for example, is very susceptible to

Table 3. The vitamin E content of fruits and vegetables[a]

	Tocopherols (mg/100 g)					Tocotrienols (mg/100 g)					Grand total	IU per
Food	**α**	**β**	**γ**	**δ**	**Total**	**α**	**β**	**γ**	**δ**	**Total**	**(mg/100 g)**	**100 g[b]**
Avocados	2.7	0.1	0.4		**3.2**					**0**	**3.2**	**4.1**
Apples	0.3				**0.3**					**0**	**0.3**	**0.5**
Pears	0.4		0.2		**0.6**	0.3				**0.3**	**0.9**	**0.8**
Peaches	0.8				**0.8**	0.1				**0.1**	**0.9**	**1.3**
Oranges	0.3				**0.3**					**0**	**0.3**	**0.4**
Mangoes	1.0				**1.0**					**0**	**1.0**	**1.5**
Spinach	1.8		0.1		**1.9**					**0**	**1.9**	**2.7**
Broccoli	1.6		0.5		**2.1**	0.1				**0.1**	**2.2**	**2.5**
Green beans				0.1	**0.1**					**0**	**0.1**	**0**
Green peas	0.3		6.4	0.6	**7.3**					**0**	**7.3**	**1.4**
Lettuce, leaf	0.3		0.7		**1.1**					**0**	**1.1**	**0.6**
Cucumbers	0.1				**0.1**	0.1				**0.1**	**0.2**	**0.1**
Tomatoes	0.4				**0.4**					**0**	**0.4**	**0.6**
Corn (maize)	0.1		0.5		**0.6**	0.4		1.0		**1.4**	**2.0**	**0.3**

[a]Data taken from Papas (1999). Blank entries indicate that the compounds were not detectable. [b]Note that in many samples the total concentration of international units (IU) of α-tocopherol equivalents is less than half of the total of all tocopherols and tocotrienols.

Table 4. The vitamin E content of animal and fish products[a]

	Tocopherols (mg/100 g)					Tocotrienols (mg/100 g)					Grand total	IU per
Food	**α**	**β**	**γ**	**δ**	**Total**	**α**	**β**	**γ**	**δ**	**Total**	**(mg/100 g)**	**100 g[b]**
Whole milk	**0.1**				0.1					0	0.1	.1
Eggs	1.9	0.1	0.6	0.4	**3.0**	0.3				**0.3**	**3.3**	**3.1**
Beef	0.3				**0.3**					**0**	**0.3**	**0.5**
Chicken	0.4				**0.4**					**0**	**0.4**	**0.6**
Pork	0.2				**0.2**					**0**	**0.2**	**0.3**
Cod, fresh	0.6				**0.6**					**0**	**0.6**	**1.0**
Shrimp, fresh	0.8				**0.8**					**0**	**0.8**	**1.2**
Tuna, canned	0.5		0.1		**0.6**	0.1				**0.1**	**0.7**	**0.8**

[a]Data taken from Papas (1999). Blank entries indicate that the compounds were not detectable. [b]Note that in many samples the total concentration of international units (IU) of α-tocopherol equivalents is less than half of the total of all tocopherols and tocotrienols.

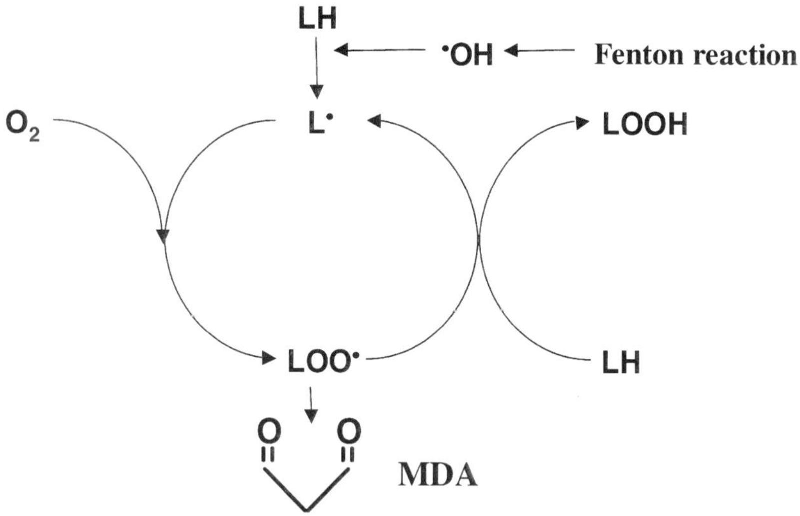

Figure 2. Lipid peroxidation. Lipids containing polyunsaturated fatty acids (LH) are oxidized to form lipid peroxyl radicals (LOO•), which can generate mutagenic aldehydes such as malondialdehyde (MDA). Lipid hydroperoxides (LOOH) accumulate during lipid peroxidation. Vitamin E inhibits lipid peroxidation by quenching LOO• radicals.

in vitro lipid peroxidation due to its very high PUFA content and its relatively low vitamin E content (see Table 4). Rats fed a fish oil diet show evidence of increased *in vivo* lipid peroxidation and decreased levels of plasma and cardiac vitamin E (Jokela *et al.*, 1998). Vitamin E can quench peroxyl radicals, and is thereby effective in preventing the chain-propagating reactions of lipid peroxidation.

Oils that have undergone lipid peroxidation may have adverse physiological effects if consumed. For example, Eder (1999) found that oxidized oils in a diet could diminish the ability of rat heart and liver tissues to desaturate fatty acids. Work by Sheehy *et al.* in 1994 showed that the consumption of thermally oxidized sunflower oil by chicks reduces the tissue content of α-tocopherol and increases the susceptibility of tissues to lipid peroxidation. The addition of vitamin E to refined olive oil increases the stability of the oil under pro-oxidant conditions and decreases the oxidative damage caused by adriamycin in rats (Quiles *et al.*, 1999).

The biological membranes of subcellular organelles are primary sites for lipid peroxidation damage (Sevanian and Ursini, 2000). Lipid hydroperoxides and aldehydic lipid peroxidation can also induce strand breaks in DNA (Yang and Schaich, 1996). 4-HNE accumulates rapidly following spinal cord injury

Table 5. Effects of vitamin E on *in vivo* lipid peroxidation in humans, as measured by the rate of formation of breath pentane[a]

Diet	Number of volunteers	Pentane formation rate (pmol/kg·min)	Plasma vitamin E (μM)
Normal	10	6.3±1.0	17.6±1.6
Parenteral nutrition	5	15.4±1.9	6.2±0.5
Normal + 10 days of 1000 IU vitamin E	5	4.8±0.5	33.9±3.8

[a]During lipid peroxidation, *n*-3 fatty acids decompose to produce ethane, whereas *n*-6 fatty acids produce pentane, which can be detected in exhaled breath. A higher breath pentane formation rate therefore indicates more peroxidation of *n*-6 fatty acids. Breath pentane and plasma vitamin E levels were measured in 10 normal subjects, and in 5 patients receiving parenteral nutrition diets who were deficient in vitamin E. Half of the normal subjects then received 1000 IU of *R,R,R*,-α-tocopheryl acetate for 10 days, and breath pentane and plasma vitamin E levels were re-measured. Data taken from Lemoyne *et al.* (1987). IU, international units of *R,R,R*-α-tocopherol equivalents.

and inhibits glutamine uptake, which may potentiate neurone cell dysfunction and death by excitotoxic mechanisms (Springer *et al.*, 1997). The destruction of rat pancreatic islet β-cells by cytokines also appears to involve the production of cytotoxic aldehydes generated during lipid peroxidation (Suarez Pinzon *et al.*, 1996).

Lipid peroxidation is well documented *in vitro* but is more difficult to verify in intact organisms. The pioneering work of Lemoyne *et al.* (Lemoyne *et al.*, 1988; Van Gossum *et al.*, 1988; Lemoyne *et al.*, 1987) demonstrated, however, that vitamin E is an effective *in vivo* antioxidant. Pentane is formed by the peroxidation of *n*-6 fatty acids and can be collected from human breath and analysed by gas chromatography. As shown in Table 5, Lemoyne *et al.* (1987) showed that the rate of breath pentane formation was inversely correlated with plasma vitamin E levels, and could be suppressed by taking a vitamin E supplement.

Since α-tocopherol has the highest activity in protecting the fetuses of pregnant female rats from death and resorption, it is reasonable to suspect that this superior biological activity is correlated with a superior ability to inhibit lipid peroxidation. Figure 3 shows, however, that *R,R,R*-γ-tocopherol and *R,R,R*-α-tocopherol have almost identical abilities to inhibit lipid peroxidation in large unilamellar liposomes. However, *R,R,R*-γ-tocopherol has about one-tenth the biological activity of *R,R,R*-α-tocopherol. Similarly, *R,R,R*-α-tocopherol and all-*rac*-α-tocopherol have very similar antioxidant activities, although all-*rac*-α-tocopherol has about half the biological potency of *R,R,R*-α-tocopherol. Work by Suarna *et al.* (1993) has shown that dietary tocotrienols are incorporated into circulating human lipoproteins and that they react with peroxyl radicals as efficiently as the corresponding tocopherol isomers. As

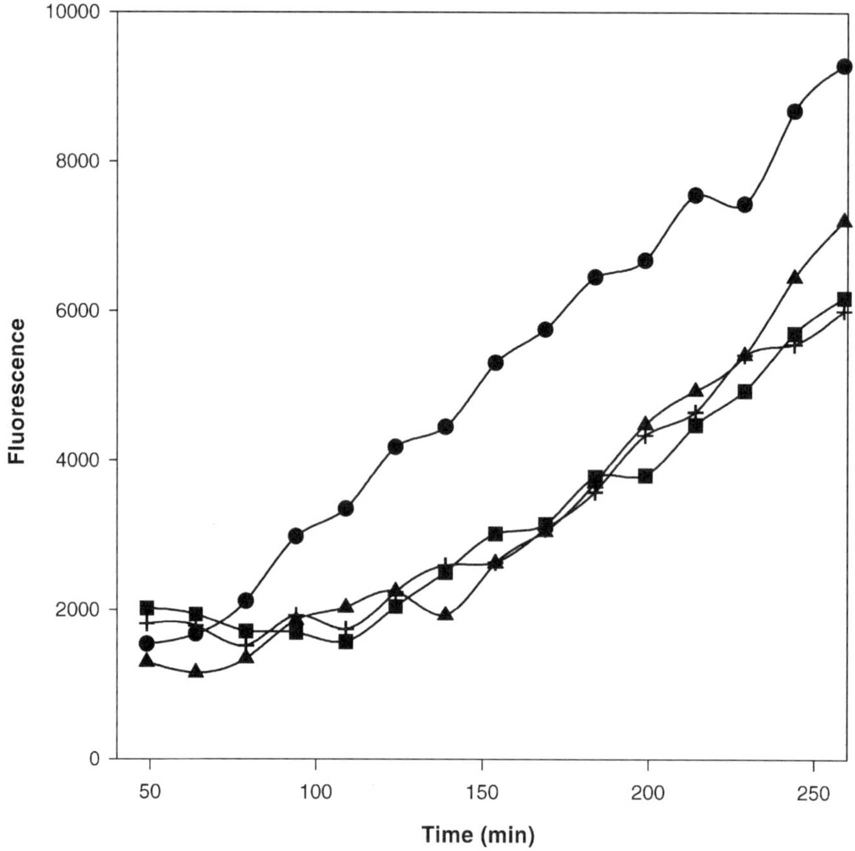

Figure 3. Both α-tocopherol and γ-tocopherol have similar abilities to quench *in vitro* lipid peroxidation. The addition of 16 μM *R,R,R*-α-tocopherol (■) or 16 μM *R,R,R*-γ-tocopherol (+) separately, or a mixture of both 8 μM *R,R,R*-α-tocopherol and 8 μM *R,R,R*-γ-tocopherol (▲), equally inhibited the formation of fluorescent lipid peroxidation products (assayed as described by Shimasaki, 1994) compared with levels seen in control samples (●) of large unilamellar liposomes oxidized in the presence of 7.4 mM 2,2′-azobis(2-amidinopropane) dihydrochloride.

discussed below, the varying biological activities of different forms of vitamin E may be primarily a function of biodiscrimination rather than major differences in their intrinsic antioxidant abilities.

D. Vitamin E compounds differ in their reactivities towards reactive nitrogen species

Although the reactivities of α- and γ-tocopherol towards reactive oxygen species are similar, this is not the case for reactive nitrogen oxide species. As

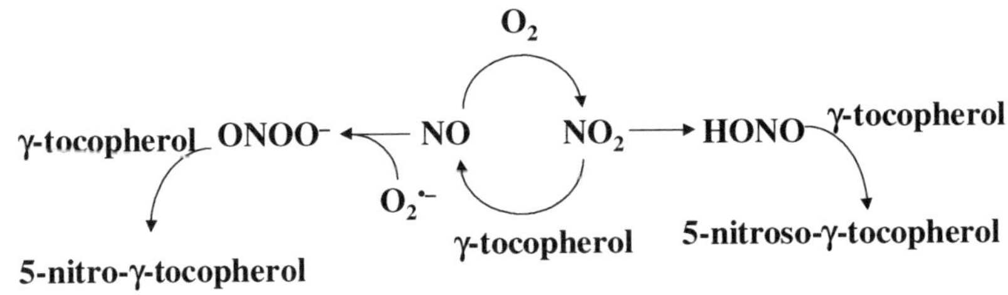

Figure 4. γ-Tocopherol plays a unique role in preventing damage caused by reactive nitrogen species; see text for details.

shown in Figure 4, nitric oxide (NO) rapidly reacts with the superoxide radical ($O_2^{\bullet -}$) to yield peroxynitrite ($ONOO^-$), a potent mutagenic agent generated during chronic inflammation. Christen *et al.* (1997) found that γ-tocopherol, by virtue of not having a methyl group at R1 (see Figure 1), could inhibit lipid peroxidation induced by peroxynitrite more effectively than α-tocopherol (which does have a methyl group at R1). In addition, the nitration of γ-tocopherol at the nucleophilic 5-position (i.e. at R1) was not affected by the presence of α-tocopherol (Christen *et al.*, 1997). As shown in Figure 4, nitric oxide can also react with oxygen to yield nitrogen dioxide (NO_2). Cooney *et al.* (1993) found that γ-tocopherol can react with NO_2 to yield nitric oxide, but that α-tocopherol is much less effective at this reduction. In addition, γ-tocopherol can detoxify nitrous acid (HONO) formed from NO_2 to yield 5-nitroso-γ-tocopherol (Figure 4). Cooney *et al.* (1993) also found that γ-tocopherol is a more potent inhibitor of neoplastic transformation during the post-initiation phase in 3-methylcholanthrene-treated C3H/10T1/2 murine fibroblasts than α-tocopherol.

E. Vitamin E in animal fat and plant fat, and its health-related significance

In general, fats derived from animal sources (e.g. butter and lard; see Table 1) contain much less vitamin E than fats from plant sources (e.g. soybean oil and corn oil). Coconut oil is an exception among the plant oils in that it contains very little vitamin E. In addition, animal fats (see Tables 1 and 4) contain primarily α-tocopherol, whereas plant fats (see Tables 1–3) often contain high levels of non-α-tocopherols such as γ- or δ-tocopherol. Butter, for example, contains about 2 mg α-tocopherol per 100 g and only trace amounts of other tocopherols (Tomasch *et al.*, 2001). In contrast, corn oil contains 11 mg α-tocopherol and 60 mg γ-tocopherol per 100 g. Some plant oils contain significant quantities of tocotrienols. Cranberry seed oil is the richest known source of tocotrienols, containing 173 mg/100 g (Nawar, 2001). These differences in the levels and types of vitamin E in animal fats versus plant fats may have profound health-related significance. For example, epidemiological studies suggest that the consumption of animal fats, but not vegetable fats, is associated with an increased risk of colon cancer (Caygill *et al.*, 1996) and prostate cancer (Giovannucci *et al.*, 1993).

It is also interesting that γ-tocopherol was recently found to be more effective than α-tocopherol in inhibiting the *in vitro* growth of a human prostate cancer cell line (Moyad *et al.*, 1999). Helzlsouer *et al.* (2000) recently looked at the association between α-tocopherol, γ-tocopherol and selenium and the subsequent development of prostate cancer. Men whose plasma γ-tocopherol concentrations were in the top quintile had a remarkable 5 fold lower risk of developing colon cancer than men in the lowest quintile (Helzlsouer *et al.*, 2000). Jiang *et al.* (2001) recently reviewed the data suggesting that γ-tocopherol could play a significant role in reducing the risks of cancer and cardiovascular disease, and concluded: "Gamma-tocopherol, the major form of vitamin E in the US diet, deserves more attention."

In an animal model, it was recently found that γ-tocopherol, but not α-tocopherol, could suppress the expression of *ras*-p21 in colon epithelial cells (Stone *et al.*, 2002b). *Ras*-p21 is a small 21 kDa G-protein with GTPase activity that plays a key role in regulating cell growth and is known to be an early biomarker for the development of colon cancer. The *ras* gene is often overexpressed in colon cancer tissue. The superoxide radical ($O_2^{\bullet -}$) functions as a signal-transduction messenger mediating the downstream effects of *ras* (Irani *et al.*, 1997; Irani and Goldschmidt-Clermont, 1998). Fibroblasts transformed with constitutively active forms of *ras*-p21 produce large amounts of $O_2^{\bullet -}$ (Irani *et al.*, 1997). It is likely that factors influencing the expression of wild-type *ras* could be extremely important in understanding the etiology of cancer.

There is also clinical evidence that γ-tocopherol plays a unique role in preventing coronary heart disease (CHD). Ohrvall *et al.* (1996) and Kontush *et al.* (1999) found that patients with CHD have lower plasma levels of γ-tocopherol (but not α-tocopherol) than healthy controls. In a follow-up on this work, it was found that the levels of α- and γ-tocopherol in serum are influenced by the type of fat in the diet (Ohrvall *et al.*, 2001). In particular, the lipid-adjusted serum γ-tocopherol concentration decreased by 37% when subjects were fed a diet rich in saturated fat (butter), to levels similar to those found in CHD patients. It has become increasingly clear that it is important to understand the dietary factors that can potentially influence serum γ-tocopherol levels. Nevertheless, this area is only just beginning to receive the appropriate research focus and attention.

Tocotrienols also appear to have properties that are distinct from those of α-tocopherol. In a very interesting paper, Ikeda *et al.* (2000) examined the tissue distribution of tocotrienols and tocopherols after feeding nude mice (BALB/c Slc-nu) and hairless mice (SKH1) a diet containing a tocotrienol-rich vitamin E extract from palm oil. α-Tocopherol was found in the skin, liver, kidney and plasma. In most tissues, α-tocotrienol and γ-tocotrienol were present at only very low levels; however, high levels of α-tocotrienol and γ-tocotrienol were detected in the skin, suggesting that this tissue has a unique ability to discriminate between various vitamin E analogs.

Although the results of various studies have not been uniform, tocotrienols have long been considered to be potentially important in the treatment of hypercholesterolemia and cardiovascular disease. Tomeo *et al.* (1995) investigated the ability of 'Palm Vitee', a fraction of palm oil enriched in γ-tocotrienol and α-tocopherol, to reduce carotid atherosclerosis and lower serum lipid levels. They found evidence that Palm Vitee reduced carotid atherosclerosis, but no evidence that it lowered serum total cholesterol, low-density lipoprotein (LDL) cholesterol, or triacylglycerol levels. In contrast, Qureshi *et al.* (1995) found that Palm Vitee (containing 40 mg α-tocopherol and 220 mg tocotrienols) or γ-tocotrienol lowered plasma cholesterol levels in subjects who had been acclimated to an American Heart Association (AHA) "Step 1" diet. Interestingly, α-tocopherol was found to attenuate the cholesterol-lowering effects of tocotrienols (Qureshi *et al.*, 1995).

More recently, Mensink *et al.* (1999) found that consuming a tocol blend (yielding 80 mg α-tocopherols and 140 mg tocotrienols per day) did not lower serum total cholesterol or LDL cholesterol levels in mildly hypercholesterolemic subjects. In this study the ratio of tocotrienols to α-tocopherol in the supplement was 1.7, whereas in the Qureshi *et al.* (1995) study it was 5.5. It seems likely, therefore, that a tocol supplement must have a high ratio of tocotrienols to α-tocopherol in order to obtain a cholesterol-lowering effect. Alternatively, a low-fat diet (such as the AHA "Step 1" diet) may act synergistically with tocotrienols to lower serum cholesterol levels.

Rice bran tocotrienols have not been as well studied as Palm Vitee, but recent work with hereditary hypercholesterolemic swine has shown promising results (Qureshi *et al.*, 2001). Administration of rice bran tocotrienols for 6 weeks markedly lowered serum levels of total cholesterol (by 32–38%), LDL cholesterol (by 35–43%), and apolipoprotein B (by 20–28%) in the treated swine. Watkins *et al.* (1999) showed that adults (aged 49–83) taking rice bran oil supplements for one year had 14% lower plasma total cholesterol levels, 20% lower LDL cholesterol levels, and higher high-density lipoprotein (HDL) C levels than individuals taking a placebo. No data are currently available for adolescents and young adults.

F. Dietary factors influencing α- and γ-tocopherol levels

In an animal model, increasing the level of dietary *R,R,R*-γ-tocopherol in a diet containing a constant level of *R,R,R*-α-tocopherol was found to increase the levels of α-tocopherol in serum and most tissues (Clement and Bourre, 1997). This remarkable result has yet to be reproduced in humans, but certainly suggests that the γ-tocopherol content of foods may play an important role in enhancing the bioavailability of α-tocopherol. The potential health-related significance of dietary γ-tocopherol is being increasingly recognized. It is relevant, therefore, to note that vitamin E supplements containing only α-tocopherol have been found to decrease plasma γ-tocopherol levels in humans. For example, work from our laboratory has shown that subjects taking an *R*,*R*,*R*-α-tocopherol supplement show an increased level of plasma α-tocopherol but a decreased level of plasma γ-tocopherol (see Figure 5). Subjects taking a placebo show no significant changes, but switching the placebo-assigned subjects to the *R*,*R*,*R*-α-tocopherol supplement results in the expected decrease in plasma γ-tocopherol levels. Switching the *R*,*R*,*R*-α-tocopherol group to the placebo also results in the expected increase in plasma γ-tocopherol levels. Similar results have been reported in the literature (Handelman *et al.*, 1985).

Cooney *et al.* (2001) found that γ-tocopherol from sesame seeds, but not from walnuts or soy oil, significantly increased serum levels of γ-tocopherol. This effect may be due to the presence of sesamolin, a lignan present in sesame seeds. Sesamolin was also present in the serum of individuals who consumed the sesame seeds, but not of those who consumed the walnuts or soy oil (Cooney *et al.*, 2001). Lemcke-Norojarvi *et al.* (2001) have studied how linola, corn (maize) and sesame oils, which are all good sources of γ-tocopherol, influence the relative and absolute serum concentrations of α- and γ-tocopherol. Both corn oil and sesame oil could increase the serum levels of γ-tocopherol, but serum α-tocopherol levels did not change. It is not yet known if diets that increase serum γ-tocopherol levels will decrease the incidences of either CHD or cancer. Clearly this is an area worthy of continued investigation.

G. Cellular and biochemical differences in vitamin E compounds

Recent work by Jiang *et al.* (2000) has shown that γ-tocopherol is far superior to α-tocopherol in suppressing the activity of cyclooxygenase-2 activity in cultured RAW264.7 macrophages. This is extremely significant because cyclooxygenase-2 plays a key role in inflammatory processes and is associated with diseases such as cancer and cardiovascular disease (Katori and Majima, 2000). Using the same cell line, Gao and Stone (unpublished work) found that γ-tocopherol was taken up to a much greater extent than α-tocopherol. Moreover, the presence of γ-tocopherol in the culture media dramatically increased the degree to which α-tocopherol was taken up by the cells.

Tocotrienols, which are found at particularly high levels in rice bran oil, cranberry oil, palm oil and barley oil, also have unique biological properties distinct from those of tocopherols. Goh *et al.* (1994) used an *in vitro* assay of the activation of Epstein–Barr virus early antigen expression in human lymphoblastoid cells to study the ability of tocotrienols and tocopherols to prevent tumor promotion. Their results showed that γ- and δ-tocotrienols, but

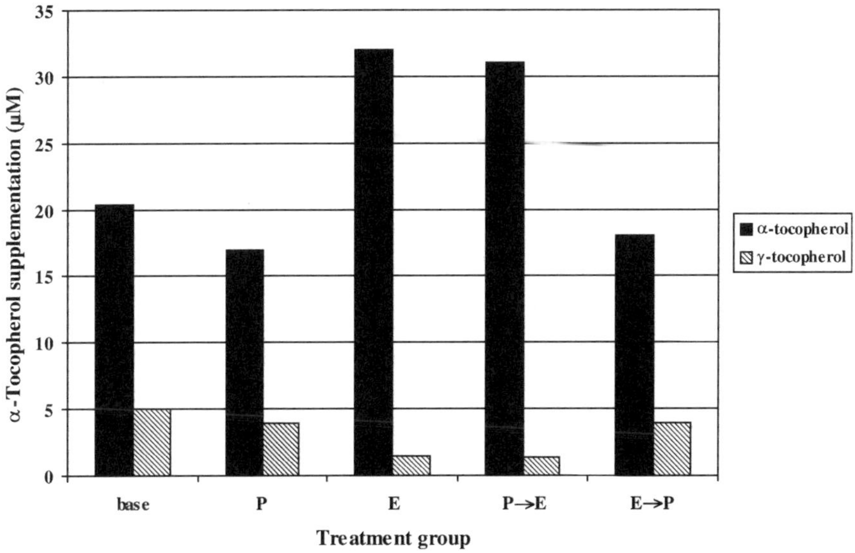

Figure 5. Dietary supplementation with α-tocopherol decreases plasma levels of γ-tocopherol. Baseline plasma levels of α-tocopherol and γ-tocopherol were measured in 14 volunteers ("base"). Half the subjects then consumed either placebo ("P") or 400 IU of *R,R,R*-α-tocopherol ("E") per day for 6 weeks and plasma tocopherol levels were again measured. The placebo group was then switched to the vitamin E group (P→E) and the E groups switched to the P group (E→P) for an additional 6 weeks and plasma tocopherols measured again.

not α- or γ-tocopherol, strongly prevent tumor promotion. The antiproliferative effects of tocotrienols (from palm oil) and α-tocopherol were recently compared using both estrogen-responsive MCF7 human breast cancer cells and estrogen-unresponsive MDA-MB-231 human breast cancer cells (Nesaretnam *et al.*, 1998). The tocotrienols inhibited growth of the estrogen-responsive cells in both the presence and absence of estradiol; complete growth suppression was achieved with a tocotrienol concentration of 8 μg/ml. Growth of the estrogen-unresponsive cells was inhibited with 20 μg/ml tocotrienol. In marked contrast, α-tocopherol had no inhibitory effect on the growth of either cell line. These results suggest that tocotrienols can directly inhibit the growth of breast cancer cells, and do not act by an estrogen-receptor-mediated pathway.

H. Vitamin E biodiscrimination

There have been considerable advances in our understanding of vitamin E biochemistry within the past ten years (Stone and Papas, 1997; Stocker and Azzi, 2000; Brigelius Flohe and Traber, 1999). The high biological activity of α-tocopherol appears to be the result of biodiscrimination occurring in the liver. The different chemical and stereochemical forms of vitamin E appear to be absorbed from foods and secreted into chylomicrons equally well (Ferslew *et al.*, 1993; Traber *et al.*, 1990a; Traber *et al.*, 1990b; Traber and Arai, 1999). In contrast, very-low-density lipoproteins (VLDL) secreted by the liver are preferentially enriched with α-tocopherol relative to other forms of vitamin E (Traber *et al.*, 1990a). It is now clear that this biodiscrimination step is, in part, the result of an α-tocopherol transfer protein (α-TTP) in the liver that preferentially binds α-tocopherol and transfers it to VLDL. Work by Hosomi *et al.* (1997) has demonstrated that the affinity of various vitamin E homologues for α-TTP is a major determinant of their biological activities.

The hepatic 31 kDa α-TTP is a cytosolic protein that has been purified and found to exist as two charge isoforms (Kuhlenkamp *et al.*, 1993). The primary structures of α-TTP from rat liver (Sato *et al.*, 1993) and human liver (Arita *et al.*, 1995) have been determined and found to have a structural homology with retinaldehyde-binding protein (Sato *et al.*, 1993). It is now known that a defect in α-TTP is the cause of ataxia (a nervous disorder characterized by the inability to coordinate muscular movement) associated with familial isolated vitamin E deficiency (Traber *et al.*, 1990c; Schuelke *et al.*, 1999). Recent work also indicates that α-TTP is very important for the normal development of placental labyrinthine trophoblasts: the placenta of female mice deficient in α-TTP was found to be extremely impaired and the embryos died at mid-gestation (Jishage *et al.*, 2001).

I. Vitamin E metabolism

Since γ-tocopherol is not transferred from the liver to VLDL as efficiently as α-tocopherol, it is reasonable to suggest that the "excess" hepatic γ-tocopherol might be secreted via the bile back into the intestine and excreted (Traber and Kayden, 1989). It was recently found, however, that rats fed a diet containing 0.156 mmol γ-tocopherol per kg diet had lower fecal tocopherol levels than rats fed a diet containing an identical level of α-tocopherol (Stone *et al.*, 2002b). In fact, rats fed the γ-tocopherol diet had lower tocopherol levels in all tissues than rats fed the α-tocopherol diet (Stone *et al.*, 2002b). These data strongly suggest that γ-tocopherol is catabolized more rapidly than α-tocopherol. Swanson *et al.* (1999) found that a major route of γ-tocopherol catabolism is conversion by phytyl tail oxidation to 2,7,8-trimethyl-2-(β-carboxyethyl)-6-hydroxychroman (also called LLU-α or γ-CEHC) followed by glucuronide conjugation and urinary excretion. γ-CEHC is involved with sodium excretion and could, therefore, be important in controlling the body's pool of extracellular fluid – which is of clinical significance in diseases such as hypertension, congestive heart failure, and cirrhosis (Wechter *et al.*, 1996).

The mechanism for the catabolism of dietary tocopherols into their carboxychroman metabolites (see Figure 6) most likely involves cytochrome

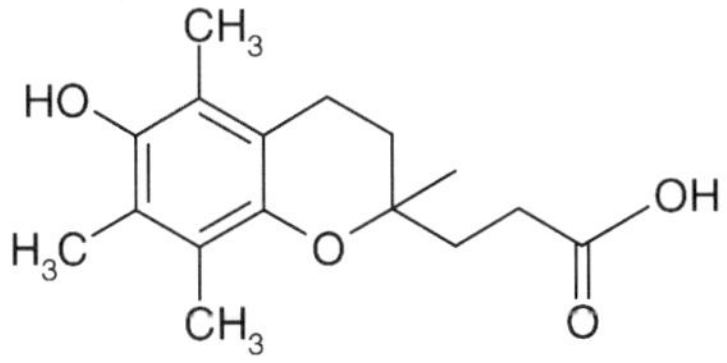

α-CEHC

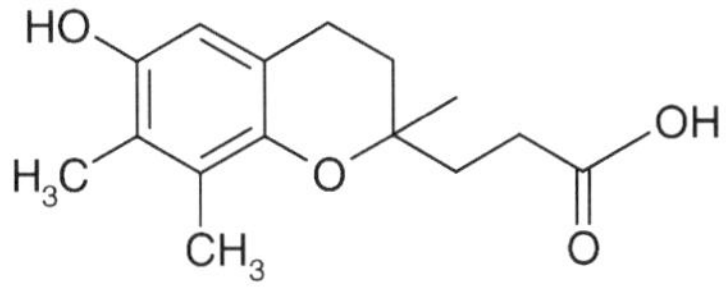

γ-CEHC

Figure 6. Newly identified carboxychroman metabolites of tocopherols. α-CEHC [2,5,7,8-tetramethyl-2-(β-carboxyethyl)-6-hydroxychroman] and γ-CEHC [2,7,8-trimethyl-2-(β-carboxyethyl)-6-hydroxychroman] are water-soluble metabolites of α-tocopherol and γ-tocopherol, respectively.

P-450 (CYP) enzymes, specifically CYP3A. Parker *et al.* (2000) found that ketoconozole, a selective inhibitor of CYP3A, inhibited the metabolism of γ- and α-tocopherol in rat primary hepatocytes, whereas sulfaphenazole and cyclosporin (inhibitors of CYP2C and CYP27, respectively) did not. It was also significant that sesamin [see discussion of sesamolin in Section F, above] strongly inhibited tocopherol catabolism, which may account for why this lignan increases tissue tocopherol concentrations (Parker *et al.*, 2000).

J. Nutritional genomics and tocopherols

It is clear from the above review that forms of vitamin E in addition to α-tocopherol are likely to be important in optimizing human health. All of these additional tocopherols and tocotrienols are found in plant oils. The problem remains, however, that the levels of these additional forms of vitamin E in oils may be too low to be useful (particularly for therapeutic purposes) without further chemical extraction and concentration by methods such as fractional distillation. It would be extremely cost effective if functional plant oils could be obtained with much higher vitamin E levels than are now available. Moreover, the chemical distribution of tocopherols in many oils is biased towards one particular chemical form. For example, almond oil contains primarily α-tocopherol, while walnut oil contains primarily γ-tocopherol. Shintani and DellaPenna (1998) have pioneered a promising approach to these problems which they have termed "nutritional genomics". This approach is based on identifying a particular metabolic pathway (e.g. α-tocopherol synthesis) and the gene targets important in both the synthesis and regulation of the species involved in that pathway. For example, three of the enzymes involved in the synthesis of α-tocopherol have been cloned: *p*-hydroxyphenol pyruvic acid dioxygenase; a phytyl transferase; and γ-tocopherol methyl transferase, which places a methyl group on γ-tocopherol thereby converting it to α-tocopherol (Shigeoka *et al.*, 1992). Shintani and DellaPenna (1998) have used the nutritional genomics approach to clone γ-tocopherol methyl transferase and overexpress it in *Arabidopsis* seeds, which shifted the oil composition of this seed in favor of α-tocopherol. Clearly this approach has considerable promise. It is conceivable that future applications of the nutritional genomics approach could result in plant oils with dramatically increased levels of tocopherols and/or tocotrienols and with a chemical distribution of tocopherols and tocotrienols optimized for human health.

Note added by the editor – the following figures were valid in 1998 (Netscher, 1999):

- The annual worldwide production of synthetic vitamin E, a mixture of the eight stereoisomers of α-tocopherol, is approximately 15 000–20 000 tonnes. It is used in feed (71%), pharmaceuticals (24%), cosmetics (3%), and foods (2%). The main producers are BASF (Germany), Rhone–Poulenc (France), Eisai (Japan), and Roche (Switzerland).
- Natural vitamin E is produced at a level of about 2000 tonnes/year from soybean deodorizer distillate. This is methylated to increase the proportion of α-tocopherol and is used mainly in human applications (largely pharmaceutical). It could be considered to be semi-synthetic α-tocopherol. The major producers are Henkel (USA), Archer Daniels Midland (USA), Eisai (Japan), and Hoffmann–La Roche/Cargill (USA).

References

Arita, M, Sato, Y, Miyata, A, Tanabe, T, Takahashi, E, Kayden, HJ, Arai, H and Inoue, K (1995) Human alpha-tocopherol transfer protein: cDNA cloning, expression and chromosomal localization. *Biochemical J.*, 306(2), 437–443.

Azzi, A and Stocker, A (2000) Vitamin E: non-antioxidant roles. *Prog. Lipid Res.*, 39, 231–255.

Bieri, JG and Evarts, RP (1974) Gamma tocopherol: metabolism, biological activity and significance in human vitamin E nutrition. *Am. J. Clin. Nutr.*, 27(9), 980–986.

Brigelius Flohe, R and Traber, MG (1999) Vitamin E: function and metabolism. *FASEB J.*, 13, 1145–1155.

Burton, GW, Traber, MG, Acuff, RV, Walters, DN, Kayden, H, Hughes, L and Ingold, KU (1998) Human plasma and tissue alpha-tocopherol concentrations in response to supplementation with deuterated natural and synthetic vitamin E. *Am. J. Clin. Nutr.*, 67, 669–684.

Caygill, CP, Charlett, A and Hill, MJ (1996) Fat, fish, fish oil and cancer. *Br. J. Cancer*, 74, 159–164.

Christen, S, Woodall, AA, Shigenaga, MK, Southwell Keely, PT, Duncan, MW and Ames, BN (1997) Gamma-tocopherol traps mutagenic electrophiles such as NO(X) and complements alpha-tocopherol: physiological implications. *Proc. Natl Acad. Sci. USA*, 94, 3217–3222.

Clement, M and Bourre, JM (1997) Graded dietary levels of RRR-gamma-tocopherol induce a marked increase in the concentrations of alpha- and gamma-tocopherol in nervous tissues, heart, liver and muscle of vitamin-E-deficient rats. *Biochim. Biophys. Acta*, 1334, 173–181.

Cooney, RV, Franke, AA, Harwood, PJ, Hatch Pigott, V, Custer, LJ and Mordan, LJ (1993) Gamma-tocopherol detoxification of nitrogen dioxide: superiority to alpha-tocopherol. *Proc. Natl Acad. Sci. USA*, 90, 1771–1775.

Cooney, RV, Custer, LJ, Okinaka, L and Franke, AA (2001) Effects of dietary sesame seeds on plasma tocopherol levels. *Nutr. Cancer*, 39, 66–71.

Eder, K (1999) The effects of a dietary oxidized oil on lipid metabolism in rats. *Lipids,* 34, 717–725.

Ferslew, KE, Acuff, RV, Daigneault, EA, Woolley, TW and Stanton, PE (1993) Pharmacokinetics and bioavailability of the RRR and all racemic stereoisomers of alpha-tocopherol in humans after single oral administration. *J. Clin. Pharmacol.*, 33, 84–88.

Giovannucci, E, Rimm, EB, Colditz, GA, Stampfer, MJ, Ascherio, A, Chute, CC and Willett, WC (1993) A prospective study of dietary fat and risk of prostate cancer. *J. Natl Cancer Inst.*, 85, 1571–1579.

Goh, SH, Hew, NF, Norhanom, AW and Yadav, M (1994) Inhibition of tumour promotion by various palm-oil tocotrienols. *Int. J. Cancer*, 57, 529–531.

Handelman, GJ, Machlin, LJ, Fitch, K, Weiter, JJ and Dratz, EA (1985) Oral alpha-tocopherol supplements decrease plasma gamma-tocopherol levels in humans. *J. Nutr.*, 115, 807–813.

Helzlsouer, KJ, Huang, HY, Alberg, AJ, Hoffman, S, Burke, A, Norkus, EP, Morris, JS and Comstock, GW (2000) Association between alpha-tocopherol, gamma-tocopherol, selenium, and subsequent prostate cancer. *J. Natl Cancer Inst.*, 92, 2018–2023.

Hosomi, A, Arita, M, Sato, Y, Kiyose, C, Ueda, T, Igarashi, O, Arai, H and Inoue, K (1997) Affinity for alpha-tocopherol transfer protein as a determinant of the biological activities of vitamin E analogs. *FEBS Lett.*, 409, 105–108.

Ikeda, S, Niwa, T and Yamashita, K (2000) Selective uptake of dietary tocotrienols into rat skin. *J. Nutr. Sci. Vitaminol. (Tokyo)*, 46, 141–143.

Irani, K and Goldschmidt-Clermont, PJ (1998) Ras, superoxide and signal transduction. *Biochem. Pharmacol.* 55, 1339–1346.

Irani, K, Xia, Y, Zweier, JL, Sollott, SJ, Der, CJ, Fearon, ER, Sundaresan, M, Finkel, T and Goldschmidt-Clermont, PJ (1997) Mitogenic signaling mediated by oxidants in Ras-transformed fibroblasts. *Science,* 275, 1649–1652.

Jiang, Q, Elson Schwab, I, Courtemanche, C and Ames, BN (2000) Gamma-tocopherol and its major metabolite, in contrast to alpha-tocopherol, inhibit cyclooxygenase activity in macrophages and epithelial cells. *Proc. Natl Acad. Sci. USA*, 97, 11494–11499.

Jiang, Q, Christen, S, Shigenaga, MK and Ames, BN (2001) Gamma-tocopherol, the major form of vitamin E in the US diet, deserves more attention. *Am. J. Clin. Nutr.*, 74, 714–722.

Jishage, K, Arita, M, Igarashi, K, Iwata, T, Watanabe, M, Ogawa, M, Ueda, O, Kamada, N, Inoue, K, Arai, H and Suzuki, H (2001) Alpha-tocopherol transfer protein is important for the normal development of placental labyrinthine trophoblasts in mice. *J. Biol. Chem.*, 276, 1669–1672.

Jokela, R, Engstrom, K, Wallin, R and Saldeen, T (1998) Effect of in vitro stability of dietary fish oil on lipid peroxidation and prostanoids in vivo. *Uppsala J. Med. Sci.*, 103, 213–221.

Katori, M and Majima, M (2000) Cyclooxygenase-2: its rich diversity of roles and possible application of its selective inhibitors. *Inflamm. Res.*, 49, 367–392.

Kontush, A, Spranger, T, Reich, A, Baum, K and Beisiegel, U (1999) Lipophilic antioxidants in blood plasma as markers of atherosclerosis: the role of alpha-carotene and gamma-tocopherol. *Atherosclerosis,* 144, 117–122.

Kuhlenkamp, J, Ronk, M, Yusin, M, Stolz, A and Kaplowitz, N (1993) Identification and purification of a human liver cytosolic tocopherol binding protein. *Prot. Express. Purific.*, 4, 382–389.

Lemcke Norojarvi, M, Kamal Eldin, A, Appelqvist, LA, Dimberg, LH, Ohrvall, M and Vessby, B (2001) Corn and sesame oils increase serum gamma-tocopherol concentrations in healthy Swedish women. *J. Nutr.*, 131, 1195–1201.

Lemoyne, M, Van Gossum, A, Kurian, R, Ostro, M, Axler, J and Jeejeebhoy, KN (1987) Breath pentane analysis as an index of lipid peroxidation: a functional test of vitamin E status. *Am. J. Clin. Nutr.,* 46, 267–272.

Lemoyne, M, Van Gossum, A, Kurian, R and Jeejeebhoy, KN (1988) Plasma vitamin E and selenium and breath pentane in home parenteral nutrition patients. *Am. J. Clin. Nutr.,* 48, 1310–1315.

Mensink, RP, van Houwelingen, AC, Kromhout, D and Hornstra, G (1999) A vitamin E concentrate rich in tocotrienols had no effect on serum lipids, lipoproteins, or platelet function in men with mildly elevated serum lipid concentrations. *Am. J. Clin. Nutr.,* 69, 213–219.

Moyad, MA, Brumfield, SK and Pienta, KJ (1999) Vitamin E, alpha- and gamma-tocopherol, and prostate cancer. *Semin. Urol. Oncol.*, 17, 85–90.

Nawar, WW (2001) Tocotrienols and omega-3 fatty acids in cranberry seed oil. *FASEB J.*, 15, A985.

Nesaretnam, K, Stephen, R, Dils, R and Darbre, P (1998) Tocotrienols inhibit the growth of human breast cancer cells irrespective of estrogen receptor status. *Lipids,* 33, 461–469.

Netscher, T (1999) Synthesis and production of vitamin E. In: *Lipid Synthesis and Manufacture* (FD Gunstone, ed.), Sheffield Academic Press, Sheffield, pp.251–267.

Ohrvall, M, Sundlof, G and Vessby, B (1996) Gamma, but not alpha, tocopherol levels in serum are reduced in coronary heart disease patients. *J. Int. Med.,* 239, 111–117.

Ohrvall, M, Gustafsson, IB and Vessby, B (2001) The alpha and gamma tocopherol levels in serum are influenced by the dietary fat quality. *J. Hum. Nutr. Diet.*, 14, 63–68.

Papas, AM (1999) *The Vitamin E Factor*, HarperCollins, New York.

Parker, RS, Sontag, TJ and Swanson, JE (2000) Cytochrome P4503A-dependent metabolism of tocopherols and inhibition by sesamin. *Biochem. Biophys. Res. Commun.*, 277, 531–534.

Quiles, JL, Ramirez Tortosa, MC, Ibanez, S, Alfonso Gonzalez, J, Duthie, GG, Huertas, JR and Mataix, J (1999) Vitamin E supplementation increases the stability and the in vivo antioxidant capacity of refined olive oil. *Free Rad. Res.*, 31(Suppl.), S129–S135.

Qureshi, AA, Bradlow, BA, Brace, L, Manganello, J, Peterson, DM, Pearce, BC, Wright, JJ, Gapor, A and Elson, CE (1995) Response of hypercholesterolemic subjects to administration of tocotrienols. *Lipids,* 30, 1171–1177.

Qureshi, AA, Peterson, DM, Hasler Rapacz, JO and Rapacz, J (2001) Novel tocotrienols of rice bran suppress cholesterogenesis in hereditary hypercholesterolemic swine. *J. Nutr.*, 131, 223–230.

Sato, Y, Arai, H, Miyata, A, Tokita, S, Yamamoto, K, Tanabe, T and Inoue, K (1993) Primary structure of alpha-tocopherol transfer protein from rat liver. Homology with cellular retinaldehyde-binding protein. *J. Biol. Chem.,* 268, 17705–17710.

Schuelke, M, Mayatepek, E, Inter, M, Becker, M, Pfeiffer, E, Speer, A, Hubner, C and Finckh, B (1999) Treatment of ataxia in isolated vitamin E deficiency caused by alpha-tocopherol transfer protein deficiency. *J. Pediatr.*, 134, 240–244.

Sevanian, A and Ursini, F (2000) Lipid peroxidation in membranes and low-density lipoproteins: similarities and differences. *Free Rad. Biol. Med.*, 29, 306–311.

Sheehy, P J, Morrissey, P A and Flynn, A (1994) Consumption of thermally-oxidized sunflower oil by chicks reduces alpha-tocopherol status and increases susceptibility of tissues to lipid oxidation. *Br. J. Nutr.*, 71, 53–65.

Shigeoka, S, Ishiko, H, Nakano, Y and Mitsunaga, T (1992) Isolation and properties of gamma-tocopherol methyltransferase in Euglena gracilis. *Biochim. Biophys. Acta*, 1128, 220–226.

Shimasaki, H (1994) Assay of fluorescent lipid peroxidation products. *Meth. Enzymol.*, 233, 338–346.

Shintani, D and DellaPenna, D (1998) Elevating the vitamin E content of plants through metabolic engineering. *Science,* 282, 2098–2100.

Springer, JE, Azbill, RD, Mark, RJ, Begley, JG, Waeg, G and Mattson, MP (1997) 4-Hydroxynonenal, a lipid peroxidation product, rapidly accumulates following traumatic spinal cord injury and inhibits glutamate uptake. *J. Neurochem.*, 68, 2469–2476.

Stocker, A and Azzi, A (2000) Tocopherol-binding proteins: their function and physiological significance. *Antiox. Redox. Signal.*, 2, 397–404.

Stone, W (1998) Oxidative stress and antioxidants in premature infants. In: *Antioxidant Status, Diet, Nutrition and Health* (A Papas, ed.), CRC Press, Boca Raton, pp.277–298.

Stone, WL and Papas, AM (1997) Tocopherols and the etiology of colon cancer. *J. Natl Cancer Inst.*, 89, 1006–1014.

Stone, W, LeClair, I, Ponder, T, Gubbi, A, Thomas, S and Barrett Reis, B (2002a) Infants metabolically discriminate between natural and synthetic vitamin E. *Am. J. Clin. Nutr.* (submitted).

Stone, WL, Papas, AM, LeClair, IO, Qui, M and Ponder, T (2002b) The influence of dietary iron and tocopherols on oxidative stress and *ras*-p21 levels in the colon. *Cancer Detect. Prevent.*, 26, 78–84.

Suarez Pinzon, WL, Strynadka, K and Rabinovitch, A (1996) Destruction of rat pancreatic islet beta-cells by cytokines involves the production of cytotoxic aldehydes. *Endocrinology,* 137, 5290–5296.

Suarna, C, Hood, RL, Dean, RT and Stocker, R (1993) Comparative antioxidant activity of tocotrienols and other natural lipid-soluble antioxidants in a homogeneous system, and in rat and human lipoproteins. *Biochim. Biophy. Acta*, 1166, 163–170.

Swanson, JE, Ben, RN, Burton, GW and Parker, RS (1999) Urinary excretion of 2,7,8-trimethyl-2-(beta-carboxyethyl)-6-hydroxychroman is a major route of elimination of gamma-tocopherol in humans. *J. Lipid Res.,* 40, 665–671.

Tomasch, R, Wagner, KH and Elmadfa, I (2001) Antioxidative power of plant oils in humans: the influence of alpha- and gamma-tocopherol. *Ann. Nutr. Metab.*, 45, 110–115.

Tomeo, AC, Geller, M, Watkins, TR, Gapor, A and Bierenbaum, ML (1995) Antioxidant effects of tocotrienols in patients with hyperlipidemia and carotid stenosis. *Lipids,* 30, 1179–1183.

Traber, MG and Arai, H (1999) Molecular mechanisms of vitamin E transport. *Annu. Rev. Nutr.*, 19, 343–355.

Traber, MG and Kayden, HJ (1989) Preferential incorporation of alpha-tocopherol vs gamma-tocopherol in human lipoproteins. *Am. J. Clin. Nutr.*, 49, 517–526.

Traber, MG, Burton, GW, Ingold, KU and Kayden, HJ (1990a) RRR- and SRR-alpha-tocopherols are secreted without discrimination in human chylomicrons, but RRR-alpha-tocopherol is preferentially secreted in very low density lipoproteins. *J. Lipid Res.*, 31, 675–685.

Traber, MG, Rudel, LL, Burton, GW, Hughes, L, Ingold, KU and Kayden, HJ (1990b) Nascent VLDL from liver perfusions of cynomolgus monkeys are preferentially enriched in RRR- compared with SRR-alpha-tocopherol: studies using deuterated tocopherols. *J. Lipid Res.*, 31, 687–694.

Traber, MG, Sokol, RJ, Burton, GW, Ingold, KU, Papas, AM, Huffaker, JE and Kayden, HJ

(1990c) Impaired ability of patients with familial isolated vitamin E deficiency to incorporate alpha-tocopherol into lipoproteins secreted by the liver. *J. Clin. Invest.*, 85, 397–407.

Van Gossum, A, Shariff, R, Lemoyne, M, Kurian, R and Jeejeebhoy, K (1988) Increased lipid peroxidation after lipid infusion as measured by breath pentane output. *Am. J. Clin. Nutr.*, 48, 1394–1399.

Watkins, T, Geller, M, Kooyenga, D and Bierenbaum, M (1999) Hypocholesterolemic and antioxidant effect of rice bran oil non-saponifiables in hypercholesterolemic subjects. *Environ. Nutr. Interact.,* 3, 115–122.

Wechter, WJ, Kantoci, D, Murray, ED, D'Amico, DC, Jung, ME and Wang, WH (1996) A new endogenous natriuretic factor: LLU-alpha. *Proc. Natl Acad. Sci. USA*, 93, 6002–6007.

Yang, MH and Schaich, KM (1996) Factors affecting DNA damage caused by lipid hydroperoxides and aldehydes. *Free Rad. Biol. Med.*, 20, 225–236.

CHAPTER 4

Other natural antioxidants – rice bran oil, sesame oil, rosemary extract, flavonoids

CLIFFORD HALL III

North Dakota State University, Fargo, North Dakota, USA

A. Introduction

Natural antioxidants can be simply defined as a group of natural compounds that exhibit antioxidant activity. This chapter will discuss the reasons for incorporating natural antioxidants into functional foods, the factors that affect antioxidant concentrations in extracts or oils, and applications of these materials in food systems. Methods for measuring natural antioxidants are not presented, but additional information is available in the referenced literature.

This chapter will focus on developments in rice bran oil and sesame seed oil, both of which have exceptional oxidative stability and diverse biological activity. In addition, rosemary antioxidants will be described, with an account of methods for extracting bioactive compounds and a discussion of the stability of these compounds in food systems. Recent literature on the stability and health benefits of flavonoids will complete the chapter.

B. Rice (*Oryza sativa*) bran oil

1. General characteristics

Rice bran is an underutilized co-product of rice processing. Based on global rice production, approximately 40 million tonnes of rice bran should be available annually (Childs, 2001). Rice bran contains 15–23% oil, representing an annual potential production of 6.0–9.2 million tonnes of crude rice bran oil. Palmitic acid (12–20%), oleic acid (34–50%) and linoleic acid (20–36%) are the predominant fatty acids (Gaydou *et al.*, 1980; Hemavathy and Prabhakar, 1987; Gupta, 1989). Rice bran oil is also a rich source of antioxidants. It contains 600–1000 mg/kg tocols (tocopherols and tocotrienols) and up to 15 000 mg/kg oryzanols (Figure 1), depending on the extraction and refining method (Rogers *et al.*, 1993; Hu *et al.*, 1996; Shin *et al.*, 1997; De and

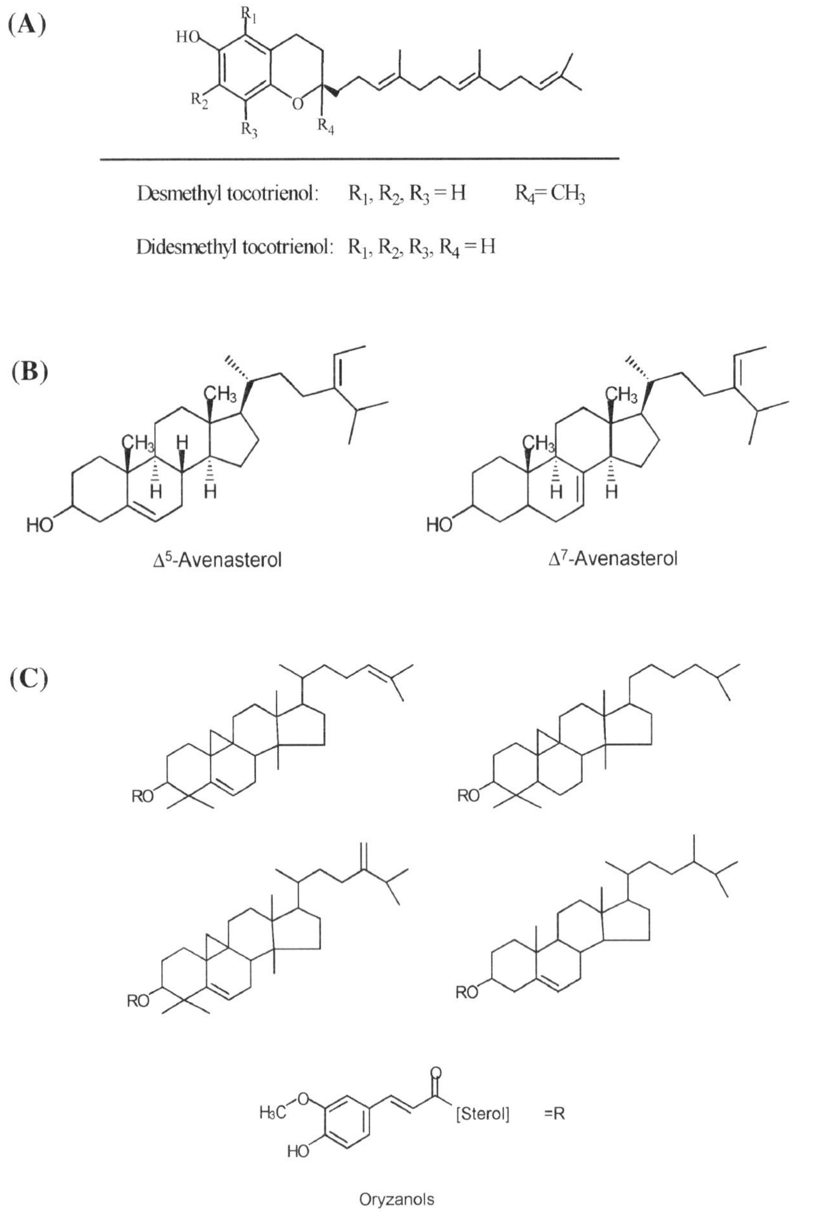

Figure 1. Selected antioxidant components from rice bran oil. (A) Two unique tocotrienols, desmethyl tocotrienol and didesmethyl tocotrienol. Unlike the more common α-, β-, γ- and δ-tocotrienols [see Chapter 3], desmethyl tocotrienol has no $-CH_3$ substituents at positions R_1, R_2 or R_3. Didesmethyl tocotrienol in addition lacks the $-CH_3$ group normally at R_4. (B) The sterols Δ^5-avenasterol and Δ^7-avenasterol. (C) Several oryzanols – sterols esterified with ferulic acid (shown here as the group R).

Bhattacharyya, 1998; Xu and Godber, 1999; Krishna *et al.,* 2001). In addition, sterols may account for up to 7500 mg/kg of crude rice bran oil (Ramsay *et al.*, 1991; Formo *et al.*, 1979). The total sterol concentration includes the sterol contribution from oryzanols, as the saponification step used in the method for determining sterols hydrolyses the ester bond of the oryzanols (Xu and Godber, 1999). Gaydou *et al.* (1980) reported that campesterol, stigmasterol and β-sitosterol were the major sterols. Δ^5- and Δ^7-avenasterol, also present, inhibit the polymerization of frying oil (Gordon and Magos, 1983). Antioxidant concentrations depend on the method of processing the oil as well as on the original quality of the rice bran (see Section B.3 below).

Lipase, phospholipase and lipoxygenase are the enzymes primarily responsible for poor-quality rice bran oil. They are activated during the bran removal process (Vetrimani *et al.*, 1992; Takano, 1993), and can cause the rate of free fatty acid (FFA) formation to be as high as 5–7% per day (Nasirullah *et al.*, 1989). Thus, inactivation of lipases is important for producing high-quality rice bran oil. The quality of rice bran oil is inversely related to the level of FFA, and this must be kept low if the oil is to be edible and acceptable in frying applications. However, the removal of FFA is not a simple process and is accompanied by the loss of important antioxidants (Krishna *et al.*, 2001). This loss must be kept to a minimum if rice bran oil is to be used as a functional food component.

2. *Health benefits*

The modification of plasma lipid profiles by rice bran oil has been extensively studied and is the major health benefit highlighted in this section. Other health benefits associated with components of rice bran oil include suppression of B16 melanoma cell proliferation by γ- and β-tocotrienols (He *et al.*, 1997; Qureshi *et al.*, 1997, 2000; Lane *et al.*, 1999) and inhibition of tumor-promoting activity by oryzanols (Yasukawa *et al.*, 1998; Akihisa *et al.*, 2000). Cicero and Gaddi (2001) have reviewed other health benefits, including modulation of pituitary secretion and anti-ulcerogenic activity.

The health benefits of rice bran oil have been examined in both animal and human studies (deDeckere and Korver, 1996; Cicero and Gaddi, 2001). In cholesterol-fed hamsters, full-fat rice bran significantly lowered cholesterol and triacylglycerol levels in plasma and liver (Kahlon *et al.*, 1992a, 1992b). Defatted rice bran did not significantly alter lipid levels, but when combined with degummed and dewaxed rice bran oil some of the cholesterol-lowering properties were restored, thus suggesting the importance of rice bran oil components in altering plasma and liver lipid levels. Kahlon *et al.* (1996) later reported that hamsters fed a diet containing an unsaponifiable fraction from rice bran had significantly lower plasma cholesterol levels than those fed a control diet, while the ratio of high-density lipoproteins (HDL) to low-density lipoproteins (LDL) was similar for both groups of animals. Rong *et al.* (1997) concluded that oryzanols were partially responsible for the cholesterol-lowering action of rice bran oil in male F1B golden hamsters. In addition, a significant reduction in aortic fatty streak formation was found when hamsters were fed a diet containing 0.5% (fat basis) oryzanols. In hamsters fed a diet containing 5% coconut oil, 0.1% cholesterol, and 1% oryzanols for 7 weeks, total cholesterol levels were 28% lower, and non-HDL cholesterol levels were 34% lower, than in animals fed diets lacking oryzanols. Hamsters fed a diet containing 10% coconut oil, 0.05% cholesterol and 0.5% oryzanols for 10 weeks had 44% lower total cholesterol levels, 57% lower non-HDL cholesterol levels, and 46% lower triacylglycerol levels than those fed diets lacking oryzanols (Rong *et al.*, 1997). In addition, cholesterol absorption was reduced by 25% in the hamsters fed an oryzanol-containing diet.

The effectiveness of α-tocopherol, a tocotrienol-rich fraction, and didesmethyl tocotrienol (see Figure 1) on atherosclerotic lesion size were evaluated in C57BL/6 apolipoprotein E-deficient mice (Qureshi *et al.*, 2001). Diets containing tocotrienols were more effective at reducing the size of the atherosclerotic lesions than diets containing α-tocopherol. Serum levels of triacylglycerols, total cholesterol, and LDL cholesterol decreased significantly in mice fed diets containing tocotrienols; however, only didesmethyl tocotrienol significantly reduced atherosclerotic lesion size in all three genotypes of mice studied (Qureshi *et al.*, 2001).

The effects of rice bran oil and its components seen in animal studies have also been noted in humans (Cicero and Gaddi, 2001). Tsai *et al.* (1992) found that full-fat rice bran decreased serum cholesterol and LDL levels in humans, while defatted rice bran did not have a significant effect. Vissers *et al.* (2000) found that feeding human volunteers 2.1 g rice bran oil sterols per day reduced total cholesterol levels by 5% and LDL cholesterol levels by 9%. HDL cholesterol and triacylglycerol concentrations did not change significantly at this level of rice bran oil supplementation.

Kooyenga *et al.* (2001) conducted a 4-year study to evaluate the effects of antioxidants in patients with carotid atherosclerosis. Patients supplemented with tocopherol and tocotrienol from rice bran oil had approximately double the serum α-tocopherol levels of those not given the supplements. In addition, cholesterol and LDL values decreased by 14% and 20%, respectively, supporting the use of antioxidants as a way to control carotid atherosclerosis. In a double-blind 12-week study, hypercholesterolemic human subjects fed 200 mg of a tocotrienol-rich fraction (obtained by molecular distillation from specially processed rice bran oil) had 12% and 16% reductions in total cholesterol and LDL cholesterol levels, respectively (Qureshi *et al.*, 1997). Subjects consuming the American Heart Association "Step 1" diet plus 200 mg of a tocotrienol-rich rice bran oil fraction had 17% and 24% reductions in total cholesterol and LDL cholesterol levels, respectively. Significant decreases in apolipoprotein B, lipoprotein (a), platelet factor 4, and thromboxane B2 (15%, 17%, 14% and 31%, respectively) were also found. The tocotrienol-rich fraction consisted of α-tocopherol (6%), α-tocotrienol (12%), γ-tocotrienol (21%), δ-tocotrienol (10%), desmethyl tocotrienol (4.5%), and didesmethyl tocotrienol (17%), with sterols, triacylglycerols, and unidentified tocotrienols making up the remaining 30% of the extract. Qureshi and colleagues (Pearce *et al.*, 1992; Qureshi *et al.*, 1999, 2000) reported that didesmethyl tocotrienol was the most potent cholesterol inhibitor, followed by desmethyl tocotrienol, then δ-, γ-, α- and β-tocotrienols. Tocotrienols inhibit cholesterol synthesis by post-transcriptional suppression of β-hydroxy-β-methylglutaryl coenzyme A reductase activity (Pearce *et al.*, 1992; Parker *et al.*, 1993).

Although tocotrienols and oryzanols have been identified as the major physiologically active components in rice bran oil, other components such as sterols could also contribute to the health-promoting characteristics of rice bran oil. If rice bran oil is to be used as a functional food component, processing methods that give maximum retention of the active components must be developed. Alternatively, processes that concentrate the active component would also be of value to the functional food market.

3. *Stability*

Rice bran oil with a high FFA content is often used as an industrial oil, because FFA removal often leads to a reduction in the content of desirable antioxidant/bioactive components. A short discussion on methods to control FFA formation is therefore warranted. Lipase inactivation becomes significant when rice bran oil cannot be extracted from rice bran immediately after the milling process. As mentioned above, the rate of FFA formation within bran lipids can reach 5–7% per day (Nasirullah *et al.*, 1989), with consequent reductions in oil yield and quality. Pretreatment of the rice bran by physical methods has been the primary means to inactivate rice bran lipase prior to oil extraction.

Extrusion is the most common method to stabilize rice bran, and the stability of rice bran obtained in this way has been thoroughly investigated (Kim *et al.*, 1985, 1987a, 1987b; Shen *et al.*, 1997). As the extrusion temperature increases from 110°C to 140°C, tocol and oryzanol levels decrease. Extrusion at 110°C resulted in 7% and 4% reductions in tocol and oryzanol levels, respectively, while extrusion at 140°C resulted in 21% and 8% reductions in the same components. The 110°C extrusion process was deemed optimal due to the better retention of antioxidants, although the FFA content of the treated bran increased by 2–6% during one year of storage.

An alternative to extrusion cooking is microwave processing (Rhee and Yoon, 1984; Tao *et al.*, 1993; Vetrimani *et al.*, 1992; Ramezanzadeh *et al.*, 1999, 2000). Tao *et al.* (1993) evaluated microwave heating as a method to stabilize rice bran against hydrolytic rancidity. The FFA content of microwave-treated rice bran increased from 4.0% to 4.9% for long-grain rice, and from 4.6% to 6.2% for medium-grain rice, during 4 weeks of storage. In contrast, the FFA content of untreated bran increased from 4.0% to 68.3% for long-grain rice, and from 4.6% to 56.8% for medium grain rice. Optimum conditions for microwave heating (340 W, 2450 MHz) of rice bran were 21% moisture, 150 g of bran per batch and 3 min heating (1.2 s/g rice bran). Ramezanzadeh *et al.* (1999, 2000) reaffirmed the observation that microwave heating was effective in inhibiting FFA formation. In addition, they noted that the FFA profile of rice bran oil remained consistent over a 16-week storage period. The type of packing (vacuum-packed or in plastic bag) did not significantly influence either the FFA levels or the fatty acid composition. We have found that a microwave (1 kW, 2450 MHz) treatment of 1.875 s/g rice bran was sufficient to inhibit FFA formation without adversely affecting oryzanol or tocopherol levels (Hall III and Proctor, unpublished data). The oil obtained from a hexane extraction of freshly milled rice bran contained 13 144 mg/kg oryzanols and 734 mg/kg tocopherols. After microwave processing, the extracted oil contained 13 071 mg/kg oryzanols and 769 mg/kg tocopherols.

Gamma-irradiation has also been studied as a method to stabilize rice bran (Shin and Godber, 1996). However, irradiation levels in the range of 5–15 kGy were detrimental to tocols and oryzanols, as demonstrated by losses of 50–82% of tocols and 12–33% of oryzanols immediately after irradiation. Further reductions in these components occurred during storage. Furthermore, FFA levels were higher in samples irradiated at 10–15 kGy than in non-irradiated control samples, thus defeating the purpose of using irradiation to inactivate lipase.

4. *Extraction using solvents*

Hexane extraction of stabilized rice bran has been the most commonly used method for obtaining rice bran oil. Hexane has the advantage of high oil recovery, but safety concerns regarding hexane have led to a search for alternative extraction methods. Less hazardous approaches such as the use of isopropyl alcohol, supercritical fluid extraction, or enzyme-assisted extraction are potential alternative processing methods.

Comparisons between hexane and isopropyl alcohol as solvents for rice bran oil extraction have been undertaken by several investigators (Proctor and Bowen, 1996; Hu *et al.*, 1996). In general, isopropyl alcohol compares favorably with hexane with regard to oil yield and quality. Hu *et al.* (1996) reported that a 3:1 solvent:bran ratio and a 60°C extraction temperature were optimal regardless of the solvent used. Preheating the solvent prior to extraction significantly enhanced oil, tocol and oryzanol yields. An increase in temperature from 40°C to 60°C resulted in 4% and 6% increases in crude oil yield for hexane and isopropyl alcohol, respectively; similar trends were found for tocols (7% and 9% increases) and oryzanols (4% and 7% increases). Increasing the solvent:bran ratio from 2:1 to 3:1 resulted in 11% and 9% increases in crude oil yield, 13% and 10% increases in tocol yield, and 12% and 8% increases in oryzanol yield for rice bran extracted with hexane or isopropyl alcohol, respectively. At optimal conditions, the hexane-extracted rice bran oil contained approximately 744 mg/kg tocols and 13 490 mg/kg oryzanols, while the isopropyl alcohol-extracted oil contained 774 mg/kg tocols and 14 577 mg/kg oryzanols (Hu *et al.*, 1996). Similar tocol and oryzanol concentrations were found in oils extracted from microwave-stabilized rice bran: the hexane-extracted oil contained 769 mg/kg tocols and 13 071 mg/kg oryzanols, while the isopropyl alcohol-extracted oil contained 800 mg/kg tocols and 14 777 mg/kg oryzanols (Hall III and Proctor, unpublished data). Xu and Godber (2000) reported that a 1:1 mixture of hexane and isopropyl alcohol extracts more oryzanols. These reports all support the use of isopropyl alcohol as an alternative solvent to hexane for isolating important bioactive components from rice bran oil.

5. *Supercritical fluid extraction*

Supercritical fluid extraction (SFE) using carbon dioxide as the supercritical fluid is a relatively new method to extract lipids from oil-bearing materials. Carbon dioxide is a gaseous material that becomes a supercritical fluid at 31°C under a pressure of 73 atm (7.4 MPa). In this supercritical state, carbon dioxide exhibits densities and solvating properties similar to those of liquid solvents and diffusion characteristics and viscosities similar to those of gases. Thus a variety of matrices can be extracted efficiently (Meyer *et al.*, 1993). In addition, supercritical carbon dioxide (SC-CO_2) is non-flammable, non-polluting, and non-toxic. Another unique characteristic of SC-CO_2 extraction is that oil quality can be influenced by changing parameters such as the pressure and temperature of the extractant (Ramsay *et al.*, 1991). The extraction and fractionation of lipid materials from rice bran and other oilseeds can be accomplished using SC-CO_2 (Stahl *et al.*, 1980; Zhao *et al.*, 1987; Ramsay *et al.*, 1991; List *et al.*, 1993; Garcia *et al.*, 1996; Xu and Godber, 2000). Using SC-CO_2 extraction, rice bran oils can be tailored to a specific application or product quality.

The first report of using SC-CO_2 extraction to produce rice bran oil occurred over a decade ago. Zhao *et al.* (1987) reported a fractionation process using pressures of 15–35 MPa and an extraction chamber temperature of 40°C. Lower oil yields were observed at 35 MPa, which may be due to the lower FFA content (4.4% at 35 MPa, versus 8.8% at 15 MPa). Hexane extraction of the same bran gave oil of 11.9% FFA content. Oils extracted at 15 MPa, 35 MPa, or with hexane had respective tocopherol contents of 284 mg/kg, 240 mg/kg and 333 mg/kg, and oryzanol contents of 1.1%, 0.9% and 1.4%. Zhao *et al.*

(1987) also noted that milling the raw material to less than 350 μm improved oil extraction. Kuk and Dowd (1998) reported rice bran oil extraction yields of 19.0–20.4% (equivalent to 93–99.5% lipid recovery) using SC-CO_2, with lower temperature (70°C) and pressure (42 MPa) giving the lowest yields. In contrast to Zhao *et al.* (1987), these researchers found that a higher temperature and pressure gave a higher FFA content: 3.0% at 100°C and 62 MPa, compared with 1.5% at 70°C and 42 MPa. SC-CO_2 processing at 35 MPa and 35°C yielded 89% of the oil and 77.5% of the sterols obtained with hexane extraction (Ramsay *et al.*, 1991). These data follow the trend observed by Kuk and Dowd (1998) that lower pressures and temperatures extract less lipids. Xu and Godber (2000) compared the extraction efficiencies of organic solvents and SC-CO_2 and found that the optimal solvent extraction system (a 1:1 hexane:isopropanol mixture at 60°C) recovered 1.68 mg oryzanols/g rice bran. The optimal SC-CO_2 extraction (50°C, 68.9 MPa, 25 min) recovered three times as much oryzanols (5.39 mg/g rice bran). The oryzanol concentration of the solvent extract was 11.8 mg/g, and that of the SC-CO_2 extract was 51 mg/g. In addition, a fraction with a particularly high oryzanol concentration (675 mg/g) was collected between 15 min and 20 min of SC-CO_2 processing.

Shen *et al.* (1996, 1997) evaluated SC-CO_2 extraction and fractionation of rice bran oil on a pilot scale. They found that 31 MPa and 40°C gave the highest oil yield (92.9%) and that increasing the temperature at constant pressure increased the FFA levels. FFA levels were twice the triacylglycerol levels during the first hour of SC-CO_2 extraction (24 MPa, 40°C or 60°C), and removed 90% of the tocols within 3 h and 90% of the oryzanols within 6 h (Shen *et al.*, 1996). A two-stage process to fractionate rice bran oil was evaluated by Shen *et al.* (1997). During stage one, the rice bran was extracted at 24.1 MPa and 40°C to obtain an oil-rich SC-CO_2 phase. This phase was then fractionated under a selected pressure and temperature in a fractionator. On pressure reduction, the SC-CO_2 separated into an oil-rich phase (raffinate) and a CO_2-rich phase. The CO_2-rich phase was sent to a separator where the CO_2 was vaporized, leaving the final extract in the separation vessel. After adjustment of the fractionator pressure to 11.2 MPa, the raffinate and extract together accounted for 72.3% of the oil, 31% of the FFA, 85.9% of the oryzanols, and 68% of the tocols from the rice bran. The fractionator column temperature and pressure affected the distribution of the oil components: higher temperatures resulted in higher raffinate yield at the expense of the extract yield (Shen *et al.*, 1997). Part of the increase in raffinate yield is due an increase in the FFA level in the raffinate (from 6.1% to 9.2% as the temperature increases from 40°C to 50°C). Oryzanols were removed most efficiently at 40°C and pressures of 11.2 MPa or 9.9 MPa, yielding total oryzanol contents of 18.1 mg/g oil or 18.6 mg/g oil, respectively. However, irrespective of the extraction conditions, the majority of the oryzanols (65–70%) remained in the raffinate fraction. In contrast, tocols were preferentially extracted from the raffinate at higher pressure (11.2 MPa, 40°C) and were retained in the raffinate at lower pressures (8.6 MPa, 40°C).

The use of SC-CO_2 processing has promise due to its flexibility and selectivity and the fact that additional refining may not be needed. Altering the extraction pressures and temperatures allows for the removal of specific components.

6. *Refining*

Rogers *et al.* (1993) found a wide range of antioxidant concentrations in commercially available rice bran oils: 115–787 ppm oryzanols, 16–452 ppm tocopherols, and 72–1157 ppm tocotrienols. The brand with the lowest oryzanol content also had a low level of tocols. These researchers speculated that the refining process was responsible for the low oryzanol levels in the commercial oils tested.

Krishna *et al.* (2001) evaluated the effects of different refining steps on the retention of oryzanols in rice bran oil from 18 different Indian rice cultivars. Extraction with chloroform and methanol yielded oils with oryzanol contents between 1.63% and 2.72%. Only small losses (1.1% and 5.9%, respectively) of oryzanols occurred in degummed and dewaxed rice bran oils, but alkali treatments removed 93.0–94.6% of oryzanols from the original crude oil. Yoon and Kim (1994) reported that 51% of the oryzanols were lost during rice bran oil processing, which included degumming, alkali refining, bleaching, and deodorization steps. After alkali, solvent, or steam refining of high-FFA rice bran oil, oryzanol levels were reduced by 42%, 7%, or 9%, respectively (Kim *et al.*, 1985). To account for the loss in oryzanols during alkali refining, Krishna *et al.* (2001) measured oryzanol levels in the soapstock and deodorizer distillate, and found oryzanol contents of 6.3–6.9% and 0.79%, respectively. These findings explain the low oryzanol contents observed in commercial rice bran oil products (Rogers *et al.*, 1993; Krishna *et al.*, 2001). However, some commercial rice bran oils have significant levels (1.74%) of oryzanols, which may be attributed to the refining method used.

Krishna *et al.* (2001) reported that the concentration of oryzanols was not affected by physical refining (deodorization), whereas an 89.5% reduction in oryzanol levels occurred in chemically refined oil. This supports the findings of De and Bhattacharyya (1998) that deodorization at temperatures between 235°C and 265°C does not significantly influence oryzanol levels. In fact, there was a slight increase in oryzanol concentration, which may have been due to the loss of other components. Physical refining caused up to a 50% reduction in the tocopherol concentration of crude rice bran oil (De and Bhattacharyya, 1998; Yoon and Kim, 1994). A low-temperature (10°C) degumming–dewaxing process followed by bleaching and deodorization (235°C) is a practical method for preparing rice bran oil with a high oryzanol concentration (De and Bhattacharyya, 1998).

7. *Antioxidant activity*

The antioxidant activity of rice bran oil can be attributed to its content of tocols, oryzanols and sterols, and rice bran oil itself has been used as a stabilizer of frying oil. Kim *et al.* (2001) found that the non-saponifiable fraction of rice bran oil was effective at preventing the oxidation of cholesterol, as measured by the loss of cholesterol and the formation of 7-ketocholesterol in the system. After 16 h incubation at 80°C, only 8% of the cholesterol was converted to 7-ketocholesterol in samples treated with 2100 ppm of the non-saponifiable materials. Treatment with 1400 ppm or 700 ppm of the materials resulted in oxidation of 18% or 36% of the cholesterol, respectively. Nearly all the cholesterol was degraded in the absence of the non-saponifiable materials (Kim *et al.*, 2001). As expected, the levels of tocols and oryzanols decreased significantly by 4 h in model systems incubated at 80°C. Significant reductions were found at all treatment levels; however, the retention of tocols and oryzanols was greatest for the treatment with 2100 ppm of the non-saponifiable materials. Similar trends in the oxidation of cholesterol were noted by Xu *et al.* (2001), who found that oryzanols were significantly more effective than tocols at preventing cholesterol oxidation. After 24 h, the level of cholesterol oxidation products in samples with added α-tocopherol or α-tocotrienol were not significantly different from the level in the control sample without additive. Cycloartenyl ferulate, 24-methylenecycloartanyl ferulate, and campesteryl ferulate were the three major oryzanols with antioxidant activity. 24-Methylenecycloartanyl ferulate was the most effective antioxidant (Xu *et al.,* 2001).

However, opposite antioxidant efficiencies were found when oryzanols and α-tocopherol were evaluated in a linoleic acid model (Xu and Godber, 2001). Oryzanols, α-tocopherol, and ferulic acid were each added to linoleic acid at a range of concentrations (80 μM, 160 μM and 400 μM), in order to assess their ability to inhibit the formation of oxidation products (9-hydroperoxy-10-*trans*,12-*cis*-octadecadienoic acid; 9-hydroperoxy-10-*trans*, 12-*trans*-octadecadienoic acid; 13-hydroperoxy-9-*cis*,11-*trans*-octadecadienoic acid; 13-hydroperoxy-9-*trans*,11-*trans*-octadecadienoic acid). Both α-tocopherol and ferulic acid were more effective than the oryzanols at inhibiting hydroperoxide formation at the 80 μM and 160 μM concentrations, although not at 400 μM. The oryzanols showed no inhibition of hydroperoxide formation at the 80 μM concentration. A higher proportion of *trans,trans* hydroperoxides was formed in the presence of the oryzanols or ferulic acid than in the presence of α-tocopherol, suggesting that the oryzanols and ferulic acid have a different antioxidant mechanism than α-tocopherol (Xu and Godber, 2001).

The addition of rice bran oil or its components as a stabilizer in frying oils has been reported (Kamal-Eldin *et al.*, 1998; Gertz *et al.*, 2000). Kamal-Eldin *et al.* (1998) reported that the addition of a blend of sesame oil and rice bran oil enhanced the frying performance of high-oleic sunflower oil. French fry quality was used as an indicator of frying performance. Addition of the blend of rice bran oil and sesame oil extended the effective frying performance of the oil to from 35 h to 65 h. Kochhar (2000) reported similar frying oil performance with 'Good-Fry', a commercial blend of sesame oil and rice bran oil. The oxidative stability of instant noodles was also improved by the addition of a blend of rice bran oil and sesame oil (Yen and Lai, 1989). These authors noted that the stored instant noodles had acceptable sensory scores and oxidative stability only when the noodles had been fried in a blend of rice bran oil and sesame oil, or in sesame oils that had not been heated for longer than 8 h at 180°C prior to frying the noodles. A blend of rice bran oil with high-α-linolenic oils has been patented for use as a stabilizer of spray oils used in snack products (Taylor *et al.*, 1996). This patent identifies the addition of 2–5% rice bran oil, processed to retain a high level of unsaponifiables, as optimal for stabilizing spraying oils.

The stabilization of milk powders and beef using rice bran oil or its constituents has been reported (Nanua *et al.*, 2000; Kim *et al.*, 2000; Kim and Godber, 2001). The addition of a high-oryzanol rice bran oil at a level of 0.1% to whole dried milk significantly reduced the extent of oxidation, as measured by a reduction in the formation of thiobarbituric acid-reactive substances (TBARS). However, increasing the rice bran oil concentration to 0.2% did not significantly further enhance the stability of the milk. Kim *et al.* (2000) noted that the addition of 2% rice bran oil or rice fiber to restructured beef roasts increased their oxidative stability and reduced 7-ketocholesterol formation. Another study using crude rice bran oil (2%, w/w) in restructured beef roasts showed a significant reduction in 7-ketocholesterol formation and an increase in vitamin E content (Kim and Godber, 2001).

Literature data support the use of rice bran oil and constituents as antioxidants in food systems. Tocols and oryzanols appear to have different degrees of activity depending on the lipid system evaluated. Oryzanols are more effective in cholesterol systems and tocols the most effective in linoleic acid model systems. The combination of these components greatly enhances the activity of rice bran oil as an antioxidant. Further study is recommended to assess other lipid systems as well as synergistic activities between the oryzanols and tocols.

C. Sesame (*Sesamum indicum*) seed oil

1. General characteristics

Bahkali *et al.* (1998) reported that the oil contents of white and dark sesame seed cultivars grown in the Gizan area of Saudi Arabia were 47% and 49%, respectively. Similar results were found for white and dark sesame cultivars grown in Turkey (51% and 55%, respectively), Egypt (45% for both cultivars),

China (53% for a dark cultivar), and Japan (52% for a white cultivar), indicating that the growing region and/or cultivar affects oil yield (Xiao *et al.*, 1992; Yoshida and Takagi, 1997; Mohamed and Awatif, 1998; Xiurong *et al.*, 2000). Xiao *et al.* (1992) reported finding eight accessions of dark sesame seeds with oil contents >57%.

Oleic and linoleic are the principal fatty acids in sesame seeds and account for 47–53% and 34–39% of the lipid, respectively. Palmitic (9–13%), stearic (5–6%), and α-linolenic (1%) acids are also present (Tashiro *et al.*, 1990; Kamal-Eldin and Appelqvist, 1994; Bahkali *et al.*, 1998; Mohamed and Awatif, 1998; Ayorinde *et al.*, 1999). The total sterol content is approximately 5400 mg/kg for white varieties and 5000 mg/kg for dark varieties, with campesterol, β-sitosterol and Δ^5-avenasterol respectively accounting for 12%, 22% and 12% of the sterols in white sesame seeds, and 11%, 30% and 11% of the sterols in dark sesame seeds (Kamal-Eldin and Appelqvist, 1994; Mohamed and Awatif, 1998). The high level of Δ^5-avenasterol (approximately 710 mg/kg) is indicative of the potential for sesame seed oil to act an antioxidant in frying applications (Gordon and Magos, 1983).

The total tocopherol content of sesame seed oil is approximately 400 mg/kg for white varieties and 540 mg/kg for dark varieties (Mohamed and Awatif, 1998; Yoshida *et al.*, 2001). Others (Kamal-Eldin and Appelqvist, 1994; Yoshida and Kajimoto, 1994; Shahidi *et al.*, 1997) have reported tocopherol levels in the range of 330–680 mg/kg oil, with γ-tocopherol being the only tocopherol detected. Yoshida *et al.* (2001) reported that γ-tocopherol accounted for 96% of the tocopherols, with the δ compound being the only other tocopherol present. Interestingly, wild species (*Sesamum radiatum* and *S. angustifolium*) had higher γ-tocopherol contents (750 mg/kg oil and 800 mg/kg oil, respectively) than cultivated varieties.

Kang *et al.* (2000) evaluated the lignan (see Figure 2) content of Korean sesame cultivars. In general, the lignan content differs between sesame cultivars. The average sesamin and sesamolin contents were 370 mg/100 g and 202 mg/100 g, respectively, for white sesame seeds, and 247 mg/100 g and 133 mg/100 g for dark sesame seeds. Mohamed and Awatif (1998) reported sesamin levels of 392 mg/100 g in oil from white seeds and 302 mg/100g in oil from dark sesame seeds. The sesamolin contents of the same oils were 242 mg/100 g and 227 mg/100 g, respectively. The variability of sesame seed lignan profiles is illustrated by the findings of Yoshida and Kajimoto (1994) and Shahidi *et al.* (1997), who reported sesamin contents in the range of 580–682 mg/100 g oil and sesamolin contents in the range of 168–564 mg/100 g oil. Wild sesame (*S. radiatum*) was reported to have a high sesamin content of 2400 mg/100 g oil, but a low sesamolin content of only 20 mg/100 g oil (Kamal-Eldin and Appelqvist, 1994). Yasumoto *et al.* (2001) reported that two experimental sesame seed lines (0730 and 0732) had higher levels of sesame lignans than the commercial cultivar 'Masekin'. The experimental lines contained approximately

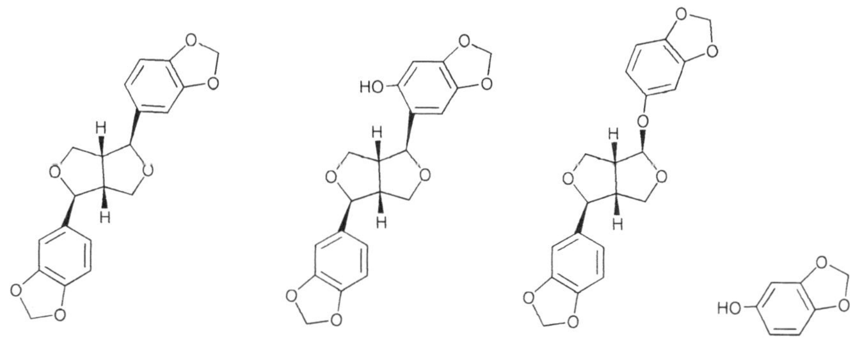

Sesamin Sesaminol Sesamolin Sesamol

Figure 2. Lignans important for the antioxidant activity of sesame seed oil.

820–1010 mg sesamin and 430–440 mg sesamolin per 100 g seed: twice the levels in the commercial cultivar (410 mg sesamin and 220 mg sesamolin per 100 g seed).

Katsuzaki *et al.* (1994) identified three sesaminol glucosides in sesame seeds as sesaminol 2′-*O*-β-D-glucopyranoside (sesaminol monoglucoside), sesaminol 2′-*O*-β-D-glucopyranosyl(1→2)*O*-β-D-glucopyranoside (sesaminol diglucoside), and sesaminol 2′-*O*-β-D-glucopyranosyl(1→2)*O*-[β-D-glucopyranosyl(1,6)]-β-D-glucopyranoside (sesaminol triglucoside). Ryu *et al.* (1998a) reported a total sesaminol glucoside content of 88.3 mg/100 g sesame seed, with the tri-, di-, and mono-glucosides accounting for 77%, 13%, and 10% of the sesaminol glucosides, respectively. In addition, an average sesaminol content of 20.5 mg/100 g sesame seed was reported by these authors. The sesaminol glucoside content was greatest in dark sesame seeds (Ryu *et al.*, 1998a). The contents of sesaminol tri-, di-, and mono-glucosides in a Korean sesame cultivar ('Danbaekggae') were 56.4 mg, 9.6 mg, and 7.5 mg per 100 g seed, respectively (Ryu *et al.*, 1998b).

2. *Health benefits*

Sesame seed oil and its constituents, particularly the tocopherols and lignans, exhibit a number of health benefits including:

- *in vivo* regulation of lipid metabolism (Hirose *et al.*, 1991; Ogawa *et al.*, 1995; Hirata *et al.*, 1996; Ashakumary *et al.*, 1999; Ide *et al.*, 2001);
- anti-inflammatory activity (Fujiwara *et al.*, 1995; Chavali *et al.*, 1998; Utsunomiya *et al.*, 2000);
- anti-hypertensive activity (Matumura *et al.*, 1995, 1998);
- antimutagenicity/anticarcinogenicity (Hirose *et al.*, 1992; Ryu and Lee, 1998; Kaur and Saini, 2000; Miyahara *et al.*, 2000, 2001a, 2001b).

Sesamin has been associated with the reduction of serum and liver lipids (Hirose *et al.*, 1991, 1996; Ide *et al.*, 2001; Yasumoto *et al.*, 2001). Hirose *et al.* (1991) reported that sesamin decreased lymphatic absorption of cholesterol. Sesamin was found to be a promoter of hepatic fatty acid oxidation in rats (Ashakumary *et al.*, 1999). Addition of a 1:1 mixture of sesamin and episesamin to the diet of rats (0.1–0.5% of total diet) promoted mitochondrial and peroxisomal palmitoyl coenzyme A (CoA) oxidation rates in a dose-dependent manner. Rats fed diets containing 0.5% sesamin had 2 fold higher mitochondrial activity and 10 fold higher peroxisomal activity than rats fed non-sesamin diets. Yasumoto *et al.* (2001) reported similar findings when a 20% sesame seed diet was fed to rats. These authors compared two experimental sesame seed lines (0730 and 0732) with the commercial cultivar Masekin. Both experimental seed lines enhanced mitochondrial and peroxisomal fatty acid oxidation rates, and were more effective than the commercial cultivar at reducing plasma triacylglycerol levels (Yasumoto *et al.*, 2001). A study of the effects of dietary sesamin on the hepatic metabolism of arachidonic and eicosapentaenoic acids in rats showed similar increases in hepatic mitochondrial and peroxisomal fatty acid oxidation (Sawada *et al.*, 2001).

Various studies on rats fed diets containing sesamin have found increases in the activity of hepatic fatty acid oxidation enzymes (acyl-CoA oxidase, carnitine palmitoyltranferase, 3-hydroxyacyl-CoA dehydrogenase and 3-ketoacyl-CoA thiolase), and decreases in the activity of enzymes responsible for fatty acid synthesis (fatty acid synthase, glucose-6-phosphate dehydrogenase, ATP-citrate lyase, and pyruvate kinase) (Hirose *et al.*, 1991; Hirata *et al.*, 1996; Ide *et al.*, 2001; Sawada *et al.*, 2001; Yasumoto *et al.*, 2001). Using specific cDNA probes to measure hepatic mRNA levels, two groups have shown that sesamin induces increases in the gene expression of hepatic mitochondrial and peroxisomal fatty acid oxidation enzymes, and decreases the gene expression of sterol regulatory element binding protein-1 (Ashakumary *et al.*, 1999; Ide *et al.*, 2001).

Sesame seed lignans also alter *in vivo* tocopherol levels (Yamashita *et al.*, 1992, 1995; Kamal-Eldin *et al.*, 1995; Izuka *et al.*, 1996; Parker *et al.*, 2000). Yamashita *et al.* (1992) reported that rats fed sesame seeds had higher *in vivo* γ-tocopherol levels than rats fed γ-tocopherol-fortified diets without sesame seeds. In addition, rats fed sesaminol- or sesamin-fortified diets had comparable *in vivo* γ-tocopherol levels. A further study by these authors showed that sesame seed and its lignans increased the level of α-tocopherol in the blood and tissues of rats fed low-tocopherol diets (Yamashita *et al.*, 1995). Yamashita *et al.* (2000) evaluated the effects of lignans on lipid peroxide formation in the plasma and liver of rats fed diets low in α-tocopherol and high in docosahexaenoic acid (DHA). Rats fed the high-DHA diet had significantly higher plasma and liver TBARS concentrations than those fed the control diet; these increases in TBARS concentrations were completely suppressed by sesamin. The addition of DHA to the rats' diet also led to reductions in plasma and liver α-tocopherol concentrations; the addition of sesamin prevented the loss of α-tocopherol and enhanced DHA concentrations in plasma triacylglycerols. Parker *et al.* (2000) also found that sesamin promoted tocopherol retention by inhibiting the cytochrome P-450 3A (CYP3A)-dependent catabolism of γ-tocopherol to γ-carboxychromans.

The enzyme Δ^5-desaturase is responsible for the conversion of dihomo-γ-linolenic acid (20:3*n*-6) to arachidonic acid (20:4*n*-6) [see Chapter 9, Figure 1], and inhibition of this enzyme can lead to a reduction in the formation of pro-inflammatory products. Fujiwara *et al.* (1995) evaluated the effects of added sesamin (0.5%, w/w of total diet) on the liver fatty acid composition of rats fed diets rich in either α-linolenic acid or γ-linolenic acid. With both diets, the addition of sesamin increased the liver weight and phospholipid content. The addition of sesamin to the α-linolenic acid diet resulted in an increase in *n*-6 fatty acids and a decrease in *n*-3 fatty acids (Fujiwara *et al.*, 1995). In addition, the liver ratio of 20:5*n*-3 to 18:3*n*-3 fatty acids increased in rats fed the α-linolenic acid plus sesamin diet, whereas the ratio of 20:4*n*-6 to 20:3*n*-6 fatty acids decreased in rats fed the γ-linolenic acid plus sesamin diet. These data suggest that only the Δ^5-desaturase enzyme involved in the metabolism of *n*-6 fatty acids was inhibited by sesamin. Chavali *et al.* (1998) found similar results when mice were fed a diet containing either safflower oil or a 2:1 mixture of linseed and safflower oils, plus sesamin (0.25%, w/w of diet). In general, they found elevated levels of α-linolenic acid, eicosapentaenoic acid and DHA, but lower levels of arachidonic acid, in the liver membrane phospholipids. Prostaglandin E_2 levels were significantly lower and tumor necrosis factor α (TNF-α) levels were higher in mice fed sesamin-containing diets, suggesting that TNF-α production could be regulated by prostaglandin E_2 synthesis (Chavali *et al.*, 1998) and that sesamin could thus serve as an anti-inflammatory agent. Utsunomiya *et al.* (2000) reported that the anti-inflammatory effects of sesamin were magnified by the concurrent consumption of linseed oil.

Lignans have been found to exhibit anti-tumor activity. Sesaminol was cytostatic at concentrations in the range of 60–100 μg/ml and cytotoxic at concentrations ≥200 μg/ml to human leukemia HL-60 cells (Ryu and Lee, 1998). A sesaminol concentration of 6 μg/ml inhibited RNA synthesis by 6% and DNA synthesis by 35%; 50μg/ml sesaminol inhibited RNA synthesis by 76% and DNA synthesis by 83%.

Miyahara *et al.* (2000) reported that sesamin and episesamin were both effective inducers of apoptosis in leukemia Molt 4B cells. Apoptosis occurred earlier in the sesamin-treated cells than in the episesamin-treated cells. In addition, sesamolin and sesaminol were found to inhibit lymphoid leukemia Molt 4B cell growth by inducing apoptosis (Miyahara *et al.*, 2001a; Miyahara *et al.*, 2001b). Kaur and Saini (2000) reported that sesamol had strong antimutagenic activity in the Ames tester strains TA100 and TA102, and that this activity may be related to its antioxidant properties.

3. *Effect of processing on sesame seed oil components*

The effectiveness of sesame seed oil as a functional food component is dependent on retention of the lignans and tocopherols in the oil. The seeds are often roasted to enhance the flavor of the oil, but heat processing can reduce the antioxidant concentration of the expressed sesame seed oil.

Seed roasting is a critical step in the processing of sesame seed oil, as documented by Fukuda *et al.* (1986a, 1986b). Yoshida (1994) evaluated the quality characteristics of oils obtained from roasted sesame seeds. Roasting at 150–180°C did not significantly affect the oxidative status of the oil. The γ-tocopherol and sesamolin contents were not significantly affected by roasting the seeds at 180°C but were depleted at 250°C. Further study was conducted to assess the relationship between sesame seed oil quality and the temperature and time of the roasting process (Yoshida and Takagi, 1997). For a given roasting temperature, the general trend was that oxidative indices increased with increased roasting time. Little difference in oxidative status was observed in oils from seeds roasted between 160°C and 250°C as long as the roasting time did not exceed 15 min. The retention of γ-tocopherol after a 25 min roasting of the seeds at 160°C, 180°C, 200°C, 220°C and 250°C was 88%, 87%, 85%, 84% and 83%, respectively. In contrast, the retention of δ-tocopherol for the same treatments was 39%, 34%, 32%, 32% and 30%, respectively. Sesamin retention under the harshest conditions (25 min at 250°C) was 94% while only 0.5% retention was observed for sesamolin under the same conditions. The concentration of sesamol in the non-roasted samples was 5.1 mg/kg, and after 25 min roasting of the seeds at 160°C, 180°C, 200°C, 220°C or 250°C the sesamol content increased by a factor of 1.7, 13.2, 23.9, 69 or 661, respectively. Yoshida and Takagi (1997) noted that a high-quality product could be obtained by roasting sesame seeds for 25 min at 160°C or 180°C, for 15 min at 200°C, or for 5 min at 220°C. Based on sensory analysis, roasting at 250°C did not produce acceptable oils. Heating at 220°C for 10 min was sufficient to alter the triacylglycerol and phospholipid levels (Yoshida *et al.*, 2000). Later studies by Yoshida *et al.* (2001) supported their earlier findings, and showed that unsaturated fatty acids located at the *sn*-2 position were significantly protected against degradation. Taking into consideration both antioxidant and lipid stability and sensory evaluation, roasting at 180°C for 25 min produced the most acceptable oil. Yen (1990) had found similar results but considered that sesame oil prepared from seeds roasted at 200°C had the best flavor score.

Yoshida and Kajimoto (1994) used microwave heating (500 W, 2350 MHz) as a sesame roasting method. All of the α-tocopherol in the seeds was lost after

8 min of microwave heating, and all of the δ-tocopherol was lost after 16 min, whereas only 6.5% and 20% reductions in γ-tocopherol levels were found after 20 min and 30 min of heating, respectively. The original contents of sesamin, sesamolin and sesamol were 6824 mg/kg, 5642 mg/kg and 54 mg/kg, respectively. The sesamin and sesamolin concentrations decreased by approximately 20% after 30 min of microwave heating; in contrast, the sesamol concentration increased 10 fold. Lipids in general were not significantly affected by microwave heating for up to 30 min, by which time the seeds had a burnt and bitter taste (Yoshida *et al.*, 1995). Shahidi *et al.* (1997) also noted a significant reduction in the antioxidant levels of sesame seeds processed using microwave heating. Under appropriate conditions, microwave energy has similar effects on oil quality as the oven-roasting process.

The stability of processed sesame seeds is dependent on the processing method. Shahidi *et al.* (1997) showed that the loss of antioxidant components was caused by a combination of factors, including seed variety, processing method, and whether or not the seed was dehulled. Irrespective of treatment, the Sudanese variety 'Gabaly Abiad' had higher retentions of sesamin (94%), sesamolin (88%) and γ-tocopherol (86%) than the Egyptian variety 'Giza 24', which had retentions of 91%, 74% and 82%, respectively. Dehulling of the Giza 24 seeds had little influence on the retention of antioxidant components. Sesamin was the only component affected by the dehulling process: a 5% greater loss of sesamin was found in dehulled samples than in non-dehulled samples (Shahidi *et al.,* 1997). Different heat treatments affected the antioxidant components in different ways. For hulled seeds, roasting at 200°C for 20 min caused the greatest reduction in sesamin and γ-tocopherol levels, while microwave heating caused the greatest reduction in sesamolin content. For dehulled seeds, steam treatment (100°C) promoted the greatest loss of sesamolin. Over the course of 35 days of storage, all samples had significant reductions in antioxidant components. Raw samples were more oxidatively stable than processed samples. Seeds treated with steam followed by roasting had the greatest oxidative stability and retention of antioxidants (Shahidi *et al.,* 1997).

4. *Antioxidant activity*

Blending oils with high levels of monounsaturated lipids with oils that have high levels of polyunsaturated fatty acids is a unique method of enhancing stability without hydrogenation. Chu and Kung (1998) found that crude sesame seed oil dramatically improved the oxidative stability of refined soybean oil. As previously highlighted, blends of rice bran oil and sesame seed oil are effective antioxidants in frying applications (Kamal-Eldin *et al.,* 1998; Kochhar, 2000). However, the antioxidant effectiveness of a hexane–isopropanol extract of crude sesame seed oil did not show a clear-cut relationship between sesame seed antioxidant levels and oxidative stability in a 'Rancimat' test system

(Kamal-Eldin and Appelqvist, 1994). The combination of a high degree of monounsaturation with increased antioxidant (avenasterol, γ-tocopherol and lignans) content may account for the improvements in frying oil performance following the addition of sesame seed oil.

Mohamed and Awatif (1998) used the unsaponifiable fraction of sesame oil as a natural antioxidant. The Δ^5- and Δ^7-avenasterols account for 12% and 0.6% of the unsaponifiables in non-roasted white sesame seeds, and 14% and 0.8% of the unsaponifiables in roasted white sesame seeds, respectively. Similar levels were found for dark sesame seeds. The tocopherol and lignan contents of the unsaponifiable fraction were higher in the white variety than in the dark variety, but decreased significantly during roasting. In contrast, the sesamol content increased during roasting, to 16 mg/kg for white varieties and 12 mg/kg for dark varieties. In general, addition of the unsaponifiable materials to sunflower oil enhanced the oxidative stability of the oil. The addition of 0.1% unsaponifiables from non-roasted or roasted seeds was more effective than the addition of lower concentrations of the materials (0.02% and 0.05%). Unsaponifiables from roasted sesame seeds appeared to be slightly more effective at inhibiting oxidation than those from non-roasted seeds (Mohamed and Awatif, 1998).

Yoshida and Takagi (1999) evaluated the role of sesame seed antioxidants in tocopherol-stripped oils. They found that microwave heating (500 W, 2450 MHz) significantly promoted oxidation of the stripped oil. When tested alone at a concentration of 800 ppm, sesamol had slightly but not significantly more antioxidant activity than γ-tocopherol or δ-tocopherol. Combinations of γ-tocopherol and sesamol were also found to be effective at preventing oxidation. After 25 min of microwave heating, the two most effective treatments were 400 ppm γ-tocopherol plus 50 ppm sesamol, and 400 ppm γ-tocopherol plus 400 ppm sesamol. During both combination treatments, the tocopherol content reduced by approximately 15–17% and the sesamol content reduced by approximately 9–11% (Yoshida and Takagi, 1999).

D. Rosemary (*Rosmarinus officinalis* L.) extract

1. General characteristics

The antioxidant activity of rosemary has been well documented over the past 50 years. The active constituents include phenolic acids, diterpenes, and flavonoids. Cuppett and Hall (1998), Ho *et al.* (2000) and Yanishlieva and Heinonen (2001) have completed thorough reviews on rosemary antioxidants, and readers should consult these for additional information. Three prominent compounds – carnosic acid, carnosol, and rosmarinic acid (Figure 3) – have received the most attention. Carnosic acid and carnosol account for 1.7–5.5%

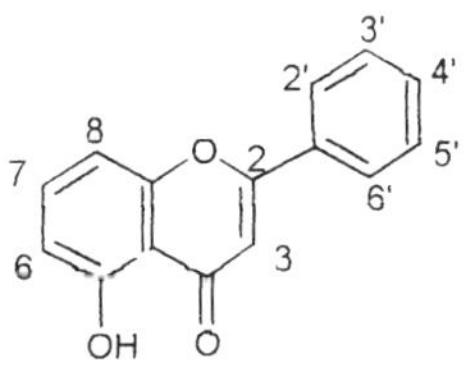

	6	7	3'	4'
Apigenin	H	OH	H	OH
Genkwanin	H	OCH_3	H	OH
Cirsimaritin	OCH_3	OCH_3	H	OH
Nepitrin	OCH_3	OGlu	OH	OH
Linaroside	OCH_3	OGlu	H	OCH_3

Selected flavones

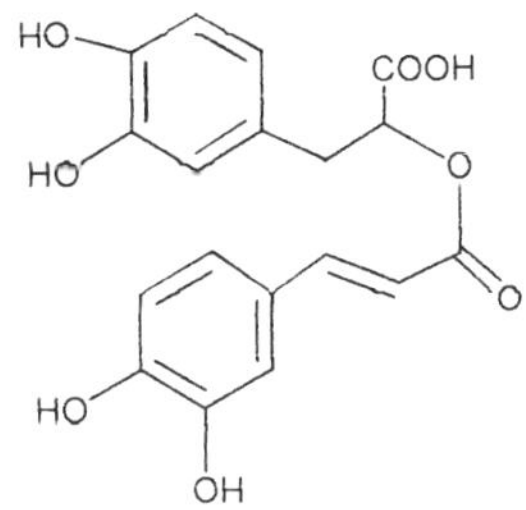

Rosmarinic acid

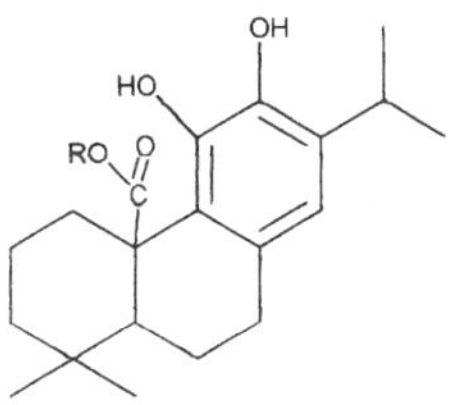

Carnosic acid: R = H
12-Methyl carnosate: R = CH_3

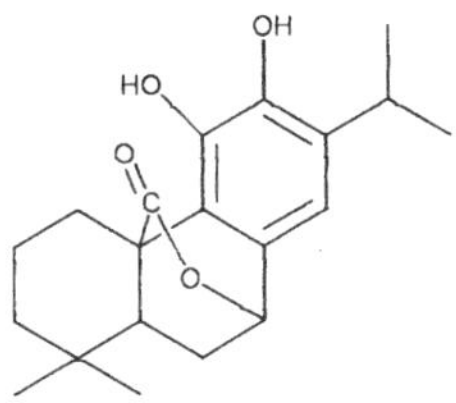

Carnosol

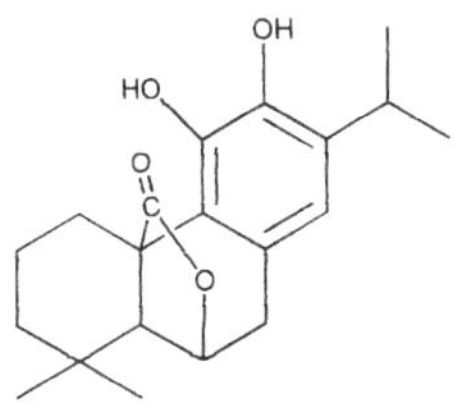

Rosmanol

Figure 3. Antioxidant components of rosemary.

and 0.2–0.4% of the dry weight of rosemary, respectively (Aeschbach and Philippossian, 1993; Richheimer *et al.*, 1996). Similar levels (0.33–3.0%) of rosmarinic acid have been isolated from dried rosemary leaves, whereas up to 10% rosmarinic acid can be obtained from cell cultures of *Rosmarinus officinalis* (Verotta, 1985; Komali and Shetty, 1998). In addition, approximately 15 flavonoids have been identified (Okamura *et al.*, 1994; Cuvelier *et al.*, 1996). Other components, such as rosmanol and 12-methyl carnosate (Figure 3), also show antioxidant activity. However, carnosic acid, carnosol, and rosmarinic acid appear to be the major compounds contributing to the health benefits of rosemary and targeted as potential functional food ingredients.

2. *Health benefits*

A variety of health benefits including cancer prevention, anti-inflammatory, antiviral, and anti-tumor activities have been claimed for carnosic acid, carnosol, and rosmarinic acid. Minnunni *et al.* (1992) found that a rosemary extract containing 8.8% carnosic acid and 1.2% carnosol had antimutagenic activity in the Ames test strain TA102. They also reported that carnosic acid had a stronger antimutagenic activity than carnosol, which these authors attributed to the higher radical-scavenging activity of carnosic acid. However, both compounds are powerful inhibitors of lipid oxidation in microsomes and liposomes and are good scavengers of peroxy, hydroxyl and superoxide radicals (Aruoma *et al.*, 1992; Haraguchi *et al.*, 1995). Haraguchi *et al.* (1995) reported that carnosic acid and carnosol were effective inhibitors of NADH- or NADPH-promoted oxidation of mitochondrial and microsomal lipids at concentrations as low as 3 μM, with complete inhibition at 9 μM. In addition, carnosic acid protected human erythrocytes against oxidative hemolysis. Rosmarinic acid has also been shown to be capable of scavenging superoxide radicals, generated by the xanthine–xanthine oxidase system, in HL-60 cells (Nakamura *et al.*, 1998).

In a human aortic endothelial cell system, carnosic acid, carnosol, and rosmarinic acid all inhibited LDL oxidation (Pearson *et al.*, 1997). Their relative activities were: carnosol > carnosic acid ≈ rosmarinic acid. The concentrations needed to achieve 50% inhibition of cell-mediated LDL oxidation were 0.33 μM, 0.74 μM, and 0.81 μM for carnosol, rosmarinic acid, and carnosic acid, respectively. The most effective concentration for maximum inhibition of LDL oxidation was 2.5 μM for all compounds.

An ethanolic extract (0.15 g/100 g body weight) of rosemary protected rats against experimentally induced hepatotoxicity (Fahim *et al.*, 1999). Amagase *et al.* (1996) reported that feeding rats diets containing 0.5% or 1% rosemary reduced 7,13-dimethylbenz[*a*]anthracene-induced DNA adduct formation by 66% and 76%, respectively. However, supplementation of rat diets with 1% carnosol did not significantly reduce the *in vivo* formation of 7,13-dimethylbenz[*a*]anthracene-induced DNA adducts, although intraperitoneal injection of carnosol (200 mg/kg body weight) was able to reduce formation of the DNA adducts (Singletary and Nelshoppen, 1991; Singletary *et al.*, 1996). In bronchial and liver cell model systems, carnosic acid and carnosol were effective in preventing DNA adduct formation by benz[*a*]pyrene or aflatoxin B_1 (Offord *et al.*, 1995, 1997). Offord *et al.* (1997) noted that these compounds prevented DNA adduct formation by two mechanisms: by inhibiting the metabolic activation of pro-carcinogens catalysed by phase 1 cytochrome P-450 enzymes, and by inducing detoxification enzymes such as glutathione *S*-transferase. Singletary (1996) also reported that carnosol was able to stimulate rat liver glutathione *S*-transferase and quinone reductase at dietary concentrations of 0.25–1%. Debersac *et al.* (2001) reported that glutathione *S*-transferase and quinone reductase were induced by an aqueous extract of rosemary but not by rosmarinic acid. They suggested that induction of the enzymes was due to flavones or monoterpenes in the extract, or to an additive effect of all the extract components.

Aruoma *et al.* (1996) reported that a deodorized rosemary extract inhibited the peroxidation of phospholipid liposomes and human immunodeficiency virus (HIV) infection. They noted that purified carnosol, at a concentration of 2 μM, also exhibited anti-HIV activity. However, Paris *et al.* (1993) found that carnosol was less effective than carnosic acid as an inhibitor of HIV protease in cell-free assay.

Rosmarinic acid has been found to have anti-inflammatory activity (Sahu *et al.*, 1999). Some anti-inflammatory activity may also be attributed to carnosol, which was found to inhibit cyclooxygenase in rat peritoneal leukocytes (Laughton *et al.*, 1991). Anti-ulcerogenic activity also was found in rats given 70% alcohol extracts of rosemary (Dias *et al.*, 2000). The anti-ulcerogenic activity was not due to an inhibition of prostaglandin production, but was related to an increase in the content of mucosal non-protein sulfhydryl groups.

3. Extraction of rosemary components

A variety of extraction technologies have been employed to produce rosemary antioxidant products, as reported in at least 15 different patents. Many use polar or non-polar solvents (Chang *et al.*, 1976; Todd, 1991; Aeschbach and Philippossian, 1993; Bailey *et al.*, 1999) or SC-CO_2 (Nguyen *et al.*, 1991). Different extraction techniques yield products with differing concentrations of antioxidant components.

Alcoholic extraction is a common practice to obtain a crude rosemary extract, but solvents such as hexane, petroleum ether, or acetone can also be used (Chang *et al.*, 1976; Todd, 1991). The extracts are generally further purified by distillation or by filtration through charcoal to remove pigments and volatile constituents. Methods to concentrate active fractions depend on the solubility traits of the antioxidants and other rosemary components. Chang *et al.* (1976) found that methanol gave the highest extract yield. However, the methanol extract had less antioxidant activity than diethyl ether or ethylene dichloride (1,2-dichloroethane) extracts. Aeschbach and Philippossian (1993) found that a 95:5 (v/v) mixture of petroleum ether and ethanol gave the highest extraction yield, but that a 98:2 (v/v) mixture of the same solvents yielded the extract with the highest carnosic acid content (16%). Another method to enhance carnosic acid levels involves extracting rosemary with acetone, followed by a concentration step and re-suspending the residue in hexane. The hexane suspension is filtered and then extracted with aqueous bicarbonate. The bicarbonate layer is adjusted to pH 2.2 and then extracted with hexane. After removal of the hexane, the sample is recrystallized to provide a creamy-colored product that is approximately 71–95% carnosic acid (Bailey *et al.*, 1999).

Tena *et al.* (1997) reported that extracts obtained with SC-CO_2 had the highest average carnosic acid concentration (35.7 mg/g), followed by acetone (26.2 mg/g), methanol (15.9 mg/g), dichloromethane (7.9 mg/g) and hexane (1.9 mg/g) extracts. An added advantage of SC-CO_2 was that the extract was free of pigments and so required less clean-up. Optimal SC-CO_2 extraction conditions were 35.5 MPa and 100°C with a CO_2 flow rate of 4 ml/min. Hidalgo *et al.* (1998) reported similar carnosic acid yields (18–35.1%) for SC-CO_2 extraction of wild and cultivated rosemary. Bicchi *et al.* (2000) also reported that SC-CO_2 extraction gave higher yields of both carnosic acid and carnosol than an acetone extract prepared via Soxhlet or sonication. A combination of SC-CO_2 and methanol extraction was effective in isolating various fractions from rosemary. SC-CO_2 extraction alone at 25 MPa and 60°C yielded an extract that was predominantly essential oils, while the addition of 5% methanol yielded an extract rich in carnosic acid and carnosol, and the addition of 10% methanol yielded an extract rich in rosmarinic acid (Bicchi *et al.*, 2000). The fractionation of essential oils from the phenolic antioxidants in rosemary was also demonstrated by Ibàñez *et al.* (1999). These authors noted that SC-CO_2 extraction at 10 MPa and 40°C gave an extract with a different essential oil composition and stronger aroma than extraction at 40 MPa and 60°C. Carnosic acid, carnosol, rosmanol, and methyl carnosate were present in rosemary extracts obtained by SC-CO_2 extraction at 35 MPa and 50°C, and at 300 MPa and 40°C; carnosic acid concentrations in the extracts were 82% and 63%, respectively (Señoráns *et al.*, 2000; Ibàñez *et al.*, 2000). As with the extraction of rice bran antioxidants, SC-CO_2 technology may provide an alternative to typical solvent extraction.

4. Antioxidant activity

Rosemary extracts have long been recognized as effective antioxidants, and attempts have been made to identify the active compound. However, conflicting evidence suggests that there may be more than one active component. The system used to evaluate the antioxidant activity is also critical in these studies.

Cuvelier *et al.* (1996) assessed the antioxidant activity of 24 pilot-plant and commercial rosemary extracts and identified 22 different compounds for investigation. These included diterpenes, flavonoids, and phenolic acids. There was no apparent correlation between antioxidant activity and extract composition but the most effective extracts contained carnosol, rosmarinic acid, and carnosic acid and, to a lesser extent, caffeic acid, rosmanol, rosmadial, cirsimaritin, and genkwanin. Carnosol was a component of all 24 extracts while rosmarinic and carnosic acids were found in 83% and 71% of the extracts, respectively (Cuvelier *et al.*, 1996). Richheimer *et al.* (1996) also evaluated the antioxidant components in a variety of plant and commercial rosemary products. In the commercial products, carnosol and carnosic acid were the predominant forms with low levels of methyl carnosate. No clear relationship between the type of commercial rosemary extract and antioxidant composition was established. These investigators found that 7-methoxy-rosmanol was present in the commercial extracts but not in extracts obtained from fresh rosemary, and speculated that this compound was a degradation product formed during processing. This observation has been supported by Schwarz and Ternes (1992) and by Cuvelier *et al.* (1994), and two quinone degradation products resulting from the oxidation process have been reported by Masuda *et al.* (2001).

Frankel and colleagues made a thorough study of the antioxidant activity of rosemary in various food systems and evaluated the effectiveness of carnosol, carnosic acid and rosmarinic acid in bulk oil and emulsion systems (Frankel *et al.*, 1996a, 1996b; Huang *et al.*, 1996; Hopia *et al.*, 1996). Frankel *et al.* (1996a) found that rosemary extract (500 ppm) was a more effective antioxidant than its purified components in bulk corn oil, as indicated by its significantly greater inhibition of hydroperoxides (81%) and hexanal (98%) formation. Rosmarinic acid (50 ppm) was the best individual antioxidant in bulk oil, resulting in greater inhibition of hydroperoxides (68%) and hexanal (78%) than carnosic acid (70% and 29% inhibition, respectively) or carnosol (6% and –19% inhibition, respectively). In bulk oil, 100 ppm tocopherol had an antioxidant activity comparable to that of carnosic acid (Frankel *et al.*, 1996a). In contrast, in an oil-in-water emulsion, tocopherol was a significantly better antioxidant than carnosic acid. In the emulsion, carnosol was the most effective rosemary antioxidant, inhibiting hydroperoxide formation by 42% and hexanal formation by 84%, but it was not significantly better than carnosic acid (37% and 83% inhibition, respectively). In the emulsion, the rosemary extract (500 ppm) was significantly less active than its purified components, inhibiting hydroperoxide formation by 12% and hexanal formation by 28%. Rosmarinic acid at 50 ppm was found to have pro-oxidant activity, but at 30 ppm it had antioxidant activity comparable to that of the rosemary extract, inhibiting hydroperoxide formation by 16% and hexanal formation by 22%.

Frankel *et al.* (1996a) also observed that carnosol and carnosic acid were significantly more active antioxidants in oil-in-water emulsions buffered at pH 4 or 5 than at pH 7. Hopia *et al.* (1996) reported that carnosic acid was a better antioxidant than carnosol in methyl linoleate, but that the reverse was true in linoleic acid. In methyl linoleate or linoleic acid, carnosic acid and carnosol were more active than α-tocopherol. However, α-tocopherol was more active in bulk corn oil triacylglycerols, followed by carnosic acid and carnosol. The oxidative stability of rapeseed oil treated with rosemary extracts correlated well with the content of carnosic acid, but not with content of carnosol (Trojakova *et al.*, 2001). In all emulsified systems, α-tocopherol exhibited more antioxidant activity than carnosol or carnosic acid (Hopia *et al.*, 1996). In bulk corn oil, the addition of carnosol to α-tocopherol had a negative impact on the antioxidant activity of α-tocopherol, whereas carnosic acid enhanced α-tocopherol activity (Hopia *et al.*, 1996). Both carnosic acid and carnosol disappeared from the test system more quickly than α-tocopherol (Hopia *et al.*, 1996).

Methyl carnosate was found to be a more effective antioxidant than carnosic acid in both bulk oil and oil-in-water emulsions (Huang *et al.*, 1996). This is in contrast to observations made by Richheimer *et al.* (1996), who found methyl carnosate to be ineffective. These differences may be due to the system used to test antioxidant activity. For example, Richheimer *et al.* (1996) used a 'Rancimat' system at 100°C, which may have promoted degradation of the methyl carnosate or may have induced volatilization due to the presence of the methyl ester. Additional work is needed to characterize the differences in the observed antioxidant activity of methyl carnosate.

Rosemary has been found to have antioxidant activity in irradiated and frying systems. The addition of ground rosemary to linoleic acid (1:10 rosemary:lipid ratio) inhibited the formation of C_{10}–C_{19} hydrocarbons by 52.5–80.5% during exposure to γ-irradiation (Lacroix *et al.*, 1997). The addition of rosemary to raw pork prior to irradiation (4.5 kGy) delayed the onset of oxidation for 3 days (Chen *et al.*, 1999), although the addition of sesamol was more effective in both irradiated raw and cooked pork during 7 days of storage. Gordon and Kourimska (1995) found that adding a rosemary extract or ascorbyl palmitate to oils retarded the formation of dimers and retarded losses of natural tocopherols during deep-fat frying. The addition of rosemary extracts to frying oil inhibited the formation of polar substances and polymers and the decomposition of polyunsaturated triacylglycerols (Reblova *et al.*, 1999). Guor *et al.* (1999) noted that thermal treatments resulted in the formation of a dark-red viscous material that upon purification was identified as a mixture of miltirone and dehydrorosmariquinone. Balogh *et al.* (2000) reported that the formation of heterocyclic aromatic amines in ground beef patties fried at 225°C was significantly lower in patties treated with either 1% or 10% (fat basis) rosemary oleoresin. Similarly, Murkovic *et al.* (1998) found that rosemary reduced the content of five heterocyclic aromatic amines to <60% of control levels.

E. Flavonoids

1. General characteristics

Flavonoids are a group of polyphenolic compounds characterized by a C_6-C_3-C_6 structural unit (Figure 4) and are ubiquitous in nature. Fruits and vegetables are major dietary sources and a typical Western diet supplies approximately 1 g of mixed flavonoids daily (Nieto *et al.*, 1993; Rodgers and Grant, 1998). The intake of flavanols, an important group of flavonoids, is in the range of 6–60 mg/day (Hollman and Arts, 2000). Examples of foods with high flavonol and flavone contents include kale(450 mg/kg), onions (300 mg/kg), broccoli (100 mg/kg), beans (50 mg/kg), apples (50 mg/kg), blackcurrants (40 mg/kg) and tea (30 mg/kg) (Hollman and Arts, 2000). A significant number of reports

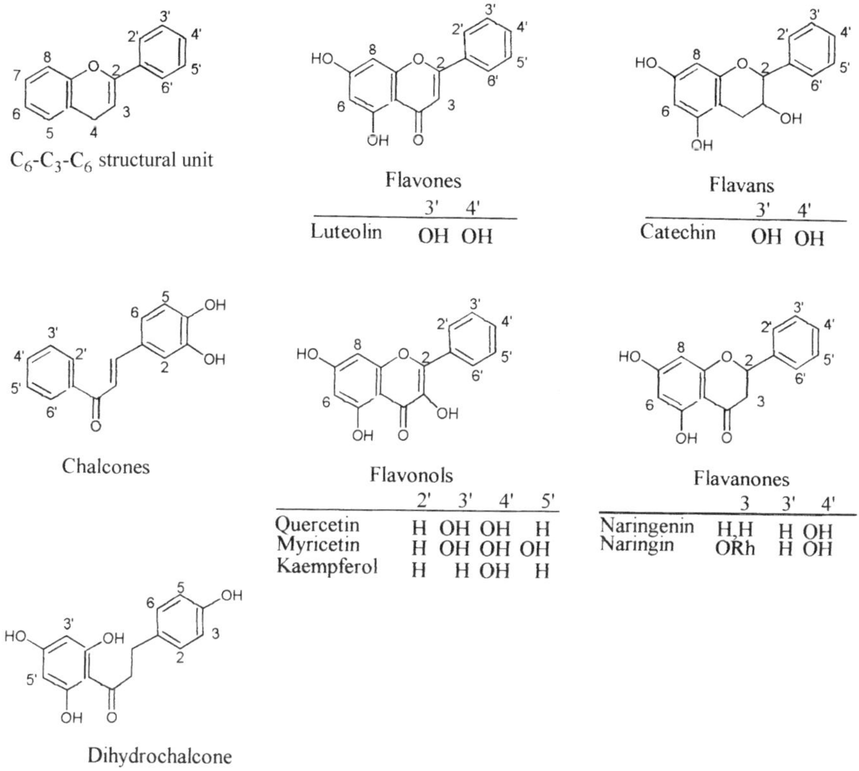

Figure 4. General categories and examples of common flavonoids, a group of polyphenolic compounds characterized by a C_6-C_3-C_6 structural unit. Rh, rhamnose.

have described the biological activity of flavonoids, which can act as radical scavengers, metal chelators, and enzyme inhibitors (Bors *et al.*, 1990; Middleton, 1998; Salah *et al.*, 1995). The role of flavonoids as food antioxidants and health promoters have recently been reviewed (Hall and Cuppett, 1997; Middleton, 1998; Pietta, 2000; Yanishlieva and Heinonen, 2001).

2. *Health benefits*

Flavonoids have anticarcinogenic and anti-tumor activity, protect against cardiovascular disease, and act as anti-inflammatory compounds (Hollman and Katan, 1998; Middleton, 1998; Waladkhani and Clemens, 1998; Agarwal, 2000). Hirano *et al.* (1994) evaluated the cytostatic activity of flavonoids on human breast carcinoma cells and found them to be effective in preventing cell growth, but failed to find a structure–activity relationship among the 21 compounds tested. Sánchez *et al.* (2001) found that flavonoids lacking a C-8 methoxy group had little cytotoxicity against Rhesus monkey kidney cells and rat glial tumor cells. However, the presence of methoxy groups at the C-2′ and C-5′ locations were also important structural features. The presence of a C-7 hydroxyl group was essential for anti-17β-hydroxysteroid dehydrogenase activity, whereas flavonoids with C-7 methoxy or C-8 hydroxyl groups showed only anti-aromatase activity (Bail *et al.*, 1998). Bagchi *et al.* (1998a) noted that mice treated with a grapeseed extract decreased 12-*O*-tetradecanoyl phorbol-13-acetate (TPA) induced production of reactive oxygen species by 70% and cytochrome *c* reduction by approximately 65%, compared with controls. DNA fragmentation decreased by 47% in hepatic tissues and by 50% in brain tissues. Grapeseed proanthocyanins effectively reduced hydrogen peroxide-induced oxidation by as much as 70% (Bagchi *et al.*, 1998b). Anti-tumor activity of grapeseed polyphenolics was observed by Bomser *et al.* (1999) and by Zhao *et al.* (1999). The most active component was procyanidin B5-3′-gallate (Zhao *et al.*, 1999). Agarwal *et al.* (2000) found that a 100 μg/ml dose of grapeseed polyphenolics inhibited the growth of DU145 human prostate carcinoma cells up to 98% by 6 days of treatment. Kamei *et al.* (1998) found that an extract of red and white wines, containing high levels of delphinidin, had greater anti-tumor activity than fractions lacking anthocyanins. Quercetin was found to be the most active flavonoid at inhibiting growth and altering enzyme activities of MCF7 human breast cancer cells, followed by myricetin and epicatechin (Rodgers and Grant, 1998).

Wang and Jiao (2000) found that blackberry juice had better radical-scavenging activity than other berries, and inhibited the formation of superoxide radicals, hydroxyl radicals and hydrogen peroxide to a greater extent. The radical-scavenging activity of the blackberry juice (100 g fresh weight) was equivalent to that of 3.78 mg ascorbic acid for H_2O_2, 11 mg of glutathione for superoxide radicals, 2.4 mg of β-carotene for singlet oxygen, and 19.1 mg of chlorogenic acid for hydroxyl radicals. A number of studies have shown that flavonoids effectively inhibit the oxidation of LDL (Meyer *et al.*, 1997, 1998a, 1998b; Brown and Rice-Evans, 1998; Heinonen *et al.*, 1998; Hwang *et al.*, 2001; Porter *et al.*, 2001). In addition, flavonoids such as hesperetin can inhibit cholesterol ester synthesis (Borradaile *et al.*, 1999). The lowering of plasma cholesterol levels by naringenin and hesperetin was due to a number of cellular activities including a reduction in acyl CoA:cholesterol acyltransferase activity, inhibition of the activity and expression of microsomal triacylglycerol transfer protein, and increased levels of LDL receptor mRNA (Wilcox *et al.*, 2001).

Other health benefits include anti-ulcer (Saito *et al.*, 1998), antiviral (Wang *et al.*, 1998) and anti-inflammatory activities. Wang *et al.* (1999a) reported that cyanidin from tart cherries had better anti-inflammatory activity than aspirin. Vitexicarpin (3′,5-dihydroxy-3,4′,6,7-tetramethoxyflavone) was shown to be a potent inhibitor of both T- and B-lymphocyte proliferation (You *et al.*, 1998). These results suggest that vitexicarpin may be a potential anti-inflammatory or immunoregulatory compound. The anti-inflammatory activity of flavonoids has been associated with the inhibition of thromboxane synthase (Skaltsa *et al.*, 2000) and of prostaglandin production (Ishiwa *et al.*, 2000).

Although not all-inclusive, these studies demonstrate that the structure of a flavonoid has a significant role in dictating its biological activity. A growing body of evidence supports the claimed health benefits of flavonoids. The development of functional foods using flavonoids will depend on the stability of these compounds in food systems.

3. *Effects of processing on flavonoid stability*

The antioxidant activity of flavonoids can be affected by a number of factors such as pH, enzymes, light and oxygen. Similar to the biological activity, the stability of flavonoids is dependent on the structural features. One critical feature is the presence of the *o*-diphenol moiety (Cheynier *et al.,* 1994; Sarni *et al.*, 1995), which tends to undergo oxidation more readily than non-*o*-diphenolic flavonoids. Kader *et al.* (1998) reported that cyanidin 3-glucoside (*o*-diphenolic anthocyanin) degradation did not occur in the presence of polyphenol oxidase (PPO) alone. However, PPO plus chlorogenic acid promoted its degradation, whereby chlorogenic acid and cyanidin 3-glucoside were involved in a coupled oxidation process. Cyanidin 3-glucoside was consumed during the process by a hydrogen-donating mechanism that regenerated the chlorogenic acid. The resulting cyanidin 3-glucoside quinone could condense with cyanidin 3-glucoside or chlorogenoquinone to form polymeric compounds (Kader *et al.*, 1998). Similar findings were reported for the degradation of cyanidin 3-glucoside in the presence of caftaric acid (Sarni *et al.*, 1995). Chlorogenoquinone will also undergo condensation with the non-*o*-diphenolic anthocyanin pelargonidin 3-glucoside, to form polymeric compounds. However, in this case limited chlorogenic acid regeneration occurred, indicating hydrogen donation from pelargonidin 3-glucoside to chlorogenic acid was not part of the oxidation mechanism (Kader *et al.*, 1999). Similar findings were observed for reactions between malvidin 3-glucoside and caftaric acid (Sarni *et al.*, 1995), and between malvidin or delphinidin and caftaric acid (Cheynier *et al.*, 1994).

Heat can also promote the degradation of flavonoids. Seeram *et al.* (2001) reported that 2,4-dihydroxybenzoic acid, 2,4,6-trihydroxybenzoic acid and protocatechuic acid were the major degradation products of cyanidin glycosides after heating under reflux or by microwave heating. Drying of red grape pomace at 100°C or 140°C significantly reduced the amount of extractable polyphenolics, but a mild heat treatment (60°C) did not (Larrauri *et al.*, 1997). Microwave processing of strawberry purées did not significantly affect the anthocyanin content (Ancos *et al.*, 1999). The short microwave processing time (<1 min) may account for the contradictory observations found by Seeram *et al.* (2001). Also, the mild heat treatment in jam processing did not significantly promote the degradation of quercetin, myricetin or kaempferol (Amakura *et al.*, 2000; Häkkinen *et al.*, 2000). Häkkinen *et al.* (2000) reported only a 15% reduction in quercetin content and 18% reduction in kaempferol content for strawberry jams. Soups made from bilberries had a 40% loss in quercetin, while steam-extracted blackcurrant juice lost 67% of the myricetin and 85% of the quercetin. Blackcurrant juice obtained from cold pressing had higher quercetin (48%) and myricetin (37%) retention. However, although a higher initial reduction in quercetin content occurred during steam extraction, additional losses were minimal over 9 months of storage. In contrast, 100% of the kaempferol in strawberry jam was lost by 3 months of storage, and 100% of the myricetin in steam-extracted blackcurrant juice was lost after 9 months (Häkkinen *et al.*, 2000).

4. *Antioxidant activity*

Yokozawa *et al.* (1997) evaluated 41 flavonoids for radical-scavenging activity and found that several structural components were needed for effective radical scavengers. These included the presence of an *o*-dihydroxy group on ring B, a 2,3-double bond conjugated with a 4-oxo group, and 3- and 5-hydroxyl units on ring A. Compounds with all three structural features tended to have the best radical-scavenging activity. Pekkarinen *et al.* (1999) noted that myricetin had the best radical-scavenging activity in methyl linoleate, followed by quercetin, (+)-catechin, kaempferol, and rutin. These relative activities were related to the presence of the structural features important for radical-scavenging activity. Myricetin had the most hydroxyl groups on the B ring, followed by quercetin and kaempferol. In addition, quercetin has the 2,3-double bond, unlike (+)-catechin. The presence of the sugar moiety on rutin greatly reduced its activity, possibly because of solubility problems. Myricetin also protected α-tocopherol from decomposition (Pekkarinen *et al.*, 1999).

The lipid system also influences the effectiveness of flavonoids as antioxidants. For example, (+)-catechin was more effective than quercetin in an aqueous-based system (Meyer *et al.*, 1998b). Wang *et al.* (1999b) found that 6,7-dimethoxy-5,8,4′-trihydroxyflavone was the most active of the polyphenols isolated from tart cherries. These researchers also noted that the pure compound was more effective than a mixture of the isolated phenols at the concentrations originally found in the extract.

The processing and handling of flavonoid-rich sources can significantly affect their antioxidant capacity. Heating of grape pomace at 100°C significantly reduced its antioxidant activity and radical-scavenging activity (Larrauri *et al.*, 1997; Larrauri *et al.*, 1998). Tsuda *et al.* (1994) found that cyanidin was a better antioxidant in liposomes than cyanidin 3-glucoside. They proposed that the activity of cyanidin 3-glucoside after 10 days was due to its conversion to cyanidin. Murakami *et al.* (1984) noted that well-fermented tempeh lacked isoflavone glucosides but retained high concentrations of isoflavones in the aglycone form. Daidzein, glycitein and genistein are the predominant isoflavones present in tempeh; daidzein and genistein are considered to be active antioxidants. Other examples of antioxidant activity of flavonoids can be found in recent reviews (Hall and Cuppett, 1997; Hall, 2001).

F. Conclusions

Rice bran oil and sesame seed oil are two underutilized oils with good antioxidant activity in a wide variety of test systems. These oils are unique in their inherent oxidative stability, which is attributed to the presence of a variety of compounds such as sterols and phenols in addition to high levels of monounsaturated fatty acids. Rosemary extracts have been used commercially to enhance the oxidative stability of oils and meats and to prevent heterocyclic amine formation in fried products. Flavonoids have been researched extensively in biological systems and have been found to be very good *in vivo* antioxidants, although these compounds tend to be less stable than the other antioxidant sources described here. In addition to their antioxidant effects, many of these sources have biological activities such as blood cholesterol-lowering, anti-carcinogenic, or anti-inflammatory properties. These materials are ideal candidates for incorporating into functional foods due to their antioxidant properties and beneficial biological activities.

References

Aeschbach, R and Philippossian, G (1993) Carnosic acid obtention and uses. US Patent 5,256,700.

Agarwal, R (2000) Cell signaling and regulators of cell cycle as molecular targets for prostate cancer prevention by dietary agents. *Biochem. Pharmacol.*, 60, 1051–1059.

Agarwal, C, Sharma, Y and Agarwal, R (2000) Anticarcinogenic effect of a polyphenolic fraction isolated from grape seeds in human prostate carcinoma DU145 cells: modulation of mitogenic signaling and cell-cycle regulators and induction of G1 arrest and apoptosis. *Mol. Carcinogen.*, 28, 129–138.

Akihisa, T, Yasukawa, K, Yamaura, M, Ukiya, M, Kimura, Y, Shimizu, N and Arai, K (2000) Triterpene alcohol and sterol ferulates from rice bran and their anti-inflammatory effects. *J. Agric. Food Chem.*, 48, 2313–2319.

Amagase, H, Sakamoto, K, Segal, E and Milner, J (1996) Dietary rosemary suppresses 7,13-dimethylbenz(a)anthracene binding to rat mammary cell DNA. *J. Nutr.*, 126, 1475–1480.

Amakura, Y, Umino, Y, Tsuji, S and Tonogai, Y (2000) Influence of jam processing on the radical scavenging activity and phenolic content in berries. *J. Agric. Food Chem.*, 48, 6292–6297.

Ancos, B, Cano, M, Hernandez, A and Monreal, M (1999) Effects of microwave heating on pigment composition and colour of fruit purees. *J. Sci. Food Agric.*, 79, 663–670.

Ashakumary, L, Rouyer, I, Takahashi, Y, Ide, T, Fukuda, N, Aoyama, T, Hashimoto, T,

Mizugaki, M and Suano, M (1999) Sesamin, a sesame lignan, is a potent inducer of hepatic fatty acid oxidation in the rat. *Metabolism*, 48, 1303–1313.

Aruoma, O, Halliwell, B, Aeschbach, R and Loliger, J (1992) Antioxidant and prooxidant properties of active rosemary constituents: carnosol and carnosic acid. *Xenobiotica*, 22, 257–268.

Aruoma, O, Spencer, J, Rossi, R, Aeschbach, R, Khan, A, Mahmood, N, Munoz, A, Murcia, A, Butler, J and Halliwell, B (1996) An evaluation of the antioxidant and antiviral action of extracts of rosemary and Provencal herbs. *Food Chem. Toxicol.*, 34, 449–456.

Ayorinde, F, Eribo, B, Balan, K, Johnson, J and Wan, L (1999) Determination of major triacylglycerol components of polyunsaturated specialty oils using matrix-assisted laser desorption/ionization time-of-flight mass spectrometry. *Rapid Commun. Mass Spectrosc.*, 13, 937–942.

Bagchi, D, Garg, A, Krohn, R, Bagchi, M, Bagchi, D, Balmoori, J and Stohs, S (1998a) Protective effects of grapeseed proanthocyanidins and selected antioxidants against TPA-induced hepatic and brain lipid peroxidation and DNA fragmentation, and peritoneal macrophage activation in mice. *Gen. Pharmacol.*, 30, 771–776.

Bagchi, D, Kuszynski, C, Balmoori, J, Bagchi, M and Stohs, S (1998b) Hydrogen peroxide-induced modulation of intracellular oxidized states in cultured macrophage J774A.1 and neuroactive PC-12 cells, and protection by a novel grapeseed proanthocyanidin extract. *Phytotherapy Res.*, 12, 568–571.

Bahkali, H, Hussain, M and Basahy, A (1998) Protein and oil composition of sesame seeds (Sesamum indicum, L.) grown in the Gizan area of Saudi Arabia. *Int. J. Food Sci. Nutr.*, 49, 409–414.

Bail, J, Laroche,T, Marre-Fournier, F and Habrioux, G (1998) Aromatase and 17β-hydroxysteroid dehydrogenase inhibition by flavonoids. *Cancer Lett.*, 133, 101-106.

Bailey, D, Richheimwr, S, Bank, V and King, B (1999) High purity carnosic acid from rosemary and sage extracts by pH-controlled precipitation. US Patent 5,859,293.

Balogh, Z, Gray, J, Gomaa, E and Booren, A (2000) Formation and inhibition of heterocyclic aromatic amines in fried ground beef patties. *Food Chem. Toxicol.*, 38, 395–401.

Bicchi, C, Binello, A and Rubiolo, P (2000) Determination of phenolic diterpene antioxidants in rosemary (Rosmarinus officinalis L.) with different methods of extraction and analysis. *Phytochem. Anal.*, 11, 236–242.

Bomser, J, Singletary, K, Wallig, M and Smith, M (1999) Inhibition of TPA-induced tumor promotion in CD-1 mouse epidermis by a polyphenolic fraction from grape seeds. *Cancer Lett.*, 135, 151–157.

Borradaile, N, Carroll, K and Kurowska, E (1999) Regulation of HepG2 cell apolipoprotein B metabolism by citrus flavonones hesperetin and naringenin. *Lipids*, 34, 591–598.

Bors, W, Heller, W, Michel, C and Saran, M (1990) Flavonoids as antioxidants: determination of radical-scavenging efficiencies. *Methods Enzymol.*, 186, 343–355.

Brown, J and Rice-Evans, C (1998) Luteolin-rich artichoke extract protects low density lipoprotein from oxidation *in vitro*. *Free Rad. Res.*, 29, 247–255.

Chang, S, Ostric-Matijasevic, B, Hang, C and Hsieh, A (1976). Method of producing an antioxidant composition from rosemary and sage. US Patent 3,950,266.

Chavali, S, Zhong, W and Forse, R (1998) Dietary α-linolenic acid increases TNF-alpha, and decreases IL-6, IL-10 in response to LPS: effects of sesamin on the delta-5 desaturation of omega-6 and omega-3 fatty acids in mice. *Prostagland. Leukotr. Essent. Fatty Acids*, 58, 185–191.

Chen, X, Jo, C, Lee, J and Ahn, D (1999) Lipid oxidation, volatiles and color changes of irradiated pork patties as affected by antioxidants. *J. Food Sci.*, 64, 16–19.

Cheynier, V, Souquet, J, Kontek, A and Moutounet, M (1994) Anthocyanin degradation in oxidising grape musts. *J. Sci. Food Agric.*, 66, 283–288.

Childs, N (2001) Rice outlook: U.S. rice market faces record supplies, low world prices. In: *Electronic Outlook Report RCS-1201*, US Department of Agriculture.

Chu, Y and Kung, Y (1998) A study on vegetable oil blends. *Food Chem.*, 62, 191–195.

Cicero, A and Gaddi, A (2001) Rice bran oil and gamma-oryzanol in the treatment of hyperlipoproteinaemias and other conditions. *Phytotherapy Res.*, 15, 277–289.

Cuppett, S and Hall, C (1998) Antioxidant activity of the Labiatae. In: *Advances in Food and Nutrition Research, Vol. 42* (S Taylor, ed.), Academic Press, San Diego, pp.245–271.

Cuvelier, M, Berset, C and Richard, H (1994) Antioxidant constituents in sage (*Salvia officinalis*). *J Agric. Food Chem.*, 42, 665–669.

Cuvelier, M, Richard, H and Berset, C (1996) Antioxidative activity and phenolic composition of pilot-plant and commercial extracts of sage and rosemary. *J Am. Oil Chem. Soc.*, 73, 645–652.

De, B and Bhattacharyya, D (1998) Physical refining of rice bran oil in relation to degumming and dewaxing. *J. Am. Oil Chem. Soc.*, 75, 1683–1686.

Debersac, P, Vernevaut, M, Amiot, M, Suschetet, M and Siess, M (2001) Effects of a water-soluble extract of rosemary and its purified component rosmarinic acid on xenobiotic-metabolizing enzymes in rat liver. *Food Chem. Toxicol.*, 39, 109–117.

deDeckere, E and Korver, O (1996) Minor constituents of rice bran oil as functional foods. *Nutr. Rev.*, 54, S120–S126.

Dias, P, Foglio, M, Possenti, A and deCarvalho, J (2000) Antiulcerogenic activity of crude hydroalcoholic extract of Rosmarinus officinalis L. *J. Ethnopharmacol.*, 69, 57–62.

Fahim, F, Esmat, A, Fadel, H and Hassan, K (1999) Allied studies on the effect of Rosmarinus officinalis L. on experimental hepatotoxicity and mutagenesis. *Int. J. Food Sci. Nutr.*, 50, 413–427.

Formo, M., Jungermann, E, Norris, F and Sonntag, N. (1979). Structure and composition of fats and oils. In: *Bailey's Industrial Oil and Fat Products* (D Swern, ed.), John Wiley & Sons, New York, p.54.

Frankel, E, Huang, S, Aeschbach, R and Prior, E (1996a) Antioxidant activity of a rosemary extract and its constituents, carnosic acid, carnosol, and rosmarinic acid, in bulk oil and oil-in-water emulsion. *J Agric Food Chem*, 44, 131–135.

Frankel, E, Huang, S, Prior, E and Aeschbach, R (1996b) Evaluation of antioxidant activity of rosemary extracts, carnosol and carnosic acid in bulk vegetable oils and fish oil and their emulsions. *J. Sci. Food Agric.*, 72, 201–208.

Fujiwara, Y, Sawada, R, Kuzuyama, M and Igarashi, O (1995) Effects of sesamin on the fatty acid composition of the liver of rats fed *n*-6 and *n*-3 fatty acid-rich diet. *J. Nutr. Sci. Vitaminol.*, 41, 217–225.

Fukuda, Y, Nagata, M, Osawa, T and Namiki, M (1986a) Contribution of lignan analogues to antioxidative activity of refined unroasted sesame seed oil. *J Am. Oil Chem. Soc.*, 63, 1027–1031.

Fukuda, Y, Nagata, M, Osawa, T and Namiki, M (1986b) Chemical aspects of the antioxidative activity of roasted sesame seed oil, and the effect of using the oil for frying. *Agric Biol Chem* 50, 857–862.

Garcia, A, Lucas, A, Rincon, J, Alvarez, A, Gracia, I and Garcia, M (1996) Supercritical carbon dioxide extraction of fatty and waxy materials from rice bran. *J. Am. Oil Chem. Soc.*, 73, 1127–1131.

Gaydou, E., Raoizafinimanana, R and Bianchini, J (1980) Quantitative analysis of fatty acids and sterols in Malagasy rice bran oil. *J. Am. Oil Chem. Soc.*, 57, 141–142.

Gertz, C, Klostermann, S and Kochhar, S (2000) Testing and comparing oxidative stability of vegetable oils and fats at frying temperature. *Eur. J. Lipid Sci. Technol.*, 102, 543–551.

Gordon, M and Kourimska, L (1995) The effects of antioxidants on changes in oils during heating and deep frying. *J. Sci. Food Agric.*, 68, 347–353.

Gordon, M and Magos, P (1983) The effect of sterols on the oxidation of edible oils. *Food Chem.*, 10, 141–147.

Guor, J, Mingfu, W, Rosen, R and Ho, CT (1999) Identification of thermal decomposition products of carnosol, an antioxidant in rosemary and sage. *J. Food Lipids,* 6, 173–179.

Gupta, H (1989) Rice bran offers India an oil source. *J. Am. Oil Chem. Soc.*, 66, 620–626.

Häkkinen, S, Kärenlampi, S, Mykkänen, H and Törrönen, R (2000) Influence of domestic processing and storage on flavonol contents in berries. *J. Agric. Food Chem.*, 48, 2960–2965.

Hall III, C (2001) Sources of natural antioxidant: oilseeds, nuts, cereals, legumes, animal products and microbial sources. In: *Antioxidants in Food: Practical Applications* (J Pokorny, N Yanishlieva and M Gordon, eds), CRC Press/Woodhead Publishing Ltd, Cambridge, pp.159–209.

Hall III, C and Cuppett, S (1997) Structure–activities of natural antioxidants. In: *Antioxidant Methodology: In Vivo* and *in Vitro* Concepts (O Aruoma and S Cuppet, eds), AOCS Press, Champaign, pp.141–172.

Haraguchi, H, Saito, T, Okamura, N and Yag, A (1995) Inhibition of lipid peroxidation and superoxide generation by diterpenoids from *Rosmarinus officinalis*. *Planta Medica*, 61, 333–336.

He, L, Mo, H, Hadisusilo, S, Qureshi, A and Elson, C (1997). Isoprenoids suppress the growth of murine B16 melanoma *in vitro* and *in vivo*. *J. Nutr.*, 127, 668–674.

Heinonen, I, Meyer, A and Frankel, E (1998) Antioxidant activity of berry phenolics on human low-density lipoprotein and liposome oxidation. *J. Agric. Food Chem.*, 46, 4107–4112.

Hemavathy, J, and Prabhakar, J (1987) Lipid composition of rice (*Oryza sativa* L.) bran. *J. Am. Oil Chem. Soc.*, 64, 1016–1019.

Hidalgo, P, Ubera, J, Tena, M and Valcarel, M (1998) Determination of the carnosic acid content in wild and cultivated *Rosmarinus officinalis*. *J. Agric. Food Chem.*, 46, 2624–2627.

Hirano, T, Gotoh, M and Oka, K (1994) Natural flavonoids and lignans are potent cytostatic agents against human leukemic HL-60 cells. *Life Sci.*, 55, 1061–1069.

Hirata, F, Fujita, K, Ishikura, Y, Hosoda, K, Ishikawa, T and Nakamura, H (1996) Hypocholesterolemic effect of sesame lignan in humans. *Atherosclerosis*, 122, 135–136.

Hirose, N, Inoue, T, Nishihara, K, Sugano, M, Abimoto, K, Shimizu, S and Yamada, H (1991) Inhibition of cholesterol absorption and synthesis in rats by sesamin. *J. Lipid Res.*, 32, 629–638.

Hirose, N, Doi, F, Ueki, T, Akazawa, K, Chijiiwa, K, Sugano, M, Akimato, K, Shimizu, S and Yamada, H (1992) Suppresion effect of sesamin against 7,12-dimethylbenz-[a]anthracene induced rat mammary carcinogenesis. *Anticancer Res.*, 12, 1259–1266.

Ho, CT, Huang, M, Lou, Y, Ma, W, Shao, Y, Wei, G, Wang, M and Chin, C (2000). Antioxidant and antitumor activity of rosemary leaves. In: *Phytochemicals and Phytopharmaceuticals* (F Shahidi and C-T Ho, eds), AOCS Press, Champaign, Illinois, pp.296–307.

Hollman, P and Arts, I (2000) Flavonols, flavones and flavanols – nature, occurrence and dietary burden. *J. Sci. Food Agric.*, 80, 1081–1093.

Hollman, P and Katan, M (1998) Bioavailability and health effects of dietary flavonols in man. *Arch. Toxicol. Suppl.*, 20, 237–248.

Hopia, A, Huang, S, Schwarz, K, German, J and Frankel, E (1996) Effect of different lipid systems on antioxidant activity of rosemary constituents carnosol and carnosic acid with and without alpha-tocopherol. *J. Agric. Food Chem.*, 44, 2030–2036.

Hu, W, Wells, J, Shin, T and Godber, J (1996) Comparison of isopropanol and hexane for extraction of vitamin E and oryzanols from stabilized rice bran. *J. Am. Oil Chem. Soc.*, 73, 1653–1656.

Huang, S, Frankel, E, Schwarz, K, Aeschbach, R and German, J (1996) Antioxidant activity of carnosic acid and methyl carnosate in bulk oils and oil-in-water emulsions. *J. Agric. Food Chem.*, 44, 2951–2956.

Hwang, J, Hodis, H and Sevanian, A (2001) Soy and alfalfa phytoestrogen extracts become potent low-density lipoprotein antioxidants in the presence of acerola cherry extract. *J. Agric. Food Chem.*, 49, 308–314.

Ibàñez, E, Oca, A, Murga, G, Lopez-Sebastian, S, Tabera, J and Reglero, G (1999) Supercritical fluid extraction and fractionation of different preprocessed rosemary plants. *J. Agric. Food Chem.*, 47, 1400–1404.

Ibàñez, E, Cifuentes, A, Crego, A, Señoráns, F, Cavero, S and Reglero, G (2000). Combined use of supercritical fluid extraction, micellar electrokinetic chromatography, and reverse phase high performance liquid chromatography for the analysis of antioxidants from rosemary (*Rosmarinus officinalis* L.). *J. Agric. Food Chem.*, 48, 4060–4065.

Ide, T, Ashakumary, L, Takahashi, Y, Kushiro, M, Fukuda, N and Sugano, M (2001) Sesamin, a sesame lignan, decreases fatty acid synthesis in rat liver accompanying the down-regulation of sterol regulatory element binding protein-1. *Biochim. Biophys. Acta*, 1534, 1–13.

Ishiwa, J, Sato, T, Mimaki, Y, Sashida, Y, Yano, M and Ito, A (2000) A citrus flavonoid, nobiletin, suppresses production and gene expression of matrix metalloproteinase 9/ gelatinase B in rabbit synovial fibroblasts. *J. Rheumatol.*, 27, 20–25.

Izuka, Y, Namiki, M and Yamashita, K (1996) Effect of lignan structure of sesame seed lignans and dietary alpha-tocopherol on alpha-tocopherol level in rats. *Nippon Eiyo Shokuryo Gakkaishi* (*J. Jpn. Soc. Nutr. Food Sci.*), 49, 149–155.

Kader, F, Haluk, J, Nicolas, J and Metche, M (1998) Degradation of cyanidin 3-glucoside by blueberry polyphenol oxidase: kinetic studies and mechanisms. *J. Agric. Food Chem.*, 46, 3060–3065.

Kader, F, Nicolas, J and Metche, M (1999) Degradation of pelargonidin 3-glucoside in the presence of chlorogenic and blueberry polyphenol oxidase. *J. Sci. Food Agric.*, 79, 517–522.

Kahlon, T, Saunders, R, Sayre, R, Chow, F, Chiu, M and Betschart, A (1992a) Cholesterol-lowering effects of rice bran and rice bran oil fractions in hypercholesterolemic hamsters. *Cereal Chem.*, 69, 485–489.

Kahlon, T, Chow, F, Sayre, R and Betschart, A (1992b) Cholesterol-lowering in hamsters fed rice bran at various levels, defatted rice bran and rice bran oil. *J. Nutr.*, 122, 513–519.

Kahlon, T, Chow, F, Chiu, M, Hudson, C and Sayre, R (1996) Cholesterol-lowering by rice bran and rice bran oil unsaponifiable matter in hamsters. *Cereal Chem.*, 73, 69–74.

Kamal-Eldin, A and Appelqvist, L (1994) Variations in the composition of sterols, tocopherols and lignans in seed oils from four Sesamum species. *J. Am. Oil Chem. Soc.*, 71, 149–156.

Kamal-Eldin, A, Petersson, D and Appelqvist, L (1995) Sesamin (a compound from sesame oil) increases tocopherol levels in rats fed ad libitum. *Lipids*, 30, 499–505.

Kamal-Eldin, A, Appelqvist, L, Gertz, C and Stier, R (1998) Enhancing the frying performance of high oleic sunflower oil using sesame oil and rice bran oil. *J. Food Serv. Syst.*, 10, 139–157.

Kamei, H, Hashimoto, Y, Koide, T, Kojima, T and Hasegawa, M (1998) Anti-tumor effect of methanol extracts from red and white wines. *Cancer Biother. Radiopharmacol.*, 13, 447–452.

Kang, M, Oh, M, Bang, J, Kim, D, Kang, C and Lee, B (2000) Varietal difference of lignan contents and fatty acids composition in Korean sesame cultivars. *Korean J. Crop Sci.*, 45, 203–206.

Katsuzaki, H, Kawakishi, S and Osawa, T (1994) Sesaminol glucosides in sesame seeds. *Phytochemistry*, 35, 733–776.

Kaur, I and Saini, A (2000) Sesamol exhibits antimutagenic activity against oxygen species mediated mutagenicity. *Mutat. Res.*, 470, 71–76.

Kim, J and Godber, J (2001) Oxidative stability and vitamin E levels increased in restructured beef roasts with added rice bran oil. *J. Food Qual.*, 24, 17–26.

Kim, S, Kim, C, Cheigh, H and Yoon, S (1985) Effect of caustic refining, solvent refining and steam refining on the deacidification and color of rice bran oil. *J. Am. Oil Chem. Soc.*, 62, 1492–1495.

Kim, C, Byun, S, Cheigh, H and Kwon, T (1987a) Comparison of solvent extraction characteristics of rice bran pretreatment by hot air drying, steam cooking and extrusion. *J. Am. Oil Chem. Soc.*, 64, 514–516.

Kim, C, Byun, S, Cheigh, H and Kwon, T (1987b) Optimization of extrusion rice bran stabilization process. *J. Food Sci.*, 52, 1355–1357.

Kim, J, Godber, J and Prinyawiwatkul, W (2000) Restructured beef roasts containing rice bran oil and fiber influences cholesterol oxidation and nutritional profile. *J. Muscle Foods*, 11, 111–127.

Kim, J, Godber, J, King, J and Prinyawiwatkul, W (2001) Inhibition of cholesterol autoxidation by the nonsaponifiable fraction in rice bran in an aqueous model system. *J. Am. Oil Chem. Soc.*, 78, 685–689.

Kochhar, P (2000) Stabilisation of frying oils with natural antioxidative components. *Eur. J. Lipid Sci. Technol.*, 102, 552–559.

Komali, A and Shetty, K (1998) Comparison of the growth pattern and rosmarinic acid production in rosemary (Rosmarinus officinalis) shoots and genetically transformed callus cultures. *Food Biotechnol.*, 12, 27–41.

Kooyenga, D, Watkins, T, Geller, M, Bierenbaum, M, Nesaretnam, K and Packer, L (2001) Antioxidants modulate the course of carotid atherosclerosis: a four-year report. In: *Micronutrients and Health: Molecular Biological Mechanism* (K Nesaretnam, ed.), AOCS Press, Champaign, pp.366–375.

Krishna, A, Khatoon, S, Shiela, P, Sarmandal, C, Indira, T and Mishra, A (2001) Effect of refining of crude rice bran oil on the retention of oryzanol in the refined oil. *J. Am. Oil Chem. Soc.*, 78, 127–131.

Kuk, M and Dowd, M (1998) Supercritical CO_2 extraction of rice bran. *J. Am. Oil Chem. Soc.*, 75, 623–628.

Lacroix, M, Smoragiewicz, W, Pazdernik, L, Kone, M and Krzystyniak, K (1997) Prevention of lipid radiolysis by natural antioxidants from rosemary (*Rosmarinus officinalis* L.) and thyme (*Thymus vulgaris* L.). *Food Res. Int.*, 30, 457–462.

Lane, R, Qureshi, A and Salser, W (1999) Tocotrienols and tocotrienol-like compounds and methods for their use. US Patent 5,919,818.

Larrauri, J, Rupérez, P and Saura-Calixto, F (1997) Effects of drying temperature on the stability of polyphenols and antioxidant activity of red grape pomace. *J. Agric. Food Chem.*, 45, 1390–1393.

Larrauri, J, Sánchez-Moreno, C and Saura-Calixto, F (1998) Effects of temperature on the free radical scavenging capacity of extracts from red and white grape pomace. *J. Agric. Food Chem.*, 46, 2694–2697

Laughton, M, Evans, P, Moroney, M, Hoult, J and Halliwell, B (1991) Inhibition of mammalian 5-lipoxygenase and cyclo-oxygenase by flavonoids and phenolic dietary additives. *Biochem. Pharmacol.*, 42, 1637–1681.

List, G, King, J, Johnson, J, Warner, K and Mounts, T (1993) Supercritical CO_2 degumming and physical refining of soybean oil. *J. Am. Oil Chem. Soc.*, 70, 473–476.

Masuda, T, Inaba, Y and Takeda, Y (2001) Antioxidant mechanism of carnosic acid: structural identification of two oxidation products. *J. Agric. Food Chem.*, 49, 5560–5565.

Matumura, Y, Kita, S, Morimoto, S, Akimoto, K, Furuya, M, Oka, N and Tanaka, T (1995)

Antihypertensive effect of sesamin, I. Protection against deoxycortiocosterone acetate-salt-induced hypertension and cardiovascular hypertrophy. *Biol. Pharmacol. Bull.*, 18, 1016–1019.

Matumura, Y, Kita, S, Tanida, Y, Taguchi, Y, Morimoto, S, Akimoto, K and Tanaka, T (1998) Antihypertensive effect of sesamin, III. Protection against development and maintenance of hypertension in stroke-prone spontaneously hypertensive rats. *Biol. Pharmacol. Bull.*, 21, 469–473.

Meyer, L, Tehrani, J, Thrall, C and Gurkin, M (1993) Spercritical fluid extraction (SFE): advantages, applications and instrumentation for sample preparation. In: *Supercritical Fluid Extraction (SFE) Instrumentation and Application*, Isco Inc, Lincoln, pp.15–19.

Meyer, A, Yi, O, Pearson, D, Waterhouse, A and Frankel, E (1997) Inhibition of human low-density lipoprotein in relation to composition of phenolic antioxidants in grapes (*Vitis vinifera*). *J. Agric. Food Chem.*, 45, 1638–1643.

Meyer, A, Jepsen, S and Sørgensen, N (1998a) Enzymatic release of antioxidants for human low-density lipoprotein from grape pomace. *J. Agric. Food Chem.*, 46, 2439–2446.

Meyer, A, Heinonen, M and Frankel, E (1998b) Antioxidant interactions of catechin, cyanidin, caffeic acid, quercetin, and ellagic acid on human LDL oxidation. *Food Chem.*, 61, 71–75.

Middleton, E (1998) Effect of plant flavonoids on immune and inflammatory cell function. *Adv. Exp. Med. Biol.*, 439, 175–182.

Minnunni, M, Wolleb, U, Mueller, O, Pfeifer, A and Aeschbacher, H (1992) Natural antioxidants as inhibitors of oxygen species induced mutagenicity. *Mutat. Res.*, 269, 193–200.

Miyahara, Y, Komiya, T, Katsuzaki, H, Imai, K, Nakagawa, M, Ishi, Y and Hibasami, H (2000) Sesamin and episesamin induce apoptosis in human lymphoid leukemia Molt 4B cells. *Int. J. Mol. Med.*, 6, 43–46.

Miyahara, Y, Hibasami, H, Katsuzaki, H, Imai, K and Komiya, T (2001a) Sesamolin from sesame seed inhibits proliferation by inducing apoptosis in human lymphoid leukemia Molt 4B cells. *Int. J. Mol. Med.*, 7, 369–371.

Miyahara, Y, Hibasami, H, Katsuzaki, H, Imai, K, Osawa, T, Ina, K and Komiya, T (2001b) Sesaminol from sesame seed induces apoptosis in human lymphoid leukemia Molt 4B cells. *Int. J. Mol. Med.*, 7, 485–488.

Mohamed, H and Awatif, I (1998) The use of sesame oil unsaponifiable matter as a natural antioxidant. *Food Chem.*, 62, 269–276.

Murakami, H, Asakawa, T, Terao, J and Matsushita, S (1984) Antioxidative stability of tempeh and liberation of isoflavones by fermentation. *Agric. Biol. Chem.*, 48, 2971–2975.

Murkovic, M, Steinberger, D and Pfannhauser, W (1998) Antioxidant spices reduce the formation of heterocyclic amines in fried meat. *Z. Lebensm. Unters. Forsch.*, 207, 477–480.

Nakamura, Y, Ohto, Y, Murakami, A and Ohigashi, H (1998) Superoxide scavenging activity of rosmarinic acid from *Perilla frutescens* Britton var. *acuta f. virdis*. *J. Agric. Food Chem.*, 46, 4545–4550.

Nanua, J, McGregor, J and Godber, J (2000) Influence of high-oryzanol rice bran oil on the oxidative stability of whole milk powder. *J. Dairy Sci.*, 83, 2426–2431.

Nasirullah, A., Krishnamurthy, M and Nagaraja, K (1989) Effect of stabilization on the quality characteristics of rice-bran oil. *J. Am. Oil Chem. Soc.*, 66, 661–663.

Nguyen, U, Frankman, G and Evans, D (1991) Process for extracting antioxidants from Labiatae herbs. US Patent 5,017,397.

Nieto, S, Garrido, A, Sahueza, J, Loyola, L, Morales, G, Leighton, F and Valenzuela, A (1993) Flavonoids as stabilizers of fish oil: an alternative to synthetic antioxidants. *J. Am. Oil Chem. Soc.*, 70, 773–778.

Offord, E, Mace, K, Ruffieux, C, Malnoe, A and Pfeifer, A (1995) Rosemary components inhibit benzo[a]pyrene-induced genotoxicity in human bronchial cells. *Carcinogenesis,* 16, 2057–2062.

Offord, E, Mace, K, Avanti, O and Pfeifer, A (1997) Mechanisms involved in the chemoprotective effects of rosemary extract studied in human liver and bronchial cells. *Cancer Lett.,* 114, 275–281.

Ogawa, H, Sasagawa, S, Murakami, T and Yoshizumi, H (1995) Sesame lignans modulate cholesterol metaboism in stroke-prone spontaneously hypertensive rat. *Clin. Exp. Pharmacol. Physiol.*, 22, S310–S312.

Okamura, N, Haraguchi, H, Hashimoto, K and Yagi, A (1994) Flavonoids in *Rosmarinus officinalis* leaves. *Phytochemistry*, 37, 1463–1466.

Paris, A, Strukelj, B, Renko, M, Turk, V, Puki, M, Umek, A and Korant, B (1993) Inhibitory effect of carnosolic acid on HIV-1 protease in cell-free assays. *J. Nat. Prod.*, 56, 1426–1430.

Parker, R, Pearce, B, Clark, R, Gordon, D and Wright, J (1993) Tocotrienols regulate cholesterol production in mammalian cells by post-transcriptional suppression of 3-hydroxy-3-methylglutarylcoenzyme A reductase. *J. Biol. Chem.*, 268, 11230–11238.

Parker, R, Sontag, T and Swanson, J (2000) Cytochrome P4503A-dependent metabolism of tocopherols and inhibition by sesamin. *Biochem. Biophys. Res. Commun.*, 277, 531–534.

Pearce, B, Parker, R, Deason, M, Qureshi, A and Wright, J (1992) Hypocholesterolemic activity of synthetic and natural tocotrienols. *J. Med. Chem.*, 35, 3595–3606.

Pearson, D, Frankel, E, Aeschbach, R and German, J (1997) Inhibition of endothelial cell-mediated oxidation of low-density lipoprotein by rosemary and plant phenolics. *J. Agric. Food Chem.*, 45, 578–582.

Pekkarinen, S, Heinonen, M and Hopia, A (1999) Flavonoids quercetin, myricetin, kaempferol and (+)-catechin as antioxidants in methyl linoleate. *J. Sci. Food Agric.*, 79, 499–506.

Pietta, P-G (2000) Flavonoids as antioxidants. *J. Nat. Prod.*, 63, 1035–1042.

Porter, M, Krueger, C, Wiebe, D, Cunningham, D and Reed, J (2001) Cranberry proanthocyanidins associate with low-density lipoprotein and inhibit in vitro Cu^{2+}-induced oxidation. *J. Sci. Food Agric.*, 81, 1306–1313.

Proctor, A and Bowen, D (1996) Ambient-temperature extraction of rice bran oil with hexane and isopropanol. *J. Am. Oil Chem. Soc.*, 73, 811–813.

Qureshi, A, Bradlow, B, Salser, W and Brace, L (1997). Novel tocotrienols of rice bran modulate cardiovascular disease risk parameters of hypercholesterolemic humans. *J. Nutr. Biochem.*, 8, 290–298.

Qureshi, A, Mo, H, Packer, L and Peterson, D (2000) Isolation and identification of novel tocotrienols from rice bran with hypocholesterolemic, antioxidant, and antitumor properties. *J. Agric. Food Chem.*, 48, 3130–3140.

Qureshi, A, Salser, W, Parmar, R and Emeson, E (2001) Novel tocotrienols of rice bran inhibit atherosclerotic lesions in C57BL/6 ApoE-deficient mice. *J. Nutr.*, 131, 2606–2618.

Ramezanzadeh, F, Rao, R, Windhauser, M, Prinyawiwatkul, W, Tulley, R and Marshall, W (1999) Prevention of hydrolytic rancidity in rice bran during storage. *J. Agric. Food Chem.*, 47, 3050–3052.

Ramezanzadeh, F, Rao, R, Prinyawiwatkul, W, Marshall, W and Windhauser, M (2000) Effects of microwave heat, packaging, and storage temperature on fatty acid and proximate compositions in rice bran. *J. Agric. Food Chem.*, 48, 464–467.

Ramsay, M, Hsu, J, Novak, R and Reighter, W (1991) Processing rice bran by supercritical fluid extraction. *Food Technol.*, 45, 98–104.

Reblova, Z, Kudrnova, J, Trojakova, L and Pokorny, J (1999) Effect of rosemary extracts on the stabilization of frying oil during deep fat frying. *J. Food Lipids,* 6,13–23.

Rhee, J and Yoon, H (1984) Stabilization of rice bran by microwave energy. *Korean J. Food Sci. Technol.*, 161, 13–19.

Richheimer, S, Bernart, M, King, G, Kent, M and Bailey, D (1996) Antioxidant activity of lipid-soluble phenolic diterpenes from rosemary. *J. Am. Oil Chem. Soc.*, 73, 507–514.

Rodgers, E and Grant, M (1998) The effect of the flavonoids, quercetin, myricetin and epicatechin on the growth and enzyme activities of MCF7 human breast cancer cells. *Chem. Biol. Interact.*, 116, 213–228.

Rogers, E, Rice, S, Nicolosi, R, Carpenter, D, McClelland, C and Romanczyk, L (1993) Identification and quantitation of γ-oryzanols components and simultaneous assessment of tocols in rice bran oil. *J. Am. Oil Chem. Soc.*, 70, 301–307.

Rong, N, Ausman, L and Nicolosi, R (1997) Oryzanol decreases cholesterol absorption and aortic fatty streaks in hamsters. *Lipids*, 32, 303–309.

Ryu, S and Lee, B (1998) Antitumor activity of crude sesaminol in sesame seed. *Korean J. Crop Sci.*, 43, 168–171.

Ryu, S, Ho, CT and Osawa, T (1998a) High performance liquid chromatographic determination of antioxidant lignan glycosides in some varieties of sesame. *J. Food Lipids*, 5, 17–28.

Ryu, S, Kim, K, Bang, J and Lee, B (1998b) Quantitative determination of sesaminol glucosides in sesame seed. *Korean J. Crop Sci.*, 43, 209–213.

Sahu, A, Rawal, N and Pangburn, M (1999) Inhibition of complement by covalent attachment of rosmarinic acid to activated C3b. *Biochem. Pharmacol.*, 57, 1439–1446.

Saito, M, Hosoyama, H, Ariga, T, Kataoka, S and Yamaji, N (1998) Antiulcer activity of grape seed extract and procyanidins. *J. Agric. Food Chem.*, 46, 1460–1464.

Salah, N, Miller, N, Paganga, G, Tijburg, L, Bolwell, G and Rice-Evans, C (1995) Polyphenolic flavanols as scavengers of aqueous phase radicals and as chain-breaking antioxidants. *Arch. Biochem. Biophys.*, 322, 339–346.

Sánchez, I, Caldrón, J, Ruiz, B, Tellez, J, Calzada, L and Taboada, J (2001) *In vitro* cytotoxicity of flavonoids against MK2 and C6 tumour cells. *Phytotherapy Res.*, 15, 290–293.

Sarni, P, Fulcrand, H, Souillol, V, Souquet, J and Cheynier, V (1995) Mechanism of anthocyanin degradation in grape must-like model solutions. *J. Sci. Food Agric.*, 69, 385–391.

Sawada, R, Ogawa, M, Nakamura, M and Igarashi, O (2001) Effect of sesamin on mitochondrial and peroxisomal beta-oxidation of arachidonic and eicosapentaenoic acids in rat liver. *Lipids,* 36, 483–489.

Schwarz, K and Ternes, W (1992) Antioxidative constituents of *Rosmarinus officinalis* and *Salvia officinalis*. *Z. Lebensm. Unters. Forsch.*, 195, 95–98.

Seeram, N, Bourquin, L and Nair, M (2001) Degradation products of cyanidin glycosides from tart cherries and their bioactivities. *J. Agric. Food Chem.*, 49, 4924–4929.

Señoráns, F, Ibanez, E, Cavero, S, Tabera, J and Reglero, G (2000) Liquid chromatographic-mass spectrometric analysis of supercritical-fluid extracts of rosemary plants. *J. Chromatogr. A*, 870, 491–499.

Shahidi, F, Amarowicz, R, Abou-Gharbia, H and Shehata, A (1997) Endogenous antioxidants and stability of sesame oil as affected by processing and storage. *J. Am. Oil Chem. Soc.*, 74, 143–148.

Shen, Z, Palmer, M, Ting, S and Fairclough, R (1996) Pilot scale extraction and fractionation of rice bran oil with dense carbon dioxide. *J. Agric. Food Chem.*, 44, 3033–3039.

Shen, Z, Palmer, M, Ting, S and Fairclough, R (1997) Pilot scale extraction and fractionation of rice bran oil using supercritical carbon dioxide. *J. Agric. Food Chem.*, 45, 4540–4544.

Shin, T and Godber, J (1996) Changes of endogenous antioxidants and fatty acid composition in irradiated rice bran during storage. *J. Agric. Food Chem.*, 44, 567–573.

Shin, T, Godber, J, Martin, D and Wells, J (1997) Hydrolytic stability and changes in E vitamers and oryzanol of extruded rice bran during storage. *J. Food Sci.*, 62, 704–708.

Singletary, K (1996) Rosemary extract and carnosol stimulate rat liver glutathione-S-transferase and quinone reductase activities. *Cancer Lett.*, 100, 139–144.

Singletary, K and Nelshoppen, J (1991) Inhibition of 7,12-dimethylbenz[a]anthracene (DMBA)-induced mammary tumorigenesis and of *in vivo* formation of mammary DMBA-DNA adducts by rosemary extract. *Cancer Lett.*, 60, 169–175.

Singletary, K, MacDonald, C and Wallig, M (1996) Inhibition by rosemary and carnosol of 7,12-dimethylbenz[a]anthracene (DMBA)-induced rat mammary tumorigenesis and in vivo DMBA-DNA adduct formation. *Cancer Lett.*, 104, 43–48.

Skaltsa, H, Bermejo, P, Lazari, D, Silvan, A, Skaltsounis, A, Sanz, A and Abad, M (2000) Inhibition of prostaglandin E_2 and leukotriene C4 in mouse peritoneal macrophages and thromboxane B2 production in human platelets by flavonoids from Stachys chrysantha and Stachys candida. *Biol. Pharmacol. Bull.*, 23, 47–53.

Stahl, E, Schutz, E and Mangold, K (1980) Extraction of seed oils with liquid and supercritical carbon dioxide. Au:*J. Agric. Food Chem.*, 28, 1153–1157.

Takano, K, (1993) Advances in cereal chemistry and technology in Japan, Mechanism of lipid hydrolysis in rice bran. *Cereal Foods World*, 38, 695–698.

Tao, J, Rao, R and Liuzzo, J (1993). Microwave heating for rice bran stabilization. *J. Microwave Power Electromagnetic Energy*, 28, 156–164.

Tashiro, T, Fukuda, Y, Osawa, T and Namiki, M (1990) Oil and minor components of sesame (*Sesamum indicum* L.) strains. *J. Am. Oil Chem. Soc.*, 67, 508–511.

Taylor, J, Richar, T, Wilhelm, C, Chrysam, M, Otterburn, M and Leveille, G (1996) Rice bran oil antioxidant. US Patent 5,552,167.

Tena, M, Valcarcel, M, Hidalgo, P and Ubera, J (1997) Supercritical fluid extraction of natural antioxidants from rosemary: comparison with liquid solvent sonication. *Anal. Chem.*, 69, 521–526.

Todd, P (1991) Stable alkaline Labiatae antioxidant extracts. US Patent 5,023,017.

Trojakova, L, Reblova, Z, Huong, T and Pokorny, J (2001) Antioxidant activity of rosemary and sage extracts in rapeseed oil. *J. Food Lipids*, 8, 1–13.

Tsai, C, Ting, H, Wang, L and Lin, T (1992) Effect of rice bran on blood lipids in man. *J. Chinese Agric. Chem. Soc.,* 30, 484–495.

Tsuda, T, Watanabe, M, Ohshima, K, Norinobu, S, Choi, S, Kawakishi, S and Osawa T (1994) Antioxidative activity of the anthocyanin pigments cyanidin 3-*O*-β-D-glucoside and cyanidin. *J. Agric. Food Chem.*, 42, 2407–2410.

Utsunomiya, T, Chavali, S, Zhong, W and Forse, R (2000) Effects of sesamin-supplemented dietary fat emulsions on the ex vivo production of lipopolysaccharide-induced prostanoids and tumor necrosis factor alpha in rats. *Am. J. Clin. Nutr.*, 72, 804–808.

Verotta, L (1985) Isolation and determination of active principles of *Rosmarinus officinalis* and *Gentiana lutea*. *Fitoterapia*, 1, 25–29.

Vetrimani, R, Jyothirmayi, N, Rao, P and Ramadoss, C (1992) Inactivation of lipase and lipoxygenase in cereal bran, germ and soybean by microwave treatment. *Lebensm. Wiss. Technol.*, 25, 532–535.

Vissers, M, Zock, P, Meijer, G and Katan, M (2000) Effect of plant sterols from rice bran oil and triterpene alcohols from sheanut oil on serum lipoprotein concentrations in humans. *Am. J. Clin. Nutr.*, 72, 1510–1515.

Waladkhani, A and Clemens, M (1998) Effect of dietary phytochemicals on cancer development (review). *Int. J. Mol. Med.*, 1, 747–753.

Wang, S and Jiao, H (2000). Scavenging capacity of berry crops on superoxide radicals, hydrogen peroxide, hydroxyl radicals, and singlet oxygen. *J. Agric. Food Chem.*, 48, 5677–5684.

Wang, H, Xia, Y, Yang, Z, Natschke, S and Lee, K (1998) Recent advances in the discovery and development of flavonoids and their analogues as antitumor and anti-HIV agents. *Adv. Exp. Med. Biol.*, 439, 191–225.

Wang, H, Nair, M, Strasburg, G, Chang, Y and Booren, A (1999a) Antioxidant and antiinflammatory activities of anthocyanins and their aglycon, cyanidin, from tart cherries. *J. Nat. Prod.*, 62, 294–296.

Wang, H, Nair, M, Strasburg, G, Booren, A and Gray, I (1999b) Antioxidant polyphenols from tart cherries (*Prunus cerasus*). *J. Agric. Food Chem.*, 47, 840–844.

Wilcox, L, Borradaile, N, de-Dreu, L and Huff, M (2001) Secretion of hepatocyte apoB is inhibited by the flavonoids, naringenin and hesperetin, via reduced activity and expression of ACAT2 and MTP. *J. Lipid Res.*, 42, 725–734.

Xiao, T, Feng, X and Zhang, X (1992) The geographical distribution of black seed-coated sesame germplasms in China and an analysis of their main characters. *Oil Crops China*, 2, 31–34.

Xiurong, Z, Yingzhong, Z, Yong, C, Xiangyun, F, Qingyuan, G, Mingde, Z and Hodgkin, T (2000) Establishment of sesame germplasm core collection in China. *Genet. Resources Crop Evol.*, 47, 273–279.

Xu, Z and Godber, S (1999) Purification and identification of components of γ-oryzanol in rice bran oil. *J. Agric. Food Chem.*, 47, 2724–2728.

Xu, Z and Godber, S. (2000) Comparison of supercritical fluid and solvent extraction methods in extracting g-oryzanol from rice bran. *J. Am. Oil Chem. Soc.*, 77, 547–551.

Xu, Z and Godber, J (2001) Antioxidant activities of major components of gamma-oryzanol from rice bran using a linoleic acid model. *J. Am. Oil Chem. Soc.*, 78, 645–649.

Xu, Z, Hua, N and Godber, J (2001) Antioxidant activity of tocopherols, tocotrienols, and γ-oryzanol components from rice bran against cholesterol oxidation accelerated by 2,2′-azobis(2-methylpropionamidine) dihydrochloride. *J. Agric. Food Chem.*, 49, 2077–2081.

Yamashita, K, Nohara,Y, Katayama, K and Namiki, M (1992) Sesame seed lignans and gamma-tocopherol act synergistically to produce vitamin E activity in rats. *J. Nutr.*, 122, 2440–2446.

Yamashita, K, Iizuka, Y, Imai, T and Namiki, M (1995) Sesame seed and its lignans produce marked enhancement of vitamin E activity in rats fed a low alpha-tocopherol diet. *Lipids*, 30, 1019–1028.

Yamashita, K, Kagaya, M, Higuti, N and Kiso, Y (2000) Sesamin and alpha-tocopherol synergistically suppress lipid-peroxide in rats fed a high docosahexaenoic acid diet. *Biofactors* 11, 11–13.

Yanishlieva, N and Heinonen, I (2001) Sources of natural antioxidants: vegetables, fruits, herbs, spices and teas. In: *Antioxidants in Food: Practical Applications* (J Pokorny, N Yanishlieva and M Gordon, eds), CRC Press/Woodhead Publishing Ltd, Cambridge, pp.210–263.

Yasukawa, K, Akihisa, T, Yasukawa, K, Tamura, T and Takido, M (1998) Inhibitory effect of cycloartenol ferulate, a component of rice bran, on tumor promotion in two-stage carcinogenesis in mouse skin. *Biol. Pharmacol. Bull.*, 21, 1072–1076.

Yasumoto, S, Katsuta, M, Okuyama, Y, Takahashi, Y and Ide, T (2001) Effect of sesame seeds rich in sesamin and sesamolin on fatty acid oxidation in rat liver. *J. Agric. Food Chem.*, 49, 2647–2651.

Yen, G (1990) Influence of seed roasting process on the changes in composition and quality of sesamum (Sesame indicum) oil. *J. Sci. Food Agric.*, 50, 563–570.

Yen, G, and Lai, S (1989) Oxidative stability of instant noodles fried with sesame oil-vegetable oil blends. *J. Chinese Agric. Chem. Soc.*, 27, 196–201.

Yokozawa, T, Dong, E, Liu, Z and Shimizu, M (1997) Antioxidative activity of flavones and flavonols. *Phytotherapy Res.*, 11, 446–449.

Yoon, S and Kim, S (1994) Oxidative stability of high-fatty acid rice bran oil at different stages of refining. *J. Am. Oil Chem. Soc.*, 71, 227–229.

Yoshida, H (1994) Composition and quality characteristics of sesame seed (Sesamum indicum) oil roasted at different temperatures in an electric oven. *J. Sci. Food Agric.*, 65, 331–336.

Yoshida, H and Kajimoto, G (1994) Microwave heating affects composition and oxidative stability of sesame (Sesamum indicum) oil. *J. Food Sci.*, 59, 613–616, 625.

Yoshida, H and Takagi, S (1997) Effects of seed roasting temperature and time on the quality characteristics of sesame (Sesamum indicum) oil. *J. Sci. Food Agric.*, 75, 19–26.

Yoshida, H and Takagi, S (1999) Antioxidant effects of sesamol and tocopherols at various concentrations in oils during microwave heating. *J. Sci. Food Agric.*, 79, 220–226.

Yoshida, H, Shigezaki, J, Takagi, S and Kajimoto, G (1995) Variations in the composition of various acyl lipids, tocopherols and lignans in sesame seed oils roasted in a microwave oven. *J. Sci. Food Agric.*, 68, 407–415.

Yoshida, H, Hirakawa, Y and Takagi, S (2000) Roasting influences on molecular species of triacylglycerols in sesame seeds (Sesamum indicum). *J. Sci. Food Agric.*, 80, 1495–1502.

Yoshida, H, Abe, S, Hirakawa, Y and Takagi, S (2001) Roasting effects on fatty acid distributions of triacylglycerols and phospholipids in sesame (Sesamum indicum) seeds. *J. Sci. Food Agric.*, 81, 620–626.

You, K, Son, K, Chang, H, Kang, S and Kim, H (1998) Vitexicarpin, a flavonoid from the fruits of Vitex rotundifolia, inhibits mouse lymphocyte proliferation and growth of cell lines in vitro. *Planta Med.*, 64, 546–550.

Zhao, W, Shishikura, A, Fujimoto, K, Arai, K and Sait, S (1987) Fractional extraction of rice bran oil with supercritical carbon dioxide. *Agric. Biol. Chem.*, 51, 1773–1777.

Zhao, J, Wang, J, Chen, Y and Agarwal, R (1999) Anti-tumor-promoting activity of a polyphenolic fraction isolated from grape seeds in the mouse skin two-stage initiation-promotion protocol and identification of procyanidin B5-3′-gallate as the most effective antioxidant constituent. *Carcinogenesis*, 20, 1737–1745.

CHAPTER 5

Diacylglycerols

HIROYUKI WATANABE AND NOBORU MATSUO

Kao Corp, Tochigi, Japan

A. Introduction

Dietary diacylglycerols (DAG) are natural components of various edible oils. We have studied the nutritional characteristics of dietary DAG in comparison with triacylglycerols (TAG) and found that DAG, particularly in the 1,3-isoform (see Figure 1), have metabolic characteristics distinct from those of TAG which may produce beneficial effects with regard to the prevention and management of postprandial lipemia and obesity.

Obesity is a growing problem worldwide, not only in industrialized countries, but also in urban areas of developing countries. According to World Health Organization (WHO) statistics, the number of obese adults increased from 200 million in 1995 to over 300 million in 2000.

Obesity is associated with a number of health risks, such as heart disease, diabetes mellitus, hypertension, gallbladder disease, and some types of cancer (Carey *et al.,* 1997; Lamon-Fava *et al.,* 1996; Rimm *et al.,* 1995; Mokdad *et al.,* 2001). The importance of reducing body fat accumulation in the prevention of lifestyle-related diseases has been noted by many experiments and surveys. Clinical studies have suggested that weight loss in the range of 5–10% of initial weight can confer significant reductions in obesity-related disorders.

Functional foods and materials that improve fat metabolism have been extensively studied. Diets for weight loss generally recommend restricting intakes of energy, total fats, and saturated fats. In contrast, we have focused on the nature of the glycerol esters rather than on the fatty acid composition of oils, and have found that DAG, particularly in the 1,3-isoform, have metabolic characteristics distinct from those of TAG and are less likely to become stored body fat. A cooking oil product containing at least 80% (w/w) DAG has been on the market in Japan as a "food for specified health use" (FOSHU) since 1999. In this chapter, the nutritional characteristics and beneficial health effects of DAG observed in several clinical studies are outlined and possible mechanisms for these effects are discussed.

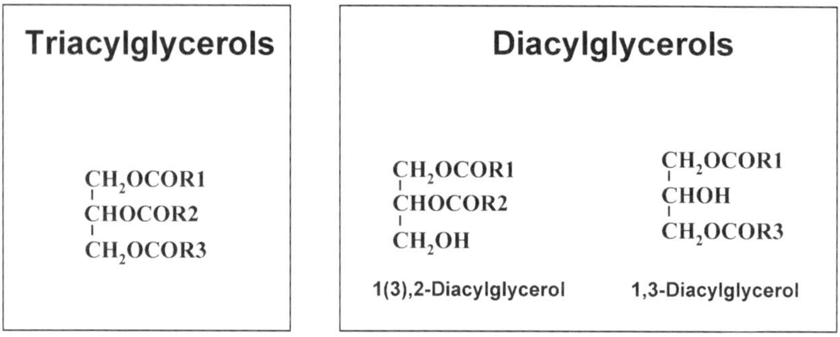

Figure 1. Schematic structures of diacylglycerols (DAG) and triacylglycerols (TAG). TAG comprise a glycerol moiety esterified with a fatty acid (R1, R2, R3) at all three carbon atoms. DAG are esterified at only two carbon atoms: either in the 1 and 3 positions, or the 1 and 2 (structurally equivalent to 2 and 3) positions.

B. Biochemical characteristics of DAG

1. Occurrence of DAG

The largest constituent of edible oils are TAG, but oils may also contain up to 10% (w/w) DAG, the level depending on the origin of the oil. Table 1 shows the content of DAG and other acylglycerols in edible oils of various origins (Abdel-Nabey *et al.,* 1992; D'alonzo *et al.*, 1982). For example, olive oil contains 5.5% DAG and crude cottonseed oil contains 9.5% DAG.

We have developed a DAG-rich oil by esterifying fatty acids derived from natural edible oils with either monoacylglycerols (MAG) or glycerol using an immobilized 1,3-regiospecific lipase (Huge-Jensen *et al.,* 1988). After esterification, the crude reaction mixture is refined in the same way as for regular edible oils, to obtain a DAG-rich oil. The DAG content of this oil is 80% (w/w) or more, and the ratio of the 1,3-isoform to the 1,2- or 2,3-isoform is approximately 7:3 as a consequence of acyl migration during the refining process.

Details of studies of the nutritional characteristics and beneficial health effects of this DAG-rich oil are detailed in Sections C and D below. A TAG-rich blend of rapeseed oil, soybean oil and safflower oil, with a similar fatty acid composition to the DAG-rich oil, was used as a comparative control in these studies.

2. Structure of DAG

DAG occur as two isoforms: 1,2- (or 2,3)-diacyl-*sn*-glycerol (1,2-DAG) and 1,3-diacyl-*sn*-glycerol (1,3-DAG); see Figure 1. The 1,2-DAG isoform is recognized as a metabolic intermediate formed after the ingestion of TAG.

Table 1. Acylglycerol composition of edible oils of various origins[a]

	Content of component (g/100 g oil)						
Component	**Soybean**	**Cottonseed**	**Palm**	**Corn (maize)**	**Safflower**	**Olive**	**Rapeseed**
TAG	97.9	87.0	93.1	95.8	96.0	93.3	96.8
DAG	1.0	9.5	5.8	2.8	2.1	5.5	0.8
MAG	<DL	0.2	<DL	<DL	<DL	0.2	0.1
Others	1.1	3.3	1.1	1.4	1.9	2.3	2.3

[a]Data taken from Abdel-Nabey *et al.* (1992); D'alonzo *et al.* (1982). <DL, below detection limit.

DAG contained in refined edible oils are mostly converted to the 1,3-isoform by migration of the acyl group during the high-temperature manufacturing process (Crossley *et al.*, 1959; Kodali *et al.*, 1990). Acyl migration is an equilibrium reaction. Although the acyl migration rate is minimal in the solid or liquid state at or below room temperature and in the absence of a catalyst, at elevated temperatures it increases considerably. The amount of the 1,3-isoform present in the equilibrium state is intrinsic to the fatty acid composition of each DAG molecule (Crossley *et al.*, 1959). For example, the equilibrium proportion of 1,3-dipalmitoyl-*sn*-glycerol is 56% (w/w) (Kodali *et al.*, 1990). In most edible oils, approximately 70% (w/w) of DAG exist as the 1,3-isoform.

3. *Energy value of DAG*

When studying the effects of a dietary fat on lipid metabolism, it is necessary to know the energy value of the fat and its digestibility. We determined the energy values of our DAG-rich and TAG-rich oils by calculation and by bomb calorimetry, and compared their absorption coefficients and digestibility, to assess whether any differences in their energy values or digestibility might contribute to the observed functional differences between in these dietary oils.

The acylglycerol and fatty acid compositions of the oils used in our studies are shown in Tables 2 and 3. The fatty acid composition of the DAG-rich oil was very similar to that of the blended TAG-rich oil. The total DAG content of the DAG-rich oil was 87.0 g/100 g and the ratio of 1(3),2-DAG to 1,3-DAG was 32:68.

The heat energy values of the oils (Table 3) were estimated using the approach outlined in the literature (Livesey, 1984; Peters *et al.*, 1991), and calculated using the equation:

$$-\Delta H_c = 0.653n - 0.166d - 0.421$$

where $-\Delta H_c$ is the heat of combustion in MJ/mol, n is the number of carbon atoms per molecule of fatty acid, and d is the number of double bonds per fatty

Table 2. Acylglycerol composition of the DAG-rich and TAG-rich oils used in our study[a]

Component	Content of component (g/100 g oil)	
	DAG-rich oil	TAG-rich oil
TAG	10.7	97.2
DAG (all isoforms)	87.0	1.1
1(3),2-DAG	27.8	ND
1,3-DAG	59.2	ND
MAG	0.82	<0.05
Free fatty acids	ND	ND

[a]Data taken from Taguchi *et al.* (2001). See Section B.1 of text for further details. ND, not determined.

Table 3. Fatty acid composition and energy value of experimental oils[a]

Fatty acid	Gross energy (kJ/g)	DAG-rich oil		TAG-rich oil	
		Weight fraction (g/g oil)	Contribution to energy value (kJ/g oil)	Weight fraction (g/g oil)	Contribution to energy value (kJ/g oil)
16:0	39.1	0.023	0.9	0.058	2.3
18:0	39.9	0.007	0.3	0.021	0.8
18:1	39.5	0.263	10.4	0.280	11.1
18:2	39.3	0.567	22.2	0.557	21.8
18:3	38.9	0.053	2.0	0.024	0.9
20:0	40.5	<DL	0.0	0.004	0.2
20:1	40.2	0.002	0.1	0.006	0.2
22:0	41.0	<DL	0.0	0.003	0.1
22:1	40.7	<DL	0.0	0.002	0.1
24:1	40.9	<DL	0.0	0.002	0.1
Total		0.913	35.9	0.956	37.6
Glycerol	18.0	0.145	2.6	0.105	1.9
Total[b]		1.000	38.5	1.000	39.5
Experimental[c]			38.9		39.6

[a]Data taken from Taguchi *et al.* (2001). See Section B.1 of text for further details. [b]Theoretical values calculated as described in the text. [c]Experimental values determined using bomb calorimetry. <DL, below detection limit (<0.0005) and regarded as 0.

acid (Livesey, 1984). The energy value of glycerol used was 18.0 kJ/g (Peters *et al.*, 1991). The heat of esterification was neglected because it amounts to only 3.8 kJ/mol of ester bond as determined with methyl stearate (Adriaanse *et al.*, 1965); in the case of tristearin this value corresponds to only 13 J/g.

The total theoretical energy values for the DAG-rich and TAG-rich oils were calculated from the energy values for each fatty acid plus glycerol, and are shown in Table 3 together with the total energy values determined experimentally using a bomb calorimeter. The theoretical and experimental values were in good agreement with each other, and indicated that the energy value of the DAG-rich oil was approximately 98% that of the TAG-rich oil. Since this energy difference is only 2%, and fats provide only part of the total energy intake of the diet, switching from using TAG-rich oils to DAG-rich oils would produce only a negligible difference in the total energy value of practical diets.

4. Absorption coefficient of DAG

Fats are digested in three different and coordinated processes: emulsification, hydrolysis of the substrate, and micellar solubilization of the hydrolysis products in the aqueous medium of the intestinal contents (Friedman and Nylund, 1980). Many determinants are therefore involved in these steps. It has been shown that intestinal fat absorption is influenced by several factors, including:

- TAG structures (Bracco, 1994);
- fatty acid chain length (Doreau and Chilliard, 1997);
- unsaturated:saturated fatty acid ratio (Apgar *et al.,* 1987);
- the esterification form of fatty acids (De Schrijver *et al.,* 1991);
- the gastrointestinal flora (Combe *et al.,* 1976; Demarne *et al.,* 1979; Demarne *et al.,* 1982).

To determine if there are any differences in bioavailability between DAG and TAG, we compared the digestibilities of our DAG-rich and TAG-rich oils in rats.

Sixteen 5 week-old male Sprague–Dawley rats were divided into two groups of eight so that the body weights of each group were approximately equal. Each animal was kept in an individual metabolic cage. Rats in one group were fed a diet containing 20% (w/w) of the DAG-rich oil for 15 days; the others were fed a diet containing 20% (w/w) of the TAG-rich oil. Feces were collected and pooled from the rats in each study group during the last 3 days of the study. The digestibility coefficient – defined as the percentage of ingested fat not excreted in the feces (Apgar *et al.,* 1987; Mitchell *et al.,* 1989; Monsma *et al.,* 1996) – was then calculated for each diet.

The overall movement of fats in the rats during the last 3 days of the experiment are shown in Table 4. No significant differences between the diet groups were seen for any of the variables: food intake, fat intake, dried feces mass, fat excretion, fat content of dry feces, or fat absorption coefficient. To see whether dietary DAG had any effects on the lipid composition of the feces, we also analysed the fecal lipid profiles of the two groups of rats using gas chromatography. No significant differences were found.

The digestibility of the DAG-rich and TAG-rich oils were also compared in

Table 4. Three-day fat intake and excretion in rats fed DAG-rich or TAG-rich diets[a]

Parameter	DAG group	TAG group
Diet intake (g/rat)	49.99 ± 1.23	50.12 ± 1.74
Fat intake (g/rat)	9.67 ± 0.24	9.67 ± 0.33
Dried feces mass (g/rat)	3.15 ± 0.17	3.35 ± 0.14
Fat excretion (g/rat)	0.354 ± 0.043	0.352 ± 0.022
Fat content of dry feces (g/100 g)	10.49 ± 0.38	11.18 ± 1.03
Fat absorption coefficient (%)	96.3 ± 0.42	96.3 ± 0.26

[a]Data taken from Taguchi *et al.* (2001). Values are means ± SD. See Section B.4 of text for further details.

mice. Mice fed the DAG-rich oil accumulated less body fat during the 5-month study than mice fed the TAG-rich oil. However, there was no evidence of a significant difference in digestibility between the two oils (Murase *et al.*, 2001).

These results suggest that the different physiological effects of DAG and TAG observed in animals and humans are not caused by differences in the energy value or absorption of the fats – but instead by their different metabolic fates after absorption into the gastrointestinal epithelial cells. Further studies of the mechanism by which dietary DAG reduces body fat accumulation in animals and humans are discussed below.

C. Nutritional functions of DAG as studied in animals

1. Identification of the digestion products in the small intestine

TAG, which form the major component of common edible oils, are hydrolysed to 2-MAG via 1(or 3),2-DAG by pancreatic lipase, and incorporated into intestinal mucosal cells. The digestion and absorption of DAG have been investigated in an experiment tracing time-course changes in the lipid composition ratio after perfusion of TAG (triolein) or DAG (a 3:7 mixture of 1,2-diolein and 1,3-diolein) in the intestinal tract of rats (Watanabe *et al.*, 1997). The 6 week-old rats (5 animals for each treatment) were fed a diet containing 10% (w/w) TAG for 7 days, deprived of food for 18 hours, then laparotomized under pentobarbital anesthesia. The portion of small intestine from 1 cm on the oral side of the opening of the common bile duct to 15 cm on the anal side of the opening was perfused according to the method of Chijiiwa and Linscheer (1987) with a lipid emulsion composed of either diolein or triolein (2%, w/w), plus bovine serum albumin (0.4%, w/w), egg yolk lecithin (0.04%, w/w), and water (97.56%, w/w). During the experiment, the body temperature of each rat was maintained at 37°C with an animal warmer. The perfusate was collected at predetermined intervals, and assayed for lipids by gas chromatography.

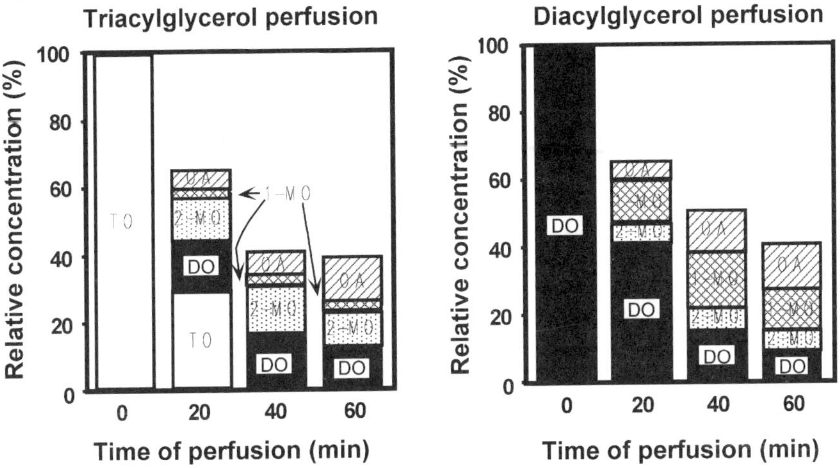

Figure 2. Time-course changes in the lipid composition ratio after the perfusion of a TAG (triolein) or DAG (diolein) in the intestinal tract of rats. TO, triolein. DO, diolein. MO, monoolein. OA, oleic acid. Data taken from Watanabe *et al.* (1997).

Figure 2 shows variations in the lipid component ratio in the small intestine perfusates. In the TAG perfusion group, the triolein was rapidly degraded to diolein, monoolein (mainly 2-monoolein) and oleic acid, and completely disappeared from the perfusate 40 minutes after initiation of perfusion. In the DAG (diolein) perfusion group, monoolein and oleic acid also formed as the diolein degraded – but, unlike the triolein perfusion, 65% (w/w) of the monoolein was 1(3)-monoolein 60 minutes after initiation of perfusion. These results thus suggest that the production of 1(3)-MAG may be a characteristic of DAG metabolism.

2. *Re-esterification rate*

The characteristics of DAG metabolism have been further clarified by studies focusing on the resynthesis (re-esterification) of TAG in gut mucosal cells. TAG resynthesis in the small intestinal epithelial cells involves the 2-MAG pathway (see Figure 3) and the glycerol-3-phosphate pathway, with the 2-MAG pathway being considered the main one (Friedman and Nylund, 1980). Whereas 2-MAG is a good substrate of the reactions in the 2-MAG pathway, the reactivity of 1(3)-MAG is lower (Lehner *et al.,* 1993). Free glycerol is a substrate of the glycerol-3-phosphate pathway, but the reaction rate is slower than that in the 2-MAG pathway, and its contribution to TAG resynthesis is small (Friedman and Nylund, 1980). It has not been quantitatively concluded

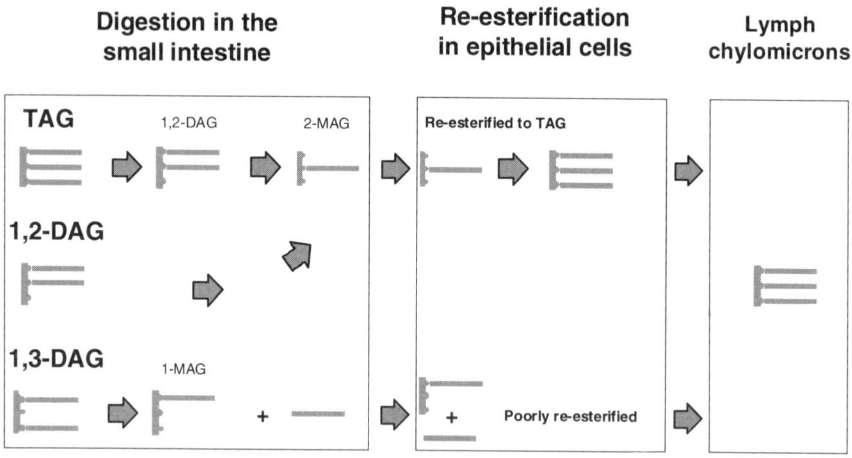

Figure 3. Metabolic characteristics of DAG in comparison with TAG. Both TAG and 1,2-DAG break down in the small intestine to form 2-MAG before being re-esterified in gut epithelial cells to re-form TAG, which circulate in lymph chylomicrons. 1,3-DAG is broken down to form 1(or 3)-MAG or glycerol, which are poorly re-esterified compared with 2-MAG.

what ratio of 1(3)-MAG produced from DAG is directly absorbed, but the amount of 2-MAG produced is smaller after the ingestion of DAG than after the ingestion of TAG, as described in the previous section. Thus the rate of resynthesis of TAG in the small intestinal epithelium may be lower following the ingestion of DAG.

Murata *et al.* (1994) reported that the release rate of resynthesized TAG into the intestinal lymph was lower after the administration of a DAG emulsion than a TAG emulsion (Murata *et al.*, 1994). Six lymph fistula rats were continuously infused with either a DAG emulsion [containing 65.6% (w/w) 1,3-DAG and 32.6% (w/w) 1(3),2-DAG] or a TAG emulsion at the rate of 1 ml/h for 1 h through a canula inserted into the stomach. The amount of test oil in each test emulsion was adjusted so that both emulsions delivered equal amounts of fatty acids (144 mg/h). Lymph fluids were collected every hour for 5 h after starting infusion of the lipid emulsions, and lymph chylomicrons were isolated, purified and analysed.

The rates of TAG transport (2–3 h after starting the infusion) and cholesterol transport (2–4 h after starting the infusion) were significantly lower in the rats infused with the DAG emulsion than in the rats infused with the TAG emulsion. The cumulative amount of TAG transport (at the end of the 5 h infusion period) and of cholesterol transport (after 3–5 h of infusion) were also significantly lower for the DAG emulsion than for the TAG emulsion.

There were no differences in the overall fatty acid composition of chylomicron

TAG between the rats receiving the TAG and DAG emulsions. However, considerable differences were observed in the fatty acid composition at the 2-position of TAG between the two groups of rats. Thus, intragastric infusion with DAG (mainly consisting of 1,3-DAG) instead of TAG not only altered the rate of lipid transport by lymph chylomicrons but also altered the structure of the chylomicron TAG molecules. These results indicate that the relative contributions of the 2-MAG pathway and the glycerol-3-phosphate pathway to TAG re-esterification may vary depending on the types of acylglycerols in the diet.

3. Anti-obesity effects of DAG in animal studies

Murase *et al.* (2001) examined the effects of dietary DAG on the development of obesity and of the resulting hyperinsulinemia and hyperleptinemia in obesity- and diabetes-prone C57BL/6J mice. C57BL/6J mice are reported to provide a good model of human obesity and of type II diabetes, which results from an interaction between environmental and genetic factors (Surwit *et al.,* 1988, 1995, 1997). These mice develop obesity, hyperinsulinemia, and hyperleptinemia when raised on a high-fat, high-sucrose diet, but remain lean on low-fat diets.

Male C57BL/6J mice at 7 weeks of age were randomly divided into 3 groups, and were allowed *ad libitum* access to water and one of three synthetic diets. Individual 'Roden CAFE' food containers were used in order to allow accurate food intakes to be calculated for each animal. The control (low-TAG) diet contained TAG (5%, w/w), protein (20%, w/w) and starch (66.5%, w/w); the high-TAG diet contained TAG (30%, w/w), protein (20%, w/w), starch (28.5%, w/w) and sucrose (13%, w/w); and the high-DAG diet was the same as the high-TAG diet but contained DAG (30%, w/w) instead of TAG. Animals were maintained on these diets for 5 months.

Their body weights during the 5 months are shown in Figure 4. Consistent with previous reports (Surwit *et al.,* 1988, 1995, 1997), feeding with the high-TAG diet resulted in significant increases in body weight compared with mice fed the control diet. On the other hand, the gain in body weight was markedly less with the high-DAG diet. In addition, the feed efficiency of the high-DAG-fed mice was lower than that of the high-TAG-fed mice, suggesting that less energy is accumulated from dietary DAG than from TAG. The average energy intakes of mice fed both the high-fat (TAG and DAG) diets were higher than that of the control mice fed a low-fat diet. However, there was no significant difference in energy intake between mice fed the high-TAG and high-DAG diets. As mentioned above, the absorption rate of dietary DAG was not significantly different from that of TAG. Therefore, it is unlikely that a difference in energy intake caused the observed differences in body fat accumulation between mice fed the high-TAG and high-DAG diets.

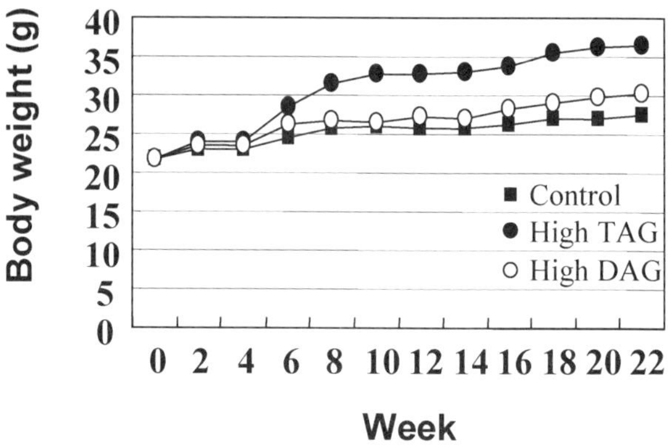

Figure 4. Comparison of the body weights of C57BL/6J mice fed one of three diets during 5 months: a control (low-fat) diet; a high-TAG diet; or a high-DAG diet. See Section C.3 of text for further details. Each point represents the mean body weight of 5 animals. Data taken from Murase *et al.* (2001).

To examine the effects of diet on regional fat accumulation, the weight distribution of four individual fat pads were analysed. After 5 months of feeding with the high-TAG diet, the weights of all the fat pads were significantly increased. Compared with the control diet, the high-TAG diet increased epididymal, mesenteric, retroperitoneal and perirenal fat pad weights by 2.8, 1.9, 3.3 and 3.0 fold, respectively (Figure 5). In contrast, the high-DAG diet resulted in significantly less fat accumulation in all four areas, resulting in epididymal, mesenteric, retroperitoneal and perirenal fat pad weights that were 73%, 92%, 76% and 89% lower, respectively, that those of mice fed the high-TAG diet. These findings parallel the results of body weight gain shown in Figure 4.

To examine the local accumulation of fat further, liver TAG and liver cholesterol contents in the mice were determined. As shown in Table 5, diet had significant effects on the liver. The weight of the liver increased by 37% when mice were fed the high-TAG diet rather than the control diet; this increase was much lower in mice fed the high-DAG diet. There was no significant difference in liver TAG concentration between mice fed control or high-TAG diets. However, the liver TAG concentration was 26% lower in mice fed the high-DAG diet than in those fed the high-TAG diet. Compared with the control diet, the total TAG content of the liver increased by 41% on the high-TAG diet, while it was unchanged on the high-DAG diet. On the other hand, there were no significant differences in liver cholesterol concentrations among the different groups of mice. These results suggest that dietary DAG might be useful for preventing TAG accumulation in the liver.

Diet had significant effects on serum insulin and leptin levels in the mice. As

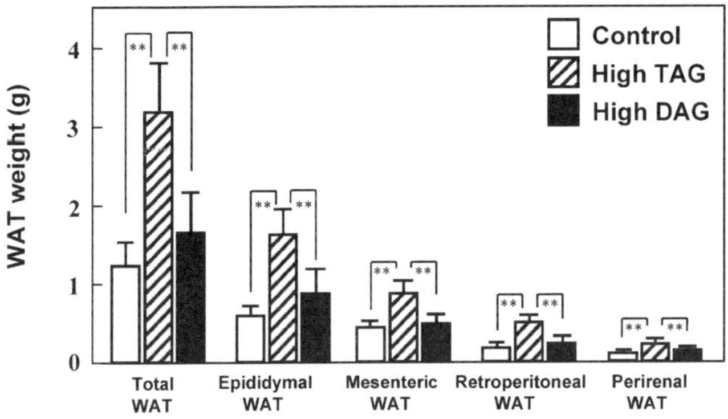

Figure 5. Fat pad white adipose tissue (WAT) weights of C57BL/6J mice after 5 months on one of three diets: a control (low-fat) diet; a high-TAG diet; or a high-DAG diet. See Section C.3 of text for further details. Animals were maintained on these diets for 5 months. Values are means±SD for 5 animals. **Significantly different, $P<0.01$. Data taken from Murase *et al.* (2001).

shown in Figure 6A, after 5 months the serum insulin concentration was 3.9 fold higher in mice fed the high-TAG diet than in the controls, indicating the development of hyperinsulinemia. In contrast, mice fed the high-DAG diet did not develop hyperinsulinemia. Serum leptin concentrations were 10 fold higher in mice fed the high-TAG diet than in the controls (Figure 6B). In contrast, mice fed the high-DAG diet did not have significantly elevated leptin levels.

In conclusion, the results of this study by Murase *et al.* (2001) indicate that structural differences between DAG and TAG (not differences in their fatty acid composition) markedly affect the nutritional behavior of lipids, including body fat accumulation, the serum lipid profile, and the development of hyperinsulinemia and hyperleptinemia.

Table 5. Effects of diet on liver weight and lipid content in mice[a]

	Diet		
Parameter	**Control**	**High-TAG**	**High-DAG**
Liver weight (g)	0.907 ± 0.057	1.246 ± 0.106[#]	1.104 ± 0.050*
[TAG] (mg/g liver)	79.4 ± 11.7	81.6 ± 13.7	60.4 ± 5.5[#,]*
[Cholesterol] (mg/g liver)	3.81 ± 0.41	3.69 ± 0.48	3.46 ± 0.35
TAG content (mg)	71.9 ± 11.4	101.6 ± 18.2[#]	66.7 ± 7.0**
Cholesterol content (mg)	3.46 ± 0.42	4.59 ± 0.61[#]	3.81 ± 0.33*

[a]Data taken from Murase *et al.* (2001). Values are means ± SD (n=5). See Section C.3 of text for further details. [#]Significantly different ($P<0.01$) from controls. *Significantly different ($P<0.05$) from high-TAG group. **Significantly different ($P<0.01$) from high-TAG group.

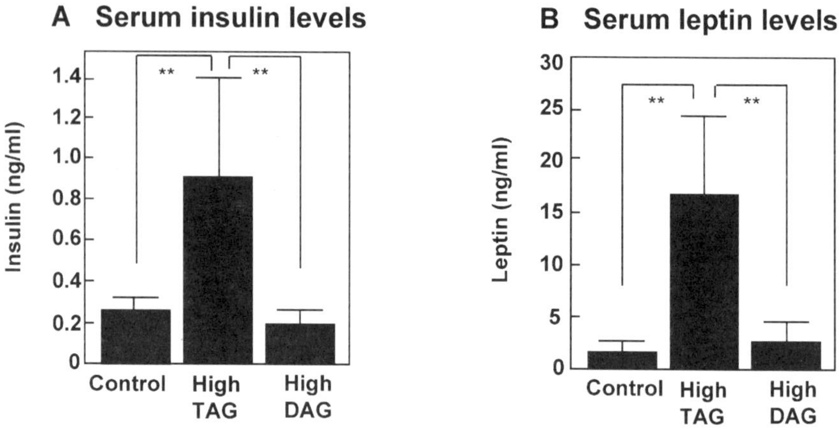

Figure 6. Fasting serum insulin (A) and leptin (B) levels in C57BL/6J mice after 5 months on one of three diets: a control (low-fat) diet; a high-TAG diet; or a high-DAG diet. See Section C.3 of text for further details. Values are means±SD for 5 animals. **Significantly different, $P<0.01$. Data taken from Murase *et al.* (2001).

4. Long-term effects of DAG ingestion on the proteins and enzymes involved in lipid metabolism

To gain insight into the effects of dietary DAG on the regulation of genes involved in obesity, lipid metabolism, and energy expenditure, Murase *et al.* (2001) analysed various mRNA levels in white adipose tissue and the liver by semi-quantitative RT-PCR (reverse transcriptase polymerase chain reaction). Leptin mRNA levels in epididymal adipose tissue were significantly higher in mice fed the high-TAG diet than that in mice fed the control diet (Figure 7). When normalized by the relative levels of β-actin mRNA, the leptin mRNA level in the mice fed the high-TAG diet was more than 4 fold higher than that in the control diet group. In contrast, no significant elevation of the leptin mRNA level occurred in mice fed the high-DAG diet. Thus, the high-DAG diet suppressed the elevation of leptin mRNA expression in adipose tissues, and resulted in lower serum leptin concentrations as discussed above (Figure 6B).

The level of tumor necrosis factor α (TNF-α), which may be involved in mediating insulin resistance, was not changed by either high-fat diet. The levels of plasminogen activator inhibitor 1 (PAI-1) and of uncoupling protein 2 (UCP-2) mRNA in adipose tissue were elevated by both high-fat diets. In these cases, there were no differences between the high-TAG and high-DAG diet groups.

In accordance with the results of a previous report (Murata *et al.*, 1997) that DAG feeding up-regulated acyl-coenzyme A (CoA) oxidase (ACO) activity in the liver of rats, Murase *et al.* (2001) found that ACO and acyl-CoA synthase

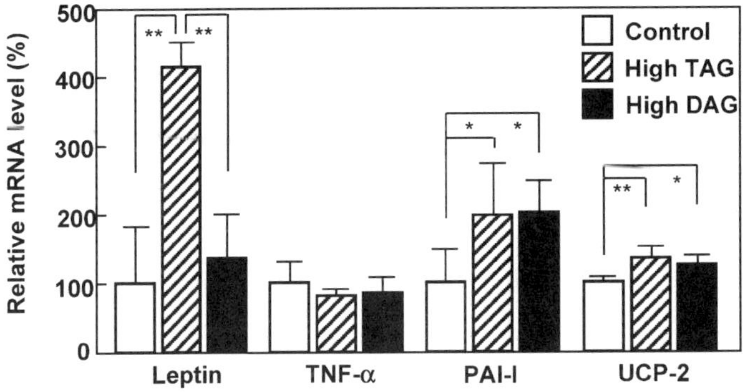

Figure 7. mRNA levels in the epididymal white adipose tissue of C57BL/6J mice fed one of three diets: a control (low-fat) diet; a high-TAG diet; or a high-DAG diet. See Section C.3 of text for further details. After 5 months of feeding with each diet, total RNA isolated from epididymal white adipose tissue was subjected to RT-PCR and visualized by ethidium bromide staining. The amounts of mRNA were quantified with a digital fluorodensitometer, and are expressed as percentages of those in the controls. Values are means±SD for 5 animals. *,**Significantly different: *$P<0.05$, **$P<0.01$. TNF-α, tumor necrosis factor α. PAI-1, plasminogen activator-1. UCP-2, uncoupling protein-2. Data taken from Murase *et al.* (2001).

(ACS) mRNA levels were significantly higher in mice fed the high-DAG diet than in mice fed the high-TAG diet (Figure 8). Hepatic UCP-2 mRNA levels in mice fed both high-fat diets were 3 fold higher than in control mice, while there was no difference between mice fed high-TAG and high-DAG diets.

Many genes related to lipid metabolism, such as ACO and UCP, are regulated by PPAR-α (peroxisome proliferator-activated receptor α) in the liver (Schoonjans *et al.*, 1996; Tsuboyama-Kasaoka *et al.*, 1999). However, PPAR-α mRNA levels did not show any significant differences among the three diet groups. It has been reported that dietary fish oil decreases levels of mature SREBP-1 (sterol regulatory element-binding protein 1) by down-regulation of mRNA in the liver, and that this may account for its beneficial effects (Kim *et al.,* 1999; Xu *et al.,* 1999). The SREBP-1 mRNA level in the liver was significantly higher in mice fed a high-TAG diet than in the mice fed a control diet. Although the SREBP-1 mRNA level in mice fed a high-DAG diet was slightly lower than in mice fed a high-TAG diet, the difference was not significant.

Although the mechanism for the suppression of body fat accumulation by dietary DAG has not been fully elucidated, it may be due to the induction of increased β-oxidation in the liver. In their 1997 study, Murata *et al.* reported that when rats were fed a diet containing about 10% (w/w) DAG for 2–3 weeks, the activities of enzymes involved in fatty acid synthesis in the liver decreased, and those involved in the β-oxidation of fatty acids increased.

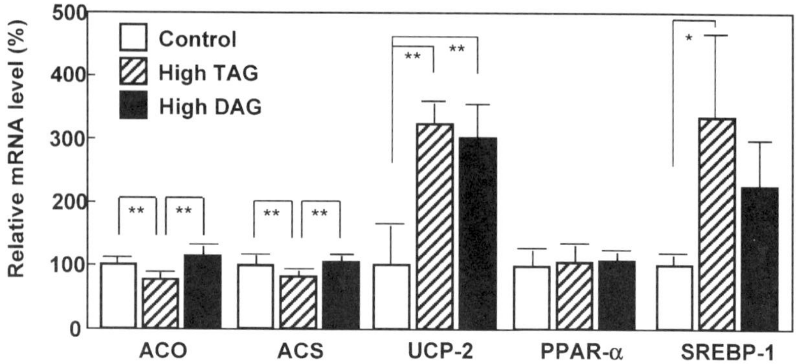

Figure 8. mRNA levels in the liver of C57BL/6J mice fed one of three diets: a control (low-fat) diet; a high-TAG diet; or a high-DAG diet. See Section C.3 of text for further details. After 5 months of feeding with each diet, total RNA isolated from the liver was subjected to RT-PCR and visualized by ethidium bromide. The amounts of mRNA were quantified with a digital fluorodensitometer, and expressed as percentages of those in the controls. Values are means±SD for 5 animals. *,**Significantly different: *$P<0.05$, **$P<0.01$. ACO, acyl-coenzyme A oxidase. ACS, acyl-coenzyme A synthase. UCP-2, uncoupling protein-2. PPAR-α, peroxisome proliferator-activated receptor α. SREBP-1, sterol regulatory element binding protein 1. Data taken from Murase *et al.* (2001).

In the first trial conducted by Murata *et al.* (1997), rats were fed diets containing either DAG or TAG for 14 days. The fatty acid compositions of the added DAG and TAG were similar, and acylglycerols were added to the experimental diets so that each provided the same total amount of fatty acids (9.39 g/100 g diet). Compared with TAG, dietary DAG resulted in significantly lower concentrations of serum and liver TAG. The activities of enzymes involved in fatty acid synthesis [fatty acid synthetase, glucose 6-phosphate dehydrogenase (EC 1.1.1.49), and malic enzyme (EC 1.1.1.40)] were significantly lower in rats fed DAG than in those fed TAG. In contrast, the rates of mitochondrial and peroxisomal oxidation of palmitoyl-CoA in liver homogenates were higher in rats fed DAG than in those fed TAG.

In the second trial conducted by Murata *et al.* (1997), varying amounts of dietary TAG were replaced by DAG while the dietary fatty acid content was maintained at 9.39 g/100 g diet. After 21 days on the new diets, significant reductions in serum and liver TAG levels were found in the groups of rats fed diets in which DAG supplied more than 6.58 g fatty acids/100 g diet. Reductions in the activities of enzymes involved in fatty acid synthesis and increases in palmitoyl-CoA oxidation rates by both mitochondrial and peroxisomal pathways were also apparent when DAG replaced TAG in diets to supply more than 6.58 g fatty acid/100 g diet. Increasing dietary levels of DAG progressively increased the activities of enzymes involved in the β-oxidation pathway in the liver, including carnitine palmitoyltransferase (EC 2.3.1.21), acyl-CoA dehydrogenase (EC 1.3.99.3), acyl-CoA oxidase (EC 1.3.3.6), enoyl-CoA hydratase (EC 4.2.1.17), 3-hydroxyacyl-CoA dehydrogenase (EC 1.1.1.35), 2,4-dienoyl-CoA reductase (EC 1.3.1.34) and Δ^3/Δ^2-enoyl-CoA isomerase (EC 5.3.3.8). These observations suggest that stimulation of fatty acid oxidation in the liver might be partially responsible for the beneficial effects of dietary DAG. In addition, the lower hepatic TAG content and lower serum cholesterol level in DAG-fed mice may be related to up-regulated lipid metabolism in the liver.

The observation that dietary DAG alters lipid metabolism in the liver led us to speculate that the mechanism might involve changes in the levels of transcription factors that regulate fatty acid metabolism. However, as mentioned earlier, we found no significant differences in PPAR-α or SREBP-1 mRNA levels in the liver between mice fed high-TAG and high-DAG diets. The precise mechanism(s) by which dietary DAG stimulates lipid metabolism in the liver remains to be elucidated. UCP-2 is a mitochondrial membrane transporter that may regulate thermogenesis through dissipation of the proton gradient across the inner mitochondrial membrane. Although no consensus has yet been reached concerning the physiological role of UCP-2, some previous studies have suggested its involvement in the regulation of thermogenesis and body weight (Ricquier and Bouillaud, 2000; Fleury *et al.*, 1997; Yanovski *et al.*, 2000). Therefore, it seems possible that stimulation of fatty acid oxidation, together with UCP-2 up-regulation, may play a role in the metabolic efficacy of dietary DAG.

D. Clinical studies

1. Double-blind controlled study on the effects of dietary DAG on postprandial serum and chylomicron TAG responses in healthy humans

Taguchi *et al.* (2000) investigated the effects of dietary DAG on postprandial lipemia in 17 healthy men by monitoring serum lipid profiles after the ingestion of test emulsions containing either DAG or TAG. None of the study participants had a history of diabetes or dyslipidemia. Each emulsion contained 10% (w/w) fat (either DAG or TAG oil), 47.5% water, 1.3% emulsifying agent (polyglycerol esters), 12.5% lemon flavoring concentrate, 12.5% orange flavoring concentrate, 3.32% erythritol, and 12.88% reduced starch hydrolysate to stabilize the mixture, which was emulsified using a high-pressure emulsifier. The study was performed in a double-blind controlled manner, and the two test emulsions were given in random order. Subjects received the DAG emulsion or TAG emulsion in doses of 44 g (n=17), 20 g (n=10) or 10 g (n=13) of experimental oil per 60 kg body weight. Blood samples were drawn every 2 h for 8 h after ingestion.

Serum TAG concentrations increased significantly relative to fasting levels

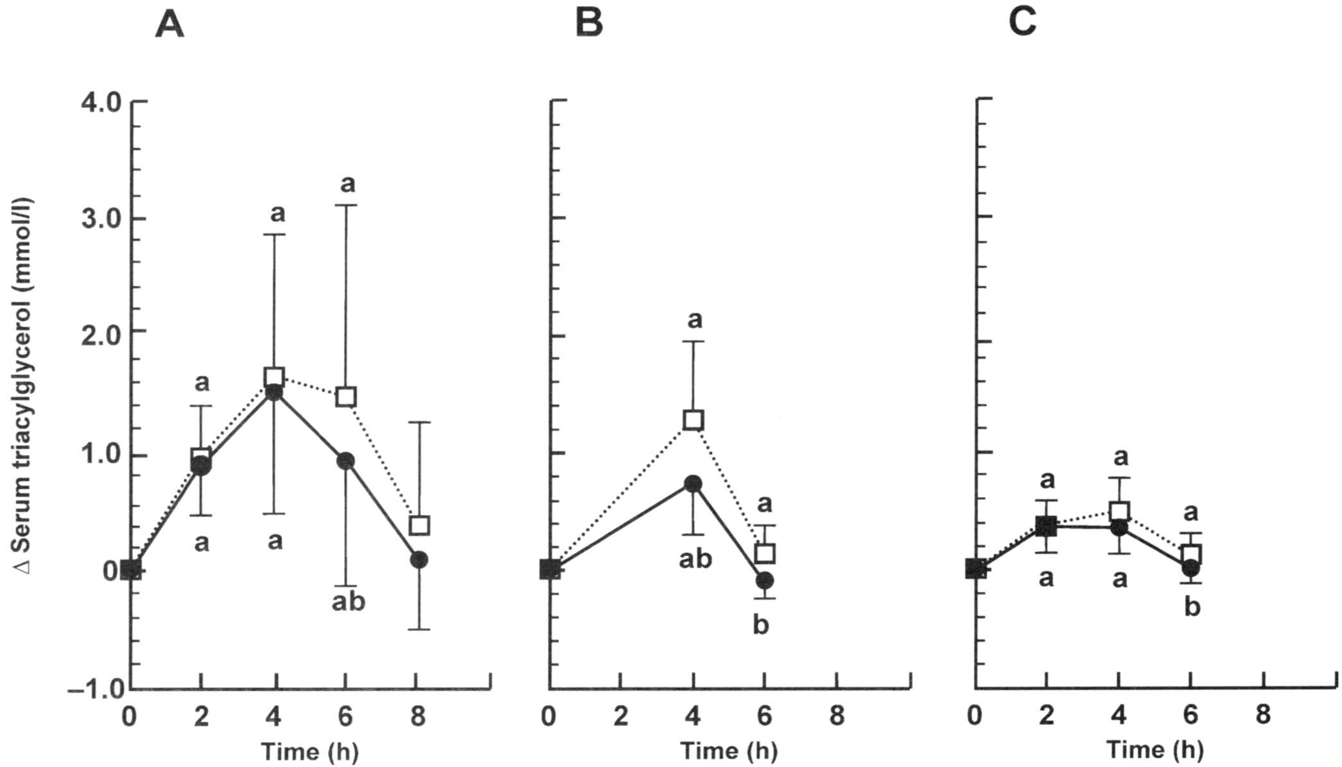

Figure 9. Mean (± SD) changes from baseline in serum TAG concentrations after the ingestion of a DAG emulsion (●) or TAG emulsion (□). (A) 44 g fat ingestion (n=17). (B) 20 g fat ingestion (n=10). (C) 10 g fat ingestion (n=13). [a]Significantly different from fasting levels, $p<0.05$ (Student's t-test for paired values). [b]Significantly different from the corresponding values for the TAG emulsion, $p<0.05$ (Student's t-test for paired values). Data taken from Taguchi *et al.* (2000).

for both the DAG and TAG emulsions at all doses (Figure 9). The postprandial serum TAG responses to the emulsions increased in a dose-dependent manner. From the data for the group ingesting the highest fat level (Figure 9A), the relative amounts of postprandial lipemia were compared by calculating the areas under the curves and above the initial fasting levels. The mean (±SD) postprandial lipemia induced by the DAG emulsion (6.54 ± 5.12 mmol·h/l) was significantly (23%, $p<0.05$) less than that induced by the TAG emulsion (8.45 ± 7.54 mmol·h/l). However, the postprandial concentrations of serum cholesterol, insulin, and plasma glucose did not differ between the TAG and DAG groups.

For a more detailed analysis of the effects of DAG on postprandial lipemia in humans, plasma lipoprotein fractions were separated and analysed following the ingestion of 20 g of DAG or TAG. The changes in plasma chylomicron lipid concentrations are shown in Figure 10. The increases in the postprandial concentrations of chylomicron TAG and cholesterol were smaller after DAG ingestion than after TAG ingestion; the difference in the concentrations of chylomicron TAG at 4 h was statistically significant (Figure 10A). The fasting and postprandial lipid concentrations of plasma very-low-density lipoproteins (VLDL), low-density lipoproteins (LDL) and high-density lipoproteins (HDL) after ingestion of the DAG and TAG emulsions are shown in Table 6. Plasma VLDL TAG levels at 4 h and 6 h were significantly lower in the DAG group than in the TAG group.

Several mechanisms might be involved in the decreased response of chylomicron TAG to dietary DAG. It is unlikely that impaired absorption of dietary fat from the small intestine is responsible, as dietary DAG has the same digestibility coefficient as TAG of similar fatty acid composition (Taguchi *et al.,* 2001). Since the calculated fatty acid content of the DAG oil (862 mg/g) was slightly lower than that of the TAG oil (899 mg/g), the theoretical amounts of fatty acid ingested differed between the emulsions; however, the differences in serum and chylomicron TAG concentrations observed in this study were far greater than would be expected from such a small difference in the amounts of fatty acid ingested. Murata *et al.* (1994) previously showed that the lymphatic transport of chylomicron TAG was slower after DAG ingestion than after TAG ingestion, using rats intragastrically infused with DAG or TAG emulsions containing equal amounts of fatty acids. We consider that a common mechanism underlies these observations of Murata *et al.* (1994) and the results of our recent human study (Taguchi *et al.*, 2000).

As previously reported by Murata *et al.* (1994), the decreased response of chylomicron TAG to dietary DAG in healthy humans may be due to a reduction of re-esterification and chylomicron assembly in the small intestine, or to a reduction of subsequent secretion of chylomicrons into the circulation. The dietary DAG used in the Taguchi *et al.* (2000) study described above was composed mainly of 1,3-isoforms. Based on the nature of gastric and pancreatic

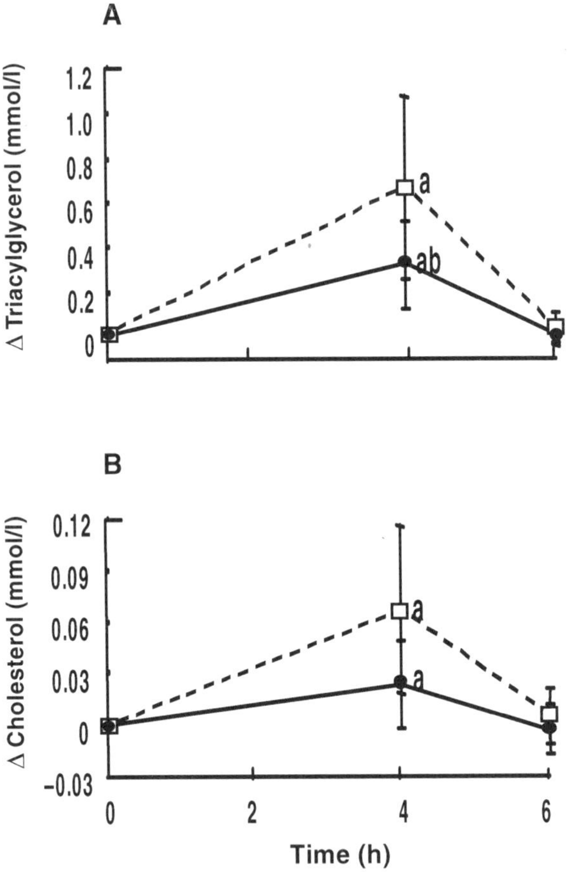

Figure 10. Mean (±SD) changes from baseline plasma chylomicron lipid concentrations after the ingestion of a DAG emulsion (●) or TAG emulsion (□) at a dose of 20 g fat (n=10). (A) Chylomicron TAG concentrations. (B) Chylomicron cholesterol concentrations. [a]Significantly different from fasting levels, p<0.05 (Student's t-test for paired values). [b]Significantly different from the corresponding values for the TAG emulsion, p<0.05 (Student's t-test for paired values). Data taken from Taguchi *et al.* (2000).

lipases, we postulate that the final digestive products of 1,3-DAG are predominantly free glycerol and free fatty acids. In contrast, the final digestion products of TAG are 2-MAG and free fatty acids. Free glycerol is readily absorbed and transported into the blood circulation, or may be phosphorylated to *sn*-glycerol-3-phosphate and utilized for the synthesis of TAG. Yang and co-workers, however, have reported that a large proportion of the TAG formed through the *sn*-glycerol-3-phosphate pathway is stored in the cytoplasm of intestinal mucosal cells rather than being directly used for chylomicron assembly (Yang and Kuksis, 1987; Yang *et al.*, 1990). Thus, the differences in the

Table 6. Fasting and postprandial concentrations of plasma lipoprotein lipids after ingesting DAG or TAG emulsions[a]

Parameter	Total cholesterol (mmol/l) DAG emulsion	Total cholesterol (mmol/l) TAG emulsion	TAG (mmol/l) DAG emulsion	TAG (mmol/l) TAG emulsion	Phospholipids (mmol/l) DAG emulsion	Phospholipids (mmol/l) TAG emulsion
VLDL						
Fasting	0.35 ± 0.22	0.36 ± 0.19	0.49 ± 0.32	0.65 ± 0.43	0.22 ± 0.14	0.24 ± 0.12
Postprandial 4 h	0.45 ± 0.25#	0.51 ± 0.21#	0.78 ± 0.44#,*	1.01 ± 0.53#	0.32 ± 0.21#	0.35 ± 0.15#
6 h	0.35 ± 0.23	0.41 ± 0.20	0.42 ± 0.27*	0.70 ± 0.44	0.21 ± 0.18	0.25 ± 0.13
LDL						
Fasting	2.88 ± 0.68	3.08 ± 0.75	0.24 ± 0.07	0.26 ± 0.05	0.79 ± 0.17	0.89 ± 0.18
Postprandial 4 h	2.86 ± 0.70	3.03 ± 0.80	0.24 ± 0.05*	0.27 ± 0.05	0.79 ± 0.19	0.88 ± 0.24
6 h	2.90 ± 0.59	3.10 ± 0.75	0.24 ± 0.05	0.25 ± 0.04	0.78 ± 0.15	0.89 ± 0.18
HDL						
Fasting	1.46 ± 0.26	1.31 ± 0.17	0.10 ± 0.03	0.09 ± 0.02	1.25 ± 0.20	1.17 ± 0.21
Postprandial 4 h	1.43 ± 0.18*	1.26 ± 0.19	0.10 ± 0.02	0.11 ± 0.02#	1.26 ± 0.21*	1.16 ± 0.24
6 h	1.50 ± 0.15*	1.34 ± 0.19	0.10 ± 0.03	0.10 ± 0.02	1.26 ± 0.21	1.18 ± 0.23

[a]Data taken from Taguchi *et al.* (2000). Subjects were given emulsions containing 20 g of fat. Values are means ± SD (n=10). See Section D.1 of text for further details. #Significantly different from fasting levels (p<0.05), assessed using paired Student's t-test. *Significantly different from the corresponding values for the TAG emulsion (p<0.05), assessed using paired Student's t-test.

metabolic fates of the digestion products of DAG and TAG in intestinal mucosal cells could be responsible for the differences in plasma chylomicron TAG concentrations.

In conclusion, the double-blind controlled study using healthy men showed that one can reduce the magnitude of postprandial lipemia by consuming DAG oil in place of TAG oil (Taguchi *et al.*, 2000). Based on the correlation between coronary artery disease and postprandial lipemia, it can be speculated that dietary DAG may be less atherogenic than TAG.

2. *Dynamics of postprandial remnant-like lipoprotein particles (RLP) in serum after DAG ingestion*

To further investigate the comparative effects of DAG and TAG on postprandial lipid metabolism, differences in the dynamics of remnant lipoproteins after consuming DAG or TAG of the same fatty acid composition were investigated in 6 male volunteers (Tada *et al.*, 2001). The subjects ate emulsified test oils (30 g lipid/m^2 of body surface area) in the early morning after fasting for at least 12 hours. Each test emulsion contained 35% (w/w) fat (either DAG or TAG oil), 61.94% water, and 3.06% emulsifying agent. This study was a double-blind controlled design. In the first examination, 3 subjects ingested

DAG and the remaining 3 subjects ingested TAG. One month later, the groups were reversed for the second examination. Changes in serum parameters were examined by analysing blood samples before eating the test emulsions, and 2 h, 3 h, 4 h, 6 h and 8 h afterwards. Serum lipid concentrations were measured by the conventional enzymatic method. Serum lipoprotein lipase (LPL) levels were measured by the method of Kobayashi *et al.* (1993) using anti-LPL antibody. Quantification of remnant-like lipoprotein particles (RLP) was carried out by the method of Nakajima *et al.* (1993, 1996).

Table 7 shows the changes in various serum lipid parameters. Serum TAG concentrations were significantly higher (two-way ANOVA, $P<0.001$) 8 h after TAG ingestion than after DAG ingestion. There were no significant differences in the serum total cholesterol levels after ingesting either DAG or TAG. Free fatty acid levels increased in both groups to a similar extent; a slight difference between the two dietary groups was detected only by two-way ANOVA. Serum HDL cholesterol concentrations were significantly higher ($P<0.01$) 3 h after DAG ingestion than 3 h after TAG ingestion. There were no significant differences in serum LPL or insulin levels between the DAG and TAG groups (Table 8). As also shown in Table 8, serum RLP cholesterol and RLP TAG levels were markedly increased in both groups from 2 h after ingesting the emulsions; these increases were significantly higher (two-way ANOVA, $P<0.001$) after TAG loading.

RLP are generated either as chylomicron remnants formed on the vascular wall by the action of LPL on secreted chylomicrons, or by the action of LPL on VLDL released from the liver and small intestine (Taskinen and Kuusi, 1987). The activation of LPL is controlled by the level of insulin (Fielding and Frayn, 1998). Therefore, it is conceivable that the differences in serum RLP cholesterol and RLP TAG levels after ingesting DAG or TAG might have been caused by altered LPL activities or serum insulin levels. However, there were no differences in serum LPL or insulin levels between the DAG and TAG diet groups, suggesting that the observed differences in serum RLP levels were unrelated to the process of RLP generation by LPL and insulin.

As mentioned earlier, foods containing DAG have been shown to reduce fatty acid synthetase activity and to increase the activities of enzymes involved in β-oxidation in the liver of rats (Murata *et al.*, 1997). The observed differences in serum RLP levels following DAG or TAG ingestion are therefore likely to be related to differences in metabolic regulation – mainly those involving the resynthesis of TAG after digestion, and the regulation of other metabolic processes in the liver. Dietary DAG may thus reduce the risk of coronary arteriosclerotic diseases by lessening the postprandial increase in RLP concentration, which is known to be closely correlated with the development of atherosclerosis.

Table 7. Changes in serum lipid concentrations following the ingestion of DAG or TAG[a]

Parameter	Treatment	Time after ingestion of lipid (h)						*P*	
		0	2	3	4	6	8	Time	Lipid
TAG	DAG	100	122.7 ± 9.7*	153.5 ± 38.2*	167.4 ± 30.7	139.0 ± 53.4	83.2 ± 23.9*	0.021	<0.001
(mmol/l)	TAG	100	171.0 ± 40.2	204.9 ± 56.0	216.6 ± 80.0	167.4 ± 58.3	96.7 ± 23.9		
Free fatty acids	DAG	100	100.6 ± 24.7	162.5 ± 45.3	193.7 ± 55.2	174.3 ± 61.7	135.7 ± 31.8		<0.001
(mmol/l)	TAG	100	119.7 ± 20.1	178.2 ± 34.8	216.8 ± 97.7	197.5 ± 111.3	119.7 ± 45.5		
Total cholesterol	DAG	100	103.5 ± 13.1	110.5 ± 27.2	116.5 ± 43.4	104.8 ± 19.1	97.1 ± 6.2		
(mmol/l)	TAG	100	100.8 ± 3.1	101.5 ± 2.9	102.3 ± 3.1	101.1 ± 2.4	101.5 ± 4.1		
Free cholesterol	DAG	100	99.5 ± 2.0**	102.5 ± 4.3*	103.6 ± 4.5*	102.0 ± 8.2	100.5 ± 6.6	0.0135	
(mmol/l)	TAG	100	104.2 ± 4.0	106.3 ± 5.7	108.0 ± 5.9	105.6 ± 6.7	102.2 ± 5.0		
HDL cholesterol	DAG	100	97.9 ± 2.9	95.1 ± 4.7**	92.4 ± 6.4	95.5 ± 7.7	103.0 ± 5.0		<0.001
(mmol/l)	TAG	100	94.9 ± 4.8	91.1 ± 5.3	90.4 ± 6.7	92.8 ± 9.3	108.6 ± 12.6		
LDL cholesterol	DAG	100	98.0 ± 3.8	99.4 ± 3.8*	98.0 ± 3.7	97.5 ± 5.8	102.4 ± 6.5		
(mmol/l)	TAG	100	99.0 ± 4.4	97.6 ± 4.0	98.0 ± 3.7	97.9 ± 4.4	102.1 ± 3.7		

[a]Data taken from Tada *et al.* (2001). Values are means ± SD (n=6) and are shown as percentages of the initial value. P values at right are given for significant main effects and interactions assessed by two-way ANOVA. See Section D.2 of text for further details. *,**Significantly different from TAG ingestion at the same time point: *P<0.05, **P<0.01.

Table 8. Changes in serum parameters following the ingestion of DAG or TAG[a]

		Time after ingestion of lipid (h)						P	
Parameter	**Treatment**	**0**	**2**	**3**	**4**	**6**	**8**	**Time**	**Lipid**
Lipoprotein lipase (μg/l)	DAG	100	100.5 ± 10.1	104.8 ± 13.4	86.5 ± 8.5	81.7 ± 8.0	75.6 ± 9.4		
	TAG	100	100.1 ± 10.2	96.5 ± 4.0	83.3 ± 2.8	82.2 ± 14.0	81.8 ± 16.4		
Insulin (μU/ml)	DAG	100	136.1 ± 84.9	92.9 ± 31.9	97.8 ± 27.4	84.6 ± 13.5	68.3 ± 42.1		
	TAG	100	113.6 ± 36.6	113.7 ± 20.0	101.0 ± 17.5	71.4 ± 15.9	53.5 ± 5.7		
RLP cholesterol (mg/dl)	DAG	100	137.1 ± 19.7**	165.0 ± 40.9*	184.9 ± 43.2	147.6 ± 52.3	81.4 ± 15.6*	0.001	<0.001
	TAG	100	198.6 ± 40.2	231.3 ± 40.5	235.9 ± 76.0	207.6 ± 91.9	109.3 ± 23.5		
RLP TAG (mg/dl)	DAG	100	286.2 ± 75.2	473.0 ± 247.6	529.7 ± 164.5	320.0 ± 135.6	88.9 ± 31.8	0.006	<0.001
	TAG	100	588.0 ± 428.1	696.8 ± 288.6	712.0 ± 337.3	461.6 ± 282.7	127.0 ± 63.3		

[a]Data taken from Tada *et al.* (2001). Values are means ± SD (n=6) and are shown as percentages of the initial value. P values at right are given for significant main effects and interactions assessed by two-way ANOVA. See Section D.2 of text for further details. *,**Significantly different from TAG ingestion at the same time point: *P<0.05, **P<0.01. RLP, remnant-like lipoprotein particles.

3. Dietary DAG suppresses the accumulation of body fat in men: a double-blind controlled trial

To investigate the effects of long-term ingestion of dietary DAG on body fat accumulation and lipid metabolism in humans, a double-blind study was conducted (Nagao *et al.*, 2000). Thirty-eight male volunteers aged 27–49 with an average body mass index (BMI) of 24.1 ± 0.4 kg/m^2 participated in the study. All subjects were generally healthy and had no history of diabetes or lipemia. For 4 weeks prior to the test period (designated as the control period), the subjects were asked to consume about 50 g of total fat daily. After the control period, all subjects underwent baseline anthropometric measurements and were randomly divided into two groups of 19. The average BMI and hepatic fat contents were not significantly different between the two groups. One group then consumed a DAG-supplemented diet and the other group consumed a TAG-supplemented diet for breakfast. Bread, mayonnaise, and shortbread containing 5 g of 1,3-DAG-rich or TAG-rich oil were prepared and subjects were allowed to choose two from these diet forms so that they consumed 10 g of one of the two test oils every day for breakfast throughout the 16-week test period. For breakfast, besides the test diet, they were asked to eat additional foods of their own choice containing about 5 g of fat. For lunch from Monday to Friday, they chose menus with known fat contents in the cafeteria, and for dinner from Monday to Friday they ate packed meals containing approximately 20 g of total fat prepared under the guidance of a nutritionist. For lunch and dinner on Saturdays, Sundays, and holidays, they ate self-selected meals containing about 50 g/day of total fat, including the test oils at breakfast. During both the control and test periods, daily alcoholic beverage intakes were restricted to an amount equivalent to about 30 ml alcohol. The subjects were also asked to maintain daily exercise at a fixed level throughout the study period, and to record the contents of their daily meals and snacks in a dietary diary throughout the entire control period and the last 4 weeks of the test period. At weeks 0 and 16, anthropometric measurements were carried out by trained investigators.

The subjects underwent computed tomography (CT) scanning within 3 days of each blood sample being taken. Transverse tomograms were taken at the umbilical level and at fixed intervals in which the spleen and liver were observed in the same plane. Total fat area, visceral fat area, and subcutaneous fat area were determined from the CT images at the umbilical level by the method of Tokunaga *et al.* (1983). Hepatic fat content was estimated by the method of Kato *et al.* (1984). Hounsfield units (HU; a measure of density) of the liver and spleen were also determined from the CT images, and the liver-to-spleen ratio of the values (L-HU/S-HU) was calculated as an index of the hepatic fat content.

Mean daily energy intakes, fat intakes, and percentages of energy from fat did not differ significantly between the two groups. Although the subjects were asked to consume 50 g/ fat per day, including the test oils, their actual average fat intake was calculated to be 43 g/day. The average daily energy intake was 7950 kJ and the percentage of energy from fat was about 21%.

The measured changes in anthropometric values in the two groups are summarized in Table 9. Body weight, BMI, and waist circumference decreased significantly in both groups after the test period. The extent of decrease in each variable, however, was significantly greater in the DAG group than in the TAG group. In both groups, the total fat area (determined by the abdominal CT images) decreased significantly after the test period, but again the extent of this decrease was significantly greater in the DAG group than in the TAG group. The visceral fat area in the DAG group decreased significantly from the baseline; in contrast, the slight reduction observed in the TAG group was not statistically significant (p=0.78). Similarly, significant reductions in the subcutaneous fat area and in the visceral to subcutaneous fat area ratio (V/S) were observed only in the DAG group.

Since we assigned the subjects into two groups randomly, except for ensuring that the average BMI and liver fat contents were not significantly different between the groups, the initial values for other variables such as total fat area and visceral fat area incidentally differed between the groups. The initial visceral fat area in the DAG group was significantly higher than that in the TAG group. An initial abundance of visceral fat has been shown to be significantly related to a larger loss of visceral fat after body weight loss (Leenen *et al.*, 1992). In fact, a weak but significant correlation between the initial values and the loss of visceral fat was observed in this study. To clarify the effect of DAG on body fat, the significance of the effect on the variables that showed significant differences between the groups at baseline (as examined by Student's t-test) were analysed by the analysis of covariance procedure. Changes from the baseline values for each group were analysed using the baseline values as the covariate. Analysis of covariance revealed that the decreases in total fat and visceral fat were significantly greater in the DAG group than in the TAG group. We therefore concluded that DAG is capable of reducing body fat.

The liver-to-spleen density ratio (L-HU/S-HU) increased slightly in the DAG group after the test period, indicating that the hepatic fat content had decreased during the test period. This ratio did not change in the TAG group. However, the differences between the two groups were not statistically significant.

It appears that the total fat and energy intakes of the subjects during the study period were less than their habitual intakes before the study. Nevertheless, the reductions in body weight, BMI, waist circumference, total fat area, visceral fat area, subcutaneous fat area, and V/S ratio that occurred during the study were significantly greater in the DAG group than in the TAG group.

Table 9. Changes of anthropometric values in human volunteers consuming DAG- or TAG-supplemented diets[a]

Parameter	Week of study	DAG group		TAG group	
		Value	Change[b]	Value	Change[b]
Body weight (kg)	0	72.1 ± 1.8		68.1 ± 1.3	
	16	69.5 ± 1.7[##]	−2.6 ± 0.3**	67.0 ± 1.5[#]	−1.1 ± 0.4
BMI (kg/cm²)	0	24.1 ± 0.4		23.5 ± 0.3	
	16	23.2 ± 0.4[##]	−1.9 ± 0.1**	23.2 ± 0.4[#]	−0.4 ± 0.1
Waist circumference (cm)	0	85.0 ± 1.4		82.0 ± 1.0	
	16	80.6 ± 1.3[##]	−4.4 ± 0.6*	79.5 ± 1.2[##]	−2.5 ± 0.6
Hip circumference (cm)	0	97.1 ± 0.9		96.1 ± 0.7	
	16	96.1 ± 0.9	−1.1 ± 0.5	95.4 ± 0.8	−0.7 ± 0.5
W/H	0	0.87 ± 0.01		0.85 ± 0.01	
	16	0.84 ± 0.01[##]	−0.04 ± 0.01	0.83 ± 0.01[##]	−0.02 ± 0.01
Body fat (g/100 g)	0	21.6 ± 1.2		20.3 ± 1.0	
	16	20.6 ± 1.3[##]	−1.1 ± 0.3	18.8 ± 1.3[#]	−1.5 ± 0.7
Total fat area (cm²)	0	227 ± 1.6*		182 ± 15	
	16	189 ± 16[##]	−38 ± 3*,[!]	164 ± 19[#]	−17 ± 8
Visceral fat area (cm²)	0	79 ± 7*		56 ± 6	
	16	63 ± 7[##]	−16 ± 2**,[!]	51 ± 6	−5 ± 3
Subcutaneous fat area (cm²)	0	148 ± 11		126 ± 10	
	16	126 ± 10[##]	−22 ± 3**	118 ± 13	−8 ± 4
V/S	0	0.55 ± 0.04		0.46 ± 0.04	
	16	0.50 ± 0.04[#]	−0.05 ± 0.02	0.46 ± 0.04	0 ± 0.02
L-HU/S-HU	0	1.24 ± 0.04		1.23 ± 0.03	
	16	1.30 ± 0.03[#]	+0.06 ± 0.03	1.24 ± 0.03	+0.01 ± 0.02

[a]Data taken from Nagao *et al.* (2000). Values are means ± SEM (n=19). See Section D.3 of text for further details. [b]16 week value minus initial (0 week) value. [#,##]Significantly different from the initial value by Student's t-test: [#]p<0.05, [##]p<0.01. *,**Significantly different from TAG group by Student's t-test for paired values: *p<0.05, **p<0.01. [!]Significantly different from TAG group by analysis of covariance, p<0.05. W/H, waist to hip circumference ratio. V/S, visceral to subcutaneous fat area ratio. L-HU/S-HU, liver to spleen ratio of Hounsfield units.

The calculated energy values of the 1,3-DAG and TAG oils given to the study participants were almost identical (Taguchi *et al.*, 2001). Moreover, the observed effects of DAG seem unlikely to be attributable to differences in digestibility, since studies on rats had shown no differences in the absorption of the two oils (Taguchi *et al.*, 2001). It therefore seems likely that the significant decreases in body weight, BMI, waist circumference, and visceral fat seen in the DAG group were caused by differences in the metabolism of DAG and TAG in the small intestine, as described above (Murata *et al.*, 1994; Taguchi *et al.*, 2000; Tada *et al.*, 2001). Although the mechanism is not fully elucidated, impaired postprandial TAG clearance has been shown to be associated with visceral obesity (Couillard *et al.*, 1998; Mekki *et al.*, 1999). Postprandial serum TAG levels increase less after DAG ingestion than after TAG ingestion, which may at least partly explain the observed reduction in visceral fat after repeated ingestion of the DAG oil.

Nelson and Tucker (1996) recently reported that total body fat increases as the intake of dietary lipids increases. Among various types of fat deposition, visceral obesity is a known risk factor for diseases such as diabetes mellitus, hyperlipidemia, hypertension, and atherosclerosis (Fujioka *et al.*, 1987; Nakamura *et al.*, 1994). Ways of reducing such risk factors are therefore of interest. The amount of test oil ingested in the Nagao *et al.* (2000) study (10 g/day) was set to match the average daily consumption of cooking oil in Japan, where the average daily fat consumption is approximately 50 g. The health benefits of the DAG-rich oil described in this study can therefore be easily achieved in daily life if normal cooking oil is replaced with DAG-rich oil. The effect of this oil for those consuming higher levels of fat now needs to be evaluated in further experiments to test the dose–response relationship in humans.

In conclusion, we have shown that DAG, in contrast to TAG, suppresses both body weight and regional fat deposition, including visceral and hepatic fat, in healthy but near-obese male volunteers. The use of DAG-rich oil in place of ordinary cooking oil may therefore be beneficial to health by suppressing visceral fat deposition.

A long-term beneficial health effect of dietary DAG has also been shown in a study of obese Americans. The Chicago Center for Clinical Research (Chicago, Illinois, USA) recently conducted a double-blind parallel study in which 127 obese men and women ingested DAG or TAG diets for 6 months. In this study, 15% of energy intake was ingested as the test oil under a mild hypocaloric condition established by subtracting 500–800 kcal from the energy requirement calculated for the subjects' body weight, activity level, and age. The decreases in body weight and body fat were significantly higher in the DAG group than in the TAG group (Maki *et al.*, in press).

4. Long-term ingestion of dietary DAG lowers serum TAG levels in type II diabetic patients with hypertriglyceridemia

Impaired postprandial TAG clearance has been shown to be associated with visceral obesity (Couillard *et al.*, 1998; Mekki *et al.*, 1999). Visceral obesity is a strong predictor of type II diabetes (Bjorntorp and Rosmond 1999). Postprandial hyperlipidemia is a serious issue in diabetic patients. Yamamoto *et al.* (2001) have suggested that consuming DAG-rich oil might be useful for reducing hypertriglyceridemia accompanying diabetes, and for relieving multiple risk factors for the development of arteriosclerosis.

A randomized, single-blind, parallel trial was conducted at the Itami City Hospital in Japan. In this study, Yamamoto *et al.* (2001) examined the influence of long-term ingestion of DAG on blood lipids in 16 patients with type II diabetes and hypertriglyceridemia whose serum TAG levels persistently increased despite continuous nutritional counseling at the outpatient clinic for a period of 14.4 ± 12.7 months. The subjects had a mean BMI of 26.3 ± 2.9. Half of the subjects (3 males and 5 females of mean age 56.8 ± 7.3 years) were randomly assigned to the DAG group, in which an ordinary (TAG-rich) cooking oil was switched to a DAG-rich oil (target dose: 10 g/day) for 3 months. The other half (4 males and 4 females of mean age 54.1 ± 18.8 years) formed the control group, in which consumption of the ordinary cooking oil was continued. In the DAG group, 5 patients were treated with nutritional counseling only, and 3 were treated with medicines to control blood glucose levels (1 with sulfonylurea, 1 with α-glucosidase inhibitor, and 1 with insulin). In the control group, 6 were treated with nutritional counseling only, and 2 with medicines (1 with sulfonylurea and 1 with α-glucosidase inhibitor).

The DAG-rich or TAG-rich oils were used for cooking, such as for frying food, and for home-made salad dressings. On weeks 0 and 12 of the study, blood samples were collected from fasting subjects. Blood serum TAG, total cholesterol, and glucose levels were measured by the conventional enzymatic method, and glycohemoglobin (HbA_{1c}) levels were measured by an automated high-pressure liquid chromatographic assay as described by Ellis *et al.* (1984).

The subjects' body weight, BMI, energy intake, percent energy from fat, total fat intake, total oil intake, and DAG and TAG intakes, measured before and at the end of the experiment, are shown in Table 10. Their serum TAG, total

Table 10. Body and dietary parameters in diabetic patients before and 3 months after replacing ordinary cooking oil with DAG-rich oil[a]

	DAG group		Control group	
Parameter	**Before**	**After**	**Before**	**After**
Body weight (kg)	60.8 ± 7.6	60.7 ± 7.2	74.5 ± 15.5	73.9 ± 15.0
BMI (kg/m^2)	24.4 ± 1.7	24.4 ± 1.6	28.1 ± 2.9	27.9 ± 2.9
Energy intake (MJ/day)	6.93 ± 0.92	6.83 ± 0.59	7.30 ± 2.57	6.88 ± 2.36
Energy from fat (%)	28 ± 5	30 ± 4	28 ± 6	27 ± 7
Total fat intake (g/day)	52.0 ± 13.7	54.9 ± 11.4	56.9 ± 30.4	51.1 ± 30.7
Total oil intake (g/day)	27.1 ± 9.4	26.4 ± 9.3	25.1 ± 15.4	23.9 ± 19.6
TAG intake (g/day)	26.8 ± 9.3	15.7 ± 8.9[#]	24.9 ± 15.2	23.5 ± 19.5
DAG intake (g/day)	0.3 ± 0.1	10.6 ± 3.9[#]	0.3 ± 0.1	0.3 ± 0.2

[a]Data taken from Yamamoto *et al.* (2001). Values are means ± SD (n=8 for each group). See Section D.4 of text for further details. [#]Significantly different from before (p<0.05).

cholesterol, HDL cholesterol levels before and at the end of the test period are shown in Table 11. The baseline levels of all parameters did not differ significantly between the DAG and control groups. No significant changes in these parameters occurred in the control group during the test period. In contrast, the serum TAG concentration in the DAG group decreased significantly ($p<0.01$) during the 3-month treatment. Fasting blood glucose and glycohemoglobin (HbA_{1c}) levels before and at the end of the test period are also shown in Table 11. There were no significant changes in blood glucose levels during the study period in either group. However, the glycohemoglobin (HbA_{1c}) content decreased significantly in the DAG group, from 6.41% ± 1.15% to 5.79% ± 0.85% of total hemoglobin (a 9.7% reduction, $p<0.05$). The normal range for HbA_{1c}, established by the Japan Diabetes Society, is 4.3–5.8% (Shima *et al.*, 1994). No significant change in HbA_{1c} level was observed in the control group.

It has been reported recently that several metabolic syndromes such as diabetes, hyperlipidemia, and hypertension are involved in the onset of arteriosclerosis (Beckman *et al.*, 2002). This implies the frequent coincidence of several serious risk factors for the development of cardiovascular disease (Reaven, 1988). The main goal of nutritional counseling for those with diabetes is to maintain good blood sugar control and to prevent complications. However, reducing the hypertriglyceridemia that accompanies diabetes might relieve multiple risk factors, preventing the development of arteriosclerosis. Thus, the use of DAG-rich oil might be useful for maintaining the quality of life of patients with diabetes. These new insights may lead to a revision of the recommended standard instructions regarding fat limitation for those with diabetes.

Table 11. Serum parameters in diabetic patients before and 3 months after replacing ordinary cooking oil with DAG-rich oil[a]

	DAG group		**Control group**	
Serum parameter	**Before**	**After**	**Before**	**After**
Total cholesterol (mmol/l)	5.82 ± 1.32	5.87 ± 0.80	6.00 ± 0.98	5.74 ± 0.70
TAG (mmol/l)	2.51 ± 0.75	1.52 ± 0.28[#,†]	3.22 ± 2.13	3.59 ± 1.70
HDL cholesterol (mmol/l)	1.27 ± 0.23	1.34 ± 0.36	1.09 ± 0.20	1.22 ± 0.36
Glucose (mmol/l)	6.72 ± 0.72	6.94 ± 0.89	7.83 ± 1.78	7.77 ± 2.50
HbA_{1c} (%)[b]	6.41 ± 1.15	5.79 ± 0.85[#]	6.88 ± 0.53	6.65 ± 0.73

[a]Data taken from Yamamoto *et al.* (2001). Values are means ± SD (n=8 for each group). See Section D.4 of text for further details. [b]Glycohemoglobin content, expressed as a percentage of the total hemoglobin concentration. [#]Significantly different from before ($p<0.05$). [†]Significantly different from control group ($p<0.05$).

5. The solubilization of phytosterols in DAG instead of TAG oil improves their serum cholesterol-lowering effect

In addition to obesity and hypertriglyceridemia, hypercholesterolemia is another health concern linked to various lifestyle-related diseases. Clinical investigations have revealed that the administration of phytosterols to human subjects reduces plasma total cholesterol and LDL cholesterol levels (see also Chapter 7). Meguro *et al.* (2001) recently investigated the relative effects on serum cholesterol and TAG levels in humans caused by phytosterols dissolved in DAG compared with phytosterols dissolved in TAG.

Twelve normocholesterolemic or moderately hypercholesterolemic (serum total cholesterol levels in the range 4.66–6.78 mmol/l) healthy men aged 30–50 years participated in this study, which was carried out using a randomized crossover design. All subjects consumed a control mayonnaise daily at supper for 2 weeks before the test period (designated as the control period) and were then randomly assigned to one of two groups. One group was given mayonnaise containing a mixture of phytosterols dissolved in DAG ("PS/DAG") and the other group was given mayonnaise containing phytosterols dispersed in TAG ("PS/TAG") daily at supper for 2 weeks. After a wash-out period, during which all subjects consumed the control mayonnaise for 4 weeks, the groups were reversed for 2 weeks.

Throughout the study, the subjects were asked to maintain their normal lifestyle and dietary pattern. Each subject recorded a food diary during the study periods. The phytosterol mixture [47% (w/w) β-sitosterol, 27% (w/w) campesterol, 25% (w/w) stigmasterol] was used without esterification or hydrogenation. The DAG-rich and TAG-rich oils used in the study were prepared by the same method as described by Taguchi *et al.* (2001); see Section B.1 above. The phytosterol mixture was dissolved or dispersed in either the DAG-rich or TAG-rich oil at a concentration of 5% (w/w). The daily PS/DAG or PS/TAG doses contained 10 g of DAG-rich or TAG-rich oils mixed with 500 mg of phytosterols. The control mayonnaise contained 10 g of the TAG-rich oil without added phytosterols. The added phytosterols did not affect the taste or flavor of the mayonnaise. Dietary fat intakes were similar when volunteers consumed the PS/DAG mayonnaise (85.84 ± 12.2 g fat per day) and PS/TAG mayonnaise (83.25 ± 14.2 g/day). Similarly, dietary energy intakes were not significantly different between the periods of ingesting the PS/DAG mayonnaise (9179.6 ± 1149.2 kJ/day) and PS/TAG mayonnaise (9217.1 ± 1250.6 kJ/day).

Overnight fasting blood samples were collected from each volunteer before breakfast at the beginning and end of each test period of the trial. Serum total cholesterol, HDL cholesterol and TAG concentrations were determined by enzymatic methods. Serum LDL cholesterol concentrations were calculated according to the method of Friedewald *et al.* (1972).

Table 12. Effects of phytosterols dissolved in DAG or TAG on serum lipid parameters in human volunteers[a]

		Value of parameter		
Serum lipid parameter	**Study group**	**Initial**	**Final**	**Change**
Total cholesterol (mmol/l)	PS/DAG	5.57 ± 0.66	5.31 ± 0.78[#]	–0.26 ± 0.39*
	PS/TAG	5.37 ± 0.64	5.40 ± 0.74	+0.03 ± 0.35
LDL cholesterol (mmol/l)	PS/DAG	3.69 ± 0.59	3.39 ± 0.66[#]	–0.30 ± 0.41
	PS/TAG	3.57 ± 0.65	3.43 ± 0.56	–0.13 ± 0.40
HDL cholesterol (mmol/l)	PS/DAG	1.09 ± 0.24	1.14 ± 0.19	+0.06 ± 0.11
	PS/TAG	1.05 ± 0.20	1.13 ± 0.23	+0.08 ± 0.11
TAG (mmol/l)	PS/DAG	1.73 ± 0.60	1.67 ± 0.52	–0.06 ± 0.44
	PS/TAG	1.65 ± 0.71	1.83 ± 0.58	+0.18 ± 0.45

[a]Data taken from Meguro *et al.* (2001). Values are means ± SD for 12 male volunteers, measured at the beginning and the end of each study period. See Section D.5 of text for further details. [#]Significantly different from the initial value by Student's t-test, $p<0.05$. *Significantly different from the PS/TAG group by Student's t-test for paired values, $p<0.05$.

Table 12 shows the serum concentrations of total, LDL, and HDL cholesterol and TAG determined at the beginning and end of each treatment. Consuming the PS/TAG mayonnaise slightly reduced LDL cholesterol levels from the initial value, but the change was not significant. However, consuming the PS/DAG mayonnaise significantly reduced total and LDL cholesterol levels from the initial values. The degree of total cholesterol reduction induced by the PS/DAG mayonnaise was significantly greater than that induced by the PS/TAG mayonnaise ($p<0.05$). In the PS/TAG group, the average serum TAG concentration increased from the initial value, while in the PS/DAG group the serum TAG concentration reduced slightly reduced from the initial value. However, this reduction was not statistically significant. The results of this study therefore suggest that phytosterols dissolved in DAG have a better serum cholesterol-lowering effect than phytosterols dissolved in TAG, and may be useful for treating hyperlipidemia.

Phytosterols have been reported to reduce serum cholesterol concentrations by competitively blocking cholesterol absorption from the intestinal lumen (Heinemann *et al.*, 1986, 1993), by displacing cholesterol from bile salt micelles (Child and Kuksis, 1986), by increasing bile salt excretion (Salen *et al.*, 1970), or by hindering the cholesterol esterification rate in the intestinal mucosa (Child and Kuksis, 1983; Ikeda and Sugano, 1983). Ikeda *et al.* (1988) also showed that the only site of action was the binding of sterols to micelles. As mentioned above, although many studies have been conducted, the precise mechanism of the inhibitory effect of phytosterols on cholesterol absorption in the intestine is not fully understood.

No significant effects of dietary DAG itself on serum cholesterol levels have been reported in humans (Nagao *et al.*, 2000). Therefore, dietary DAG appears to augment the inhibitory effect of phytosterols on cholesterol absorption in the intestine. However, it is not clear why the hypolipidemic activity of PS/DAG was greater than that of PS/TAG. One possibility is that the administration of phytosterols in soluble form may be important for obtaining a sufficient quantity in the micellar phase to inhibit cholesterol absorption (Vanhanen *et al.*, 1993; Ostlund *et al.*, 1999). However, Mattson *et al.* (1982) showed that β-sitosterol is more efficient than its esters at decreasing cholesterol absorption in humans. It has been reported that β-sitosterol esters dissolve well in dietary oils and are hydrolysed to β-sitosterol during the lipolysis of dietary fats (Miettinen and Siurala, 1971). Furthermore, no significant increase in hypocholesterolemic activity due to differences in solubility in oil were detected by Miettinen and Vanhanen (1994). These results imply that factors other than the relative solubilities of phytosterols in different oils might be responsible for their competition with cholesterol for incorporation into mixed micelles in the intestine. Further detailed experiments are required to confirm the mechanism of the hypolipidemic activity of PS/DAG observed by Meguro *et al.* (2001).

E. Applications of DAG

From the results of these studies, it has been possible to develop a phytosterol preparation that may be used as a component of daily meals to lower blood cholesterol levels without requiring alterations in daily dietary habits. A DAG-rich oil containing added phytosterols, called 'Healthy Econa Cooking Oil' (manufactured by Kao Corp), has been on the market in Japan since 2001 as a FOSHU approved by the Ministry of Public Health and Welfare. It is now sold under two major sub-brand names, one carrying a label claim that it is "less likely to become body fat", and the other having the double claim that it is "less likely to become body fat" and "lowers blood cholesterol levels".

Healthy Econa Cooking Oil has been widely accepted in the Japanese market as a cooking oil effective in the prevention of fat deposition. Other products incorporating DAG oil (but without added phytosterols) include canned tuna fish, curry, bread rolls, and margarine. There is also a 'Healthy Econa Dressings' range of salad dressings with different flavors. Further expansion of this premium oil market is expected.

Kao Corp, together with Archer Daniels Midland Co (Illinois, USA) – the largest edible oilseed processor in the USA – recently established a joint company to introduce DAG, the primary component of the Healthy Econa Cooking Oil range, throughout the world. ADM Kao LLC (Decatur, Illinois, USA) aims to supply DAG as a raw material to major food manufacturers throughout North America and Europe.

F. Conclusions

(1) DAG are minor components of most edible oils.
(2) Approximately 70% (w/w) of DAG in edible oils exist as the 1,3-isoform as a result of acyl migration.
(3) The energy value of an oil containing 87% (w/w) DAG is approximately 98% that of a TAG oil with the same fatty acid composition, so differences in their total energy values in realistic diets will be negligible.
(4) The absorption coefficient of DAG oil is the same as that of TAG oil.
(5) 1-MAG is produced during the digestion of DAG oils in the small intestine.
(6) The release rate of resynthesized TAG into the intestinal lymph is lower after administering a DAG emulsion than a TAG emulsion.
(7) DAG, unlike TAG of similar fatty acid composition, significantly suppress the body fat accumulation and elevation of circulating insulin and leptin levels induced by feeding a high-fat diet to obesity-prone and diabetes-prone C57BL/6J mice.
(8) DAG feeding affects the expression of genes related to energy homeostasis, including leptin in white adipose tissue, and acyl-CoA oxidase and acyl-CoA synthase in the liver.
(9) DAG feeding reduces fatty acid synthetase activity and elevates the activities of enzymes involved in β-oxidation in the liver of rats.
(10) The magnitude of postprandial lipemia in healthy men is lower after DAG ingestion than after TAG ingestion.
(11) The levels of postprandial RLPs in healthy men are lower after DAG ingestion than after TAG ingestion.
(12) DAG, unlike TAG, suppresses both body weight and regional fat deposition – including visceral and hepatic fat deposition – in healthy but near-obese men.
(13) Long-term ingestion of dietary DAG lowers serum TAG concentrations in type II diabetic patients with hypertriglyceridemia.
(14) The solubilization of phytosterols in DAG versus TAG improves the serum cholesterol-lowering effect.

References

Abdel-Nabey, AA, Shehata, AAY, Ragab, MH and Rossell, JB (1992) Glycerides of cottonseed oils from Egyptian and other varieties. *Riv. Ital. Sostanze Grasse*, 69, 443–447.

Adriaanse, N, Dekker, H and Coops, J (1965) Heats of combustion of normal saturated fatty acids and their methyl esters. *Recueil*, 84, 393–407.

Apgar, JL, Shively, CA and Tarka, SM, Jr (1987) Digestibility of cocoa butter and corn oil and their influence on fatty acid distribution in rats. *J. Nutr.*, 117, 660–665.

Beckman, JA, Creager, MA and Libby, P (2002) Diabetes and atherosclerosis: epidemiology, pathophysiology, and management. *JAMA*, 287(19), 2570–2581.

Bjorntorp, P and Rosmond, R (1999) Hypothalamic origin of the metabolic syndrome X. *Ann. NY Acad. Sci.*, 892, 297–307.

Bracco, U (1994) Effect of triglyceride structure on fat absorption. *Am. J. Clin. Nutr.*, 60, 1002S–1009S.

Carey, VJ, Walters, EE, Colditz, GA, Solomon, CG, Willett, WC, Rosner, BA, Speizer, FE and Manson, JE (1997) Body fat distribution and risk of non-insulin-dependent diabetes mellitus in women. The Nurses' Health Study. *Am. J. Epidemiol.*, 145, 614–619.

Chijiiwa, K and Linscheer, WG (1987) Distribution and monomer activity of cholesterol in micellar bile salt: effect of cholesterol level. *Am. J. Physiol.*, 252(3 Pt 1), G309–G314.

Child, P and Kuksis, A (1983) Critical role of ring structure in the differential uptake of cholesterol and plant sterols by membrane preparations in vitro. *J. Lipid Res.*, 24, 1196–1209.

Child, P and Kuksis, A (1986) Investigation of the role of micellar phospholipid in the preferential uptake of cholesterol over sitosterol by dispersed rat jejunal villus cells. *Biochem. Cell Biol.*, 64, 847–853.

Combe, E, Demarne, Y, Gueguen, L, Ivorec-Szylit, O, Meslin, JC and Sacquet, E (1976) Some aspects of the relationships between gastrointestinal flora and host nutrition. *World Rev. Nutr. Diet.*, 24, 1–57.

Couillard, C, Bergeron, N, Prud'homme, D, Bergeron, J, Tremblay, A, Bouchard, C, Mauriege, P and Despres, JP (1998) Postprandial triglyceride response in visceral obesity in men. *Diabetes*, 47, 953–960.

Crossley, A, Freeman, IP, Hudson, BJF and Pierce, JH (1959) Acyl migration in diglycerides. *J. Chem. Soc.*, 760–764.

D'alonzo, RP, Kozarek, WJ and Wade, RL (1982) Glyceride composition of processed fats and oils as determined by glass capillary gas chromatography. *J. Am. Oil Chem. Soc.*, 59, 292–295.

Demarne, Y, Sacquet, E, Lecourtier, MJ and Flanzy, J (1979) Comparative study of endogenous fecal fatty acids in germ-free and conventional rats. *Am. J. Clin. Nutr.*, 32, 2027–2032.

Demarne, Y, Corring, T, Pihet, A and Sacquet, E (1982) Fat absorption in germ-free and conventional rats artificially deprived of bile secretion. *Gut*, 23, 49–57.

De Schrijver, R, Vermeulen, D and Backx, S (1991) Digestion and absorption of free and esterified fish oil fatty acids in rats. *Lipids*, 26, 400–404.

Doreau, M and Chilliard, Y (1997) Digestion and metabolism of dietary fat in farm animals. *Br. J. Nutr.*, 78, S15–S35.

Ellis, G, Diamandis, EP, Giesbrecht, EE, Daneman, D and Allen, LC (1984) An automated "high-pressure" liquid-chromatographic assay for hemoglobin A1c. *Clin. Chem.*, 30, 1746–1752.

Fielding, BA and Frayn, KN (1998) Lipoprotein lipase and the disposition of dietary fatty acids. *Br. J. Nutr.*, 80, 495–502.

Fleury, C, Neverova, M, Collins, S, Raimbault, S, Champigny, O, Levi-Meyrueis, C, Bouillaud, F, Seldin, MF, Surwit, RS, Ricquier, D and Warden, CH (1997) Uncoupling protein-2: a novel gene linked to obesity and hyperinsulinemia. *Nat. Genet.*, 15, 269–272.

Friedewald, WT, Levy, RI and Fredrickson, DS (1972) Estimation of the concentration of low-density lipoprotein cholesterol in plasma, without use of the preparative ultracentrifuge. *Clin. Chem.*, 18, 499–502.

Friedman, HI and Nylund, B (1980) Intestinal fat digestion, absorption, and transport. A review. *Am. J. Clin. Nutr.*, 33, 1108–1139.

Fujioka, S, Matsuzawa, Y, Tokunaga, K and Tarui, S (1987) Contribution of intra-abdominal fat accumulation to the impairment of glucose and lipid metabolism in human obesity. *Metabolism,* 36, 54–59.

Heinemann, T, Leiss, O and von Bergmann, K (1986) Effect of low-dose sitostanol on serum cholesterol in patients with hypercholesterolemia. *Atherosclerosis*, 61, 219–223.

Heinemann, T, Axtmann, G and von Bergmann, K (1993) Comparison of intestinal absorption of cholesterol with different plant sterols in man. *Eur. J. Clin. Invest.*, 23, 827–831.

Huge-Jensen, B, Galluzzo, DR and Jensen, RG (1988) Studies on free and immobilized lipases from *Mucor miehei*. *J. Am. Oil Chem. Soc.*, 65, 905–910.

Ikeda, I and Sugano, M (1983) Some aspects of mechanism of inhibition of cholesterol absorption by β-sitosterol. *Biochim. Biophys. Acta*, 732, 651–658.

Ikeda, I, Tanaka, K, Sugano, M, Vahouny, GV and Gallo, LL (1988) Inhibition of cholesterol absorption in rats by plant sterols. *J. Lipid Res.*, 29, 1573–1582.

Kato, K, Takayama, T, Sano, H, Katada, N and Takeichi, M (1984) Clinical investigation of fatty liver by CT. *Acta Hepatol. Jpn (Kanzou)*, 25, 1097–1103.

Kim, HJ, Takahashi, M and Ezaki, O (1999) Fish oil feeding decreases mature sterol regulatory element-binding protein 1 (SREBP-1) by down-regulation of SREBP-1c mRNA in mouse liver. A possible mechanism for down-regulation of lipogenic enzyme mRNAs. *J. Biol. Chem.*, 274, 25892–25898.

Kobayashi, J, Hashimoto, H, Fukamachi, I, Tashiro, J, Shirai, K, Saito, Y and Yoshida, S (1993) Lipoprotein lipase mass and activity in severe hypertriglyceridemia. *Clin. Chim. Acta*, 216, 113–123.

Kodali, DR, Tercyak, A, Fahey, DA and Small, DM (1990) Acyl migration in 1,2-dipalmitoyl-sn-glycerol. *Chem. Phys. Lipids*, 52, 163–170.

Lamon-Fava, S, Wilson, PW and Schaefer, EJ (1996) Impact of body mass index on coronary heart disease risk factors in men and women. The Framingham Offspring Study. *Arterioscler. Thromb. Vasc. Biol.*, 16, 1509–1515.

Leenen, R, van der Kooy, K, Deurenberg, P, Seidell, JC, Weststrate, JA, Schouten, FJ and Hautvast, JG (1992) Visceral fat accumulation in obese subjects: relation to energy expenditure and response to weight loss. *Am. J. Physiol.*, 263 (5 Pt 1), E913–E919.

Lehner, R, Kuksis, A and Itabashi, Y (1993) Stereospecificity of monoacylglycerol and diacylglycerol acyltransferases from rat intestine as determined by chiral phase high-performance liquid chromatography. *Lipids*, 28, 29–34.

Livesey, G (1984) The energy equivalents of ATP and the energy values of food proteins and fats. *Br. J. Nutr.*, 51, 15–28.

Maki, KC, Davidson, MH, Tsushima, R, Matsuo, N, Tokimitsu, I, Umporowicz, DM, Dicklin, MR, Foster, GS, Ingram, KA, Anderson, BD, Frost, SD and Bell, M (2002) Consumption of diacylglycerol oil as part of a mildly reduced-energy diet enhances loss of body weight and fat compared with a triacylglycerol control oil. *Am. J. Clin. Nutr.* (in press).

Mattson, FH, Grundy, SM and Crouse, JR (1982) Optimizing the effect of plant sterols on cholesterol absorption in man. *Am. J. Clin. Nutr.*, 35, 697–700.

Meguro, S, Higashi, K, Hase, T, Honda, Y, Otsuka, A, Tokimitsu, I and Itakura, H (2001) Solubilization of phytosterols in diacylglycerol versus triacylglycerol improves the serum cholesterol-lowering effect. *Eur. J. Clin. Nutr.*, 55, 513–517.

Mekki, N, Christofilis, MA, Charbonnier, M, Atlan-Gepner, C, Defoort, C, Juhel, C, Borel, P, Portugal, H, Pauli, AM, Vialettes, B and Lairon, D (1999) Influence of obesity and body fat distribution on postprandial lipemia and triglyceride-rich lipoproteins in adult women. *J. Clin. Endocrinol. Metab.*, 84, 184–191.

Miettinen, TA and Siurala, M (1971) Bile salts, sterols, sterol esters, glycerides and fatty acids in micellar and oil phases of intestinal contents during fat digestion in man. *Z. Klin. Chem. Klin. Biochem.*, 9, 47–52.

Miettinen, TA and Vanhanen, H (1994) Dietary sitostanol related to absorption, synthesis and serum level of cholesterol in different apolipoprotein E phenotypes. *Atherosclerosis*, 105, 217–226.

Mitchell, DC, McMahon, KE, Shively, CA, Apgar, JL and Kris-Etherton, PM (1989) Digestibility of cocoa butter and corn oil in human subjects: a preliminary study. *Am. J. Clin. Nutr.*, 50, 983–986.

Mokdad, AH, Bowman, BA, Ford, ES, Vinicor, F, Marks, JS and Koplan, JP (2001) The continuing epidemics of obesity and diabetes in the United States. *JAMA*, 286, 1195–1200.

Monsma, CC, Gallaher, DD and Ney, DM (1996) Reduced digestibility of beef tallow and cocoa butter affects bile acid excretion and reduces hepatic esterified cholesterol in rats. *J. Nutr.*, 126, 2028–2035.

Murase, T, Mizuno, T, Omachi, T, Onizawa, K, Komine, Y, Kondo, H, Hase, T and Tokimitsu, I (2001) Dietary diacylglycerol suppresses high fat and high sucrose diet-induced body fat accumulation in C57BL/6J mice. *J. Lipid Res.*, 42, 372–378.

Murata, M, Hara, K and Ide, T (1994) Alteration by diacylglycerols of the transport and fatty acid composition of lymph chylomicrons in rats. *Biosci. Biotech. Biochem.*, 58, 1416–1419.

Murata, M, Ide, T and Hara, K (1997) Reciprocal responses to dietary diacylglycerol of hepatic enzymes of fatty acid synthesis and oxidation in the rat. *Br. J. Nutr.*, 77, 107–121.

Nagao, T, Watanabe, H, Goto, N, Onizawa, K, Taguchi, H, Matsuo, N, Yasukawa, T, Tsushima, R, Shimasaki, H and Itakura, H (2000) Dietary diacylglycerol suppresses accumulation of body fat compared to triacylglycerol in men in a double-blind controlled trial. *J. Nutr.*, 130, 792–797.

Nakajima, K, Saito, T, Tamura, A, Suzuki, M, Nakano, T, Adachi, M, Tanaka, A, Tada, N, Nakamura, H, Campos, E and Havel, RJ (1993) Cholesterol in remnant-like lipoproteins in human serum using monoclonal anti apo B-100 and anti apo A-I immunoaffinity mixed gels. *Clin. Chim. Acta*, 223, 53–71.

Nakajima, K, Okazaki, M, Tanaka, A, Pullinger, CR, Wang, T, Nakano, T, Adachi, M and Havel, RJ (1996) Separation and determination of remnant-like particles in human serum using monoclonal antibodies to apo B-100 and apo A-I. *J. Clin. Ligand Assay*, 19, 177–183.

Nakamura, T, Tokunaga, K, Shimomura, I, Nishida, M, Yoshida, S, Kotani, K, Islam, AH, Keno, Y, Kobatake, T, Nagai, Y, Fujioka, S, Tarui, S and Matsuzawa, Y (1994) Contribution of visceral fat accumulation to the development of coronary artery disease in non-obese men. *Atherosclerosis*, 107, 239–246.

Nelson, LH and Tucker, LA (1996) Diet composition related to body fat in a multivariate study of 203 men. *J. Am. Diet. Assoc.*, 96, 771–777.

Ostlund, RE, Jr, Spilburg, CA and Stenson, WF (1999) Sitostanol administered in lecithin micelles potently reduces cholesterol absorption in humans. *Am. J. Clin. Nutr.*, 70, 826–831.

Peters, JC, Holcombe, BN, Hiller, LK and Webb, DR (1991) Caprenin 3. Absorption and caloric value in adult humans. *J. Am. Coll. Toxicol.*, 10, 357–367.

Reaven, GM (1988) Banting lecture 1988. Role of insulin resistance in human disease. *Diabetes*, 37, 1595–1607.

Ricquier, D and Bouillaud, F (2000) The uncoupling protein homologues: UCP1, UCP2, UCP3, StUCP and AtUCP. *Biochem. J.*, 345, 161–179.

Rimm, EB, Stampfer, MJ, Giovannucci, E, Ascherio, A, Spiegelman, D, Colditz, GA and Willett, WC (1995) Body size and fat distribution as predictors of coronary heart disease among middle-aged and older US men. *Am. J. Epidemiol.*, 141, 1117–1127.

Salen, G, Ahrens, EH, Jr and Grundy, SM (1970) Metabolism of β-sitosterol in man. *J. Clin. Invest.*, 49, 952–967.

Schoonjans, K, Staels, B and Auwerx, J (1996) Role of the peroxisome proliferator-activated receptor (PPAR) in mediating the effects of fibrates and fatty acids on gene expression. *J. Lipid Res.*, 37, 907–925.

Shima, K, Endo, J, Oimoni, M, Oshima, I, Omori, Y, Katayama, Y, Kanazawa, Y, Kawai, T, Kawamori, R, Kanno, T, Kiyose, H, Nakashima, K, Nagamine, Y, Baba, S, Hoshino, T and Amoni, N (1994) Interlaboratory difference in HbA1c measurement in Japan – A report of the committee of an Interlaboratory Standardization of HbA1c Determination, the Japan Diabetes Society. *J. Jpn Diabetes Soc.*, 37, 855–864.

Surwit, RS, Kuhn, CM, Cochrane, C, McCubbin, JA and Feinglos, MN (1988) Diet-induced type II diabetes in C57BL/6J mice. *Diabetes*, 37, 1163–1167.

Surwit, RS, Feinglos, MN, Rodin, J, Sutherland, A, Petro, AE, Opara, EC, Kuhn, CM and Rebuffe-Scrive, M (1995) Differential effects of fat and sucrose on the development of obesity and diabetes in C57BL/6J and A/J mice. *Metabolism*, 44, 645–651.

Surwit, RS, Petro, AE, Parekh, P and Collins, S (1997) Low plasma leptin in response to dietary fat in diabetes- and obesity-prone mice. *Diabetes*, 46, 1516–1520.

Tada, N, Watanabe, H, Matsuo, N, Tokimitsu, I and Okazaki, M (2001) Dynamics of postprandial remnant-like lipoprotein particles in serum after loading of diacylglycerols. *Clin. Chim. Acta*, 311, 109–117.

Taguchi, H, Watanabe, H, Onizawa, K, Nagao, T, Gotoh, N, Yasukawa, T, Tsushima, R, Shimasaki, H and Itakura, H (2000) Double-blind controlled study on the effects of dietary diacylglycerol on postprandial serum and chylomicron triacylglycerol responses in healthy humans. *J. Am. Coll. Nutr.*, 19, 789–796.

Taguchi, H, Nagao, T, Watanabe, H, Onizawa, K, Matsuo, N, Tokimitsu, I and Itakura, H (2001) Energy value and digestibility of dietary oil containing mainly 1,3-diacylglycerol are similar to those of triacylglycerol. *Lipids*, 36, 379–382.

Taskinen, MR and Kuusi, T (1987) Enzymes involved in triglyceride hydrolysis. *Baillieres Clin. Endocrinol. Metab.*, 1, 639–666.

Tokunaga, K, Matsuzawa, Y, Ishikawa, K and Tarui, S (1983) A novel technique for the determination of body fat by computed tomography. *Int. J. Obes.*, 7, 437–445.

Tsuboyama-Kasaoka, N, Takahashi, M, Kim, H and Ezaki, O (1999) Up-regulation of liver uncoupling protein-2 mRNA by either fish oil feeding or fibrate administration in mice. *Biochem. Biophys. Res. Commun.*, 257, 879–885.

Vanhanen, HT, Blomqvist, S, Ehnholm, C, Hyvonen, M, Jauhiainen, M, Torstila, I and Miettinen, TA (1993) Serum cholesterol, cholesterol precursors, and plant sterols in hypercholesterolemic subjects with different apoE phenotypes during dietary sitostanol ester treatment. *J. Lipid Res.*, 34, 1535–1544.

Watanabe, H, Onizawa, K, Taguchi, H, Kobori, M, Chiba, H, Naito, S, Matsuo, N, Yasukawa, T, Hattori, M and Shimasaki, H (1997) Nutritional characterization of diacylglycerols in rats. *J. Jpn. Oil Chem. Soc.*, 46, 301–307.

Xu, J, Nakamura, MT, Cho, HP and Clarke, SD (1999) Sterol regulatory element binding protein-1 expression is suppressed by dietary polyunsaturated fatty acids. A mechanism for the coordinate suppression of lipogenic genes by polyunsaturated fats. *J. Biol. Chem.*, 274, 23577–23583.

Yamamoto, K, Asakawa, H, Tokunaga, K, Watanabe, H, Matsuo, N, Tokimitsu, I and Yagi, N (2001) Long-term ingestion of dietary diacylglycerol lowers serum triacylglycerol in type II diabetic patients with hypertriglyceridemia. *J. Nutr.*, 131, 3204–3207.

Yang, LY and Kuksis, A (1987) Size and composition of lymph chylomicrons following feeding corn oil or its fatty acid methyl esters. *Biochem. Cell. Biol.*, 65, 514–524.

Yang, LY, Kuksis, A and Myher, JJ (1990) Intestinal absorption of menhaden and rapeseed oils and their fatty acid methyl and ethyl esters in the rat. *Biochem. Cell. Biol.*, 68, 480–491.

Yanovski, JA, Diament, AL, Sovik, KN, Nguyen, TT, Li, H, Sebring, NG and Warden, CH (2000) Associations between uncoupling protein 2, body composition, and resting energy expenditure in lean and obese African American, white, and Asian children. *Am. J. Clin. Nutr.*, 71, 1405–1420.

CHAPTER 6

Lipase-catalysed synthesis of modified lipids

UWE T. BORNSCHEUER[1], MAREK ADAMCZAK[1,2] and
MOHAMED M. SOUMANOU[1,3]

[1]*University of Greifswald, Greifswald, Germany;* [2]*University of Warmia and Mazury, Olsztyn, Poland;* [3]*Abomey-Calavi University, Cotonou, Benin*

A. Introduction

Fats and oils are major nutritional compounds and the main source of energy (9 kcal/g fat). The majority of natural fats and oils are used for human nutrition directly or after appropriate modification using conventional large-scale technologies (such as hardening or hydrogenation). Natural fats and oils are not always ideal products and modification of their fatty acid composition and their regiochemical and stereochemical structure can improve their properties and nutritional value. Unbalanced fat consumption and the consumption of the "wrong" lipids can induce health problems such as cardiovascular diseases or cancer. Factors which increase the risk for coronary heart disease, such as the distribution of plasma cholesterol between low-density and high-density lipoproteins (LDL and HDL), oxidizability of LDL, and hemostasis, can be reduced or avoided by changing the source, amount, and structure of dietary fats and oils.

Some fatty acids are precursors of regulatory substances in physiological processes (Kritchevsky, 1998). For instance, polyunsaturated fatty acids (PUFA) of the *n*-3 and *n*-6 families (also known as ω-3/omega-3 and ω-6/omega-6 fatty acids) are required for the synthesis of eicosanoids (Gill and Valivety, 1997a, 1997b; Shahidi and Wanasundara, 1998). Thus, there is an increasing demand for sophisticated methods to produce triacylglycerols which meet the requirements for healthy nutrition. This can be achieved by using structured triacylglycerols (sTAG).

The term sTAG was first introduced by Babayan (1987) and can be taken to include any fat or oil that has been modified in terms of its fatty acid composition by techniques like hydrogenation, blending, interesterification, directed esterification, fatty acid enrichment, genetic engineering of plants, etc. However, a stricter definition of a "true" sTAG includes only those fats or oils

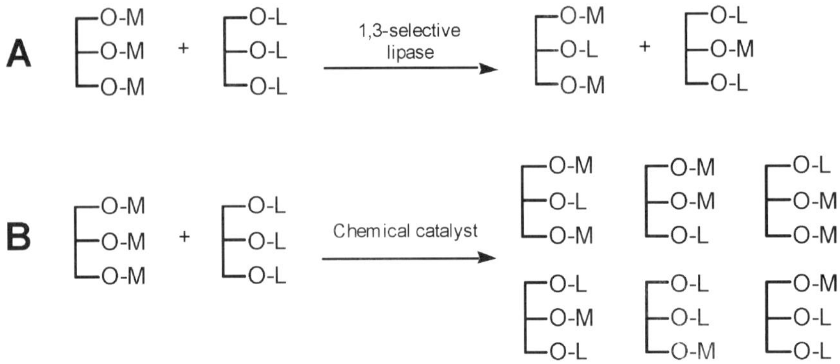

Figure 1. Synthesis of sTAG by interesterification catalysed by: (A), *sn*-1,3 selective lipase; (B), chemical catalysis. M, medium-chain fatty acid. L, long-chain fatty acid. Only the main products of the reactions are shown.

with both a defined fatty acid composition and a defined distribution of these acids along the glycerol backbone. Only then can the full important nutritional values and medical properties of sTAG be ensured.

Modified triacylglycerols can be produced by enzymatic or chemical methods, but lipases are usually needed to produce sTAG, as these biocatalysts offer the required fatty acid selectivity, positional selectivity, and stereoselectivity. Current chemical methods (unless very sophisticated) yield only mixtures of randomized triacylglycerols (Figure 1). Other advantages of lipase reactions over chemical synthesis are summarized in Table 1.

Table 1. Comparison of enzymatic and chemical methods for the synthesis of sTAG

Enzymatic process	Chemical process
• Selective	• Non-selective
• Mild reaction conditions	• High temperature, pressure and drastic pH conditions
• Biocompatible	
• Reduced environmental pollution	• Problem of environmental pollution
• Availability of lipases with different properties (possibility of screening)	
• Many possibilities for reaction engineering	• Some possibilities to change reaction conditions
• Ability to improve lipase properties (e.g. by genetic engineering and/or directed evolution and/or rational protein design and/or mutagenesis)	
• Immobilized lipases can be used many times	
• Compatible with PUFA	• PUFA can be destroyed

Palmitic acid is a major constituent of human milk fat triacylglycerols, occurring predominantly at the *sn*-2 position. Release of palmitic acid from the *sn*-1 or *sn*-3 position can lead to the malabsorption and loss of vital calcium due to the formation of calcium palmitate soaps. Loders Croklaan (Unilever, the Netherlands) currently markets a sTAG under the trade name 'Betapol' as a component of human milk fat substitutes. Betapol contains primarily 1,3-dioleoyl-2-palmitoyl-*sn*-glycerol, and is produced in a lipase-catalysed process (see Section C.2 below). Another application of sTAG is as low-calorie fats. Shorter-chain fatty acids have fewer available calories per unit weight than long-chain fatty acids. Furthermore, stearic acid is poorly absorbed, so fats containing stearic acid impart less energy. Thus, a triacylglycerol containing stearic acid in the *sn*-2 position and short-chain fatty acids in the *sn*-1,3 positions produces fewer calories. If low energy provision is the only application, then triacylglycerols bearing medium-chain fatty acids only would be sufficient (Babayan and Rosenau, 1991; Kennedy, 1991; Megremis, 1991; Fitch Haumann, 1997). However, such fats do not allow the provision of (poly)unsaturated essential fatty acids to the body. Alternatively, sTAG of the "MLM" type (where M refers to a medium-chain fatty acid at the *sn*-1 and *sn*-3 positions, and L refers to a long-chain (poly)unsaturated fatty acid at the *sn*-2 position) will be more readily adsorbed than free PUFA. Consumption of these sTAG has been reported to have positive effects on brain, liver, and adipose tissue lipid and fatty acid profiles (Gill and Valivety, 1997a, 1997b). sTAG may provide a source of γ-linolenic acid (GLA), which has physiological functions including the modulation of atopic eczema (see Chapter 9) and rheumatoid arthritis, and of arachidonic acid (AA), which is important for premature infants.

Despite the lack of selectivity, some structured lipid-like products have been synthesized by chemicals methods. The catalyst most often used for chemical interesterification of triacylglycerols or for alcoholysis between triacylglycerols and fatty acids is sodium methoxide, which is more active than other base, metal or acid catalysts. In general, sodium alkoxides are easy to use as catalysts. They are inexpensive, active at relatively low temperatures (50–90°C), and are effective at low concentrations. Although no regioselectivity or fatty acid selectivity occurs, the chemically synthesized commercial products listed in Table 2 contain sTAG with the desired nutritional function.

Among low-calorie structured fats, 'Salatrim' and 'Caprenin' are well known because they are widely applied in food products (Auerbach *et al.*, 1998). Salatrim is a mixture of short-chain (C_{2-4}) and long-chain (C_{16-22}) triacylglycerols (Smith *et al.*, 1994) produced by a random interesterification (Klemann *et al.*, 1994) followed by vacuum distillation to remove remaining short-chain fatty acids. Enzymatic syntheses of Salatrim-like products have also been reported (Fomuso and Akoh, 1997; Mangos *et al.*, 1999; Yang *et al.*, 2001). Transesterification of triacetin and hydrogenated soybean oil or stearic acid using an immobilized lipase from *Rhizomucor miehei* or from *Carica*

Table 2. Commercially available sTAG-like products[a]

Product	Manufacturer	Composition	Properties/applications	Synthesis
'Captex'	ABITEC Corp, USA	8:0, 10:0, 18:2, 8:0, 10:0, 12:0, 18:2	Clinical applications; cosmetics industry	NA
'Caprenin'	Procter & Gamble, USA	8:0, 10:0, 22:0	GRAS product; low-calorie fat (5 kcal/g); ingredient of soft candies, candy bars, and confectionery coatings for nuts, cookies, etc.	Interesterification between coconut, palm kernel and rapeseed oils
'Salatrim' ('Benefat')	Cultor Food Science, USA	2:0, 3:0, 4:0, 18:0	Low-calorie fat (5 kcal/g); intended for use in oven-baked French fries, baked and dairy products, dressings, dips, sauces, cocoa butter substitute, chocolate-flavored coatings	Interesterification of hydrogenated vegetable oils with triacylglycerols containing acetic and/or propionic and/or butyric acids
'Neobee'	Stepan Co, USA	8:0, 10:0, long-chain fatty acids	Incorporated into nutritional or medical beverages or snack bars	NA
'Impact'	Novartis Nutrition, USA	Randomized high-lauric acid oil and high-linoleic acid oil	For patients who suffered trauma, surgery, sepsis or cancer	Interesterification of oils rich in lauric and linoleic acids

[a]Modified after Fitch Haumann (1997); Akoh (1998); Kazlauskas and Bornscheuer (1998). NA, Details not available. GRAS, generally recognized as safe.

papaya gave a 85.5% yield after 6 h at 80°C (Yang *et al.*, 2001). Further improvement in yield was achieved by adsorption of triacetin on silica gel (Mangos *et al.*, 1999). Similarly, Caprenin containing caprylic (8:0), capric (10:0) and behenic (22:0) acids has been produced by both chemical (Kluesener *et al.*, 1992; Stipp and Kluesener, 1992) and enzymatic (McNeill and Sonnet, 1995; Akoh and Yee, 1997; Yankah and Akoh, 2000) methods. In one example, enzymatic esterification was catalysed by a lipase from *Geotrichum candidum* at 50°C, yielding 75% Caprenin in the final product (McNeill and Sonnet, 1995). More information about chemically synthesized low-calorie fats can be found in past reviews (Marangoni and Rousseau, 1995; Rousseau and Marangoni, 1998a).

In this chapter, we concentrate on reviewing recent achievements in the enzymatic synthesis of modified lipids, with special emphasis on sTAG. In addition, information about lipases and suitable starting materials, as well as some chemically synthesized sTAG, are included. Readers are also referred to a number of excellent reviews, books, and book chapters for further information on this topic (Akoh, 1998; Bornscheuer, 1999, 2000, 2001; Knauf and Del Vecchio, 1998; Lee and Akoh, 1998; Mukherjee, 1998; Parkin, 1998; Iwasaki and Yamane, 2000; Xu, 2000; Gunstone, 2001; Bednarski and Adamczak, 2002).

B. Lipases

Lipases (EC 3.1.1.3; systematic name: triacylglycerol hydrolases) are a versatile group of biocatalysts. Their natural function is the hydrolysis of triacylglycerols to partial glycerols and fatty acids during digestion. In contrast to carboxyl esterases, catalysis by lipases occurs at the lipid–water interface and most lipases have the property of "interfacial activation": high catalytic activity is only observed in the presence of a hydrophobic phase – a triacylglycerol dispersed in water or in an organic solvent. This phenomenon is related to the presence of a hydrophobic oligopeptide region (the "lid" or "flap") covering the entrance to the active site of the enzyme. In a hydrophobic environment, the "lid" moves aside and the substrate can enter the binding pocket (Anthonsen *et al.*, 1995). Exceptions to interfacial activation are lipases from *Pseudomonas glumae*, *Pseudomonas aeruginosa* and *Candida antarctica* B, but all of these lipases still have a (smaller) lid (Verger, 1997; Schmid and Verger, 1998).

In all lipases, a catalytic triad composed of serine, histidine, and aspartate (or glutamate in a few enzymes such as lipases from *Geotrichum candidum* and *Candida rugosa*) is present. The tetrahedral intermediate formed during the transition state is stabilized by an oxyanion hole. All lipases show a characteristic α/β-hydrolase fold, which is also found in esterases, proteases, and haloperoxidases (Ollis *et al.*, 1992). Lipases do not need cofactors, although for some enzymes the addition of Ca^{2+} ions or bile salts (in the case of gastric lipases) is necessary or at least recommended. Lipases are active over a broad range of pH values and temperatures, and most microbial enzymes show maximum activity at pH 7–9 and 30–40°C. Important exceptions are the lipase from *Rhizomucor miehei* (sold in immobilized form as 'Lipozyme RM IM' from Novozymes, Denmark) and lipase B from *Candida antarctica* (sold as 'Novozymes SP435', and also as 'Chirazyme L2' from Roche Diagnostics, Germany). Both these biocatalysts are highly active and are stable at temperatures in the range of 60–80°C.

The main interest in the application of lipases stems from their high activity and stability in non-aqueous systems (organic solvents or supercritical fluids), their broad substrate range, and high degree of stereoselectivity. In the absence of water, they may be used to catalyse the reverse reactions of esterification or transesterification, and therefore the full range of reactions which can be catalysed by lipases is considerably broadened.

Further information about general properties, structural features, and applications of lipases in organic synthesis is available in a review (Schmid and Verger, 1998), a book chapter (Kazlauskas and Bornscheuer, 1998), and a book (Bornscheuer and Kazlauskas, 1999).

Besides the fact that triacylglycerols are the natural substrates of lipases, the application of these enzymes in lipid modification is attractive because they show three main types of selectivity (Stadler *et al.*, 1995; Diks and Bosley, 2000):

(1) regioselectivity: either *sn*-1,3 regioselectivity or non-selectivity;
(2) fatty acid selectivity for long-chain PUFA, saturated fatty acids, *cis-n*-9 unsaturated fatty acids, or short-chain fatty acids;
(3) acylglycerol selectivity for monoacylglycerols (MAG), both MAG and diacylglycerols, or triacylglycerols.

These selectivities are influenced by reaction parameters such as solvent polarity (log*P*) (Laane *et al.*, 1987), water activity (a_w) (Halling *et al.*, 1994), the carrier used for immobilization, and the immobilization procedure (Catoni *et al.*, 1996). Consequently, engineering of the reaction medium can considerably change the outcome of a reaction.

Table 3 lists some information concerning lipase selectivity important for the synthesis of sTAG. In addition to the selectivities listed, discrimination between lipid enantiomers by lipases is possible. Chandler *et al.* (1998) used lipases from *Rhizomucor miehei* and *Rhizopus delemar* for the synthesis of

Table 3. Selectivity of lipases used in sTAG synthesis

Lipase	Fatty acid selectivity	Regio-selectivity	Properties/applications	References
Rhizopus delemar	M,L>>S	1,3>>2	Acidolysis: acts strongly on 12:0, 14:0, 16:0, 17:0, 18:0, 18:1, 18:3 Selectivity can differ between hydrolysis and acidolysis	Shimada *et al.* (1997b)
Rhizomucor miehei	S>M,L	1>3>>2	Acidolysis of tripalmitin with oleic acid results in a high level of monosubstituted oleoyldipalmitoyl glycerol	Chandler *et al.* (1998)
Porcine pancreas	S>M,L	1,3	Hydrolysis: high activity for 4:0, 10:0, 18:1	Berger and Schneider (1991)
Humicola lanuginosa	S,M,L	1,3>>2	Hydrolysis: high activity for 8:0–14:0	Berger and Schneider (1991)
Pseudomonas sp.	S,M,L	1,3>2	Complete hydrolysis of borage oil	Shimada *et al.* (1997c)

M, medium-chain fatty acid. L, long-chain fatty acid. S, short-chain fatty acid.

chiral triacylglycerols. Stereoselectivity measurements performed by ^{13}C-NMR (nuclear magnetic resonance) spectroscopy confirmed that immobilized *Rhizomucor miehei* lipase is strongly stereoselective for the *sn*-1 position of a triacylglycerol: under optimal conditions (in hexane, at 60°C, a_w=0.11), over 80% of the synthesized stereoisomers was found to be 1-stearoyl-2-oleoyl-3-palmitoyl-*sn*-glycerol. Only a racemic mixture was produced using immobilized *Rhizopus delemar* lipase under the same conditions.

The majority of lipases used in biocatalysis are of microbial origin, and most enzymes come from bacteria, yeast or fungi (Jaeger *et al.*, 1994, 1999; Benjamin and Pandey, 1998; Haas *et al.*, 1999). Many microbial lipase genes have been cloned and some recombinant lipases are commercially available (Shimada *et al.*, 1994b; Kazlauskas and Bornscheuer, 1998; Schmidt-Dannert, 1999). In addition, several plant lipases have been described, some of which are useful for the synthesis of sTAG (Mukherjee, 1994; Mukherjee and Hills, 1994) (Table 4). Plant lipases often show greater selectivity than microbial enzymes, especially if the seeds contain a high proportion of saturated or unusual fatty acids (Hellyer *et al.*, 1999).

Table 4. Properties and possible applications of plant lipases for sTAG synthesis

Lipase	Properties/applications	References
Papaya latex lipase (*Carica papaya*)	Selective for *sn*-1 and *sn*-3 positions and for 2:0–14:0 acyl donors (increased selectivity for longer chain length). Synthesis of sTAG for infant nutrition and dietetics products by interesterification of tripalmitin with 2:0–14:0 ethyl esters	Gandhi and Mukherjee (2001b)
Papaya latex lipase (*Carica papaya*)	Interesterification of tributyrin and hydrogenated soy oil. Synthesis of low-calorie sTAG (65–70%, w/w SLS and SLL)	Foglia and Valivety (1997)
Papaya latex lipase (*Carica papaya*)	Synthesis of low-calorie products	Gandhi and Mukherjee (2001a)
Rapeseed lipase (*Brassica napus* L.)	Enrichment of GLA in evening primrose oil and of DHA in cod liver oil by esterification with butanol	Hills *et al.* (1990)
Potato tuber lipase	Preferential partitioning of fatty acids: 4:0 – TAG, DAG, 2-MAG 8:0 – DAG 14:0 – 1-MAG, 2-MAG, 3-MAG 18:0 – 1-MAG, 2-MAG, 3-MAG, TAG Synthesis of MAG and intermediates in sTAG synthesis	Pinsirodom and Parkin (1999)

SLS, sTAG with short-chain fatty acid in *sn*-1 and *sn*-3 positions, and long-chain fatty acid in *sn*-2 position. SLL, sTAG with short-chain fatty acid in *sn*-1 position, and long-chain fatty acid in remaining positions. EPA, eicosapentaenoic acid. GLA, γ-linolenic acid. MAG, monoacylglycerol. DAG, diacylglycerol. TAG, triacylglycerol.

To achieve an efficient and economic process, lipases are usually used in immobilized form. This improves their stability at elevated temperatures, varying pH, and in organic solvents. Usually, carriers are used which allow easy recovery and re-use of the biocatalyst, with high operational stability for continuous or repeated-batch processes. However, the choice of an appropriate immobilization method and type of carrier material is crucial for success. The most popular commercial immobilized lipase used for the synthesis of sTAG is the *Rhizomucor miehei* lipase (Lipozyme RM IM) mentioned above, which is adsorbed onto an anionic resin. More recently, Novozymes introduced another lipase designed for lipid modification. This enzyme originates from *Thermomyces lanuginosa* and is sold as a silica-granulated catalyst under the trade name 'Lipozyme TL IM'. Both enzymes show very good performance in the synthesis of sTAG.

C. Strategies for sTAG synthesis

In principle, sTAG can be enzymatically produced by a one-step or a two-step synthesis.

1. One-step process

In the one-step approach, either two triacylglycerols bearing the fatty acids of interest are interesterified using a 1,3-regiospecific lipase, or one triacylglycerol is subjected to acidolysis using two equivalents of a second fatty acid (or its ester) which is to be located at the *sn*-1 and *sn*-3 positions of the final sTAG (Figure 2). The first method yields products close to those obtained by chemical interesterification, as the desired sTAG is difficult to separate from the other triacylglycerols. The major advantage thus resides only in the mild reaction conditions and the environmentally friendly process. In principle, acidolysis of a triacylglycerol with fatty acids can yield purer sTAG, as fewer by-products are formed.

Despite the fact that the yield of desired sTAG is considerably lower in a one-step process than in a two-step process, this approach has been used by most researchers due to its simplicity. As can be seen from Table 5, a wide range of starting materials, reaction conditions, and lipases have been used. Most of the work has focused on acidolysis or transesterification using starting materials based on (poly)unsaturated long-chain fatty acids and medium-chain fatty acids (Xu, 2000).

The best-known example of a one-step process is the synthesis of a cocoa butter equivalent (Quinlan and Moore, 1993). Cocoa butter is predominantly 1,3-disaturated-2-oleoyl-glycerol, where palmitic, stearic and oleic acids account for more than 95% of the total fatty acids (Table 6). Cocoa butter is crystalline and melts between 25°C and 35°C, thus imparting the desirable "mouth-feel" at body temperature. Unilever (Coleman and Macrae, 1977) and

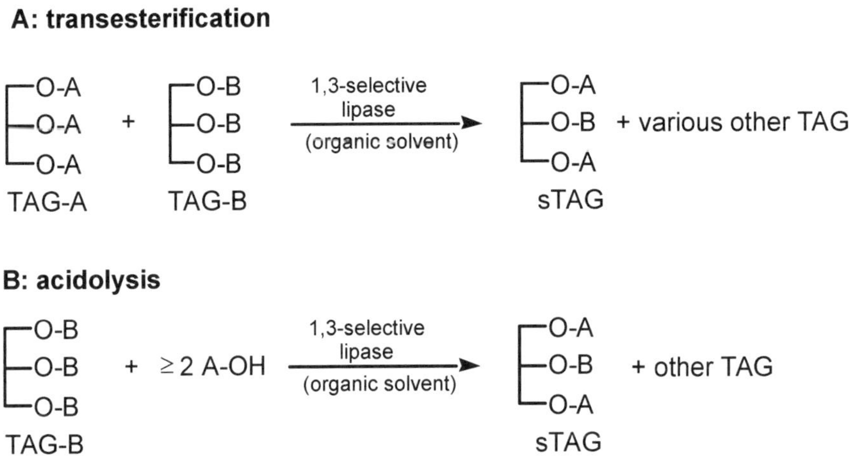

Figure 2. Principles of the one-step synthesis to obtain the sTAG ABA catalysed by a 1,3-selective lipase: (A), transesterification between two triacylglycerols (TAG-A and TAG-B); (B), acidolysis between a triacylglycerol (TAG-B) and at least 2 equivalents of fatty acid A (A-OH). A and B represent different acyl chains.

Fuji Oil (Matsuo *et al.*, 1981) filed the first patents for the lipase-catalysed synthesis of a cocoa butter equivalent from palm oil and stearic acid. Both companies currently manufacture it using 1,3-selective lipases to replace palmitic acid with stearic acid at the *sn*-1 and *sn*-3 positions. The reactions usually performed are transesterification or acidolysis of cheap oils using tristearin or stearic acid as the acyl donors and a 1,3-specific lipase (Figure 3). Suitable starting oils besides palm oil mid-fractions are sunflower, rapeseed (Adlercreutz, 1994) and olive (Chang *et al.*, 1990) oils.

Among many regioselective lipases tested for their ability to incorporate medium-chain fatty acids into triacylglycerols, the best results have been found with lipases from *Rhizomucor miehei* (Lipozyme RM IM), *Rhizopus delemar*, and *Candida antarctica* B (Huang and Akoh, 1996; Soumanou *et al.*, 1999). The maximum incorporation reported for most studies did not exceed 40–65 mol%. Beside free medium-chain fatty acids, the corresponding ethyl esters have also been used. These have higher solubilities, the negative effect on enzyme performance caused by reduced pH in the presence of free fatty acids is avoided, and reaction rates are usually higher.

Beside pure (model) triacylglycerols, natural vegetable oils such as peanut, cottonseed, soybean, safflower and rapeseed oils, as well as chicken fat and polyunsaturated fish oils, have been used as starting materials (Table 5). Although considerable efforts have been made to optimize the reactions – focusing on the effects of the molar ratio of substrates, water activity or water content, amount of enzyme, and temperature – pure sTAG have not been obtained. It has been suggested that column chromatography be used (Miura *et*

Table 5. Some enzymatic processes in the one-step synthesis of sTAG

Lipase	Reaction	Incorporation rate or amount of sTAG formed	Reaction conditions	References
Candida antarctica B	Transesterification of triolein and ethyl caprylate	62 mol% of sTAG	55°C in *n*-hexane	Huang and Akoh (1996)
Rhizopus delemar	Acidolysis of safflower or linseed oil with caprylic acid	45–50 mol% incorporated 8:0	30°C	Shimada *et al.* (1996)
Rhizomucor miehei, *Candida rugosa*, *Chromobacterium viscosum*	Transesterification between peanut oil and tricaprylin	35 mol% of sTAG	40°C in *n*-hexane	Soumanou *et al.* (1997)
Rhizomucor miehei	Acidolysis of rapeseed oil with capric acid	65 mol% incorporated 10:0	60°C	Xu *et al.* (1998a)
Rhizomucor miehei, *Candida antarctica*	Transesterification of tricaproin and trilinolein	53.5 mol% of sTAG	45°C in *n*-hexane	Fomuso and Akoh (1998)
Rhizomucor miehei	Acidolysis of triolein with lauric acid	70 mol% of sTAG	70°C	Miura *et al.* (1999)
Rhizomucor miehei	Transesterification between tricaprylin and EPA ethyl ester	87.0 mol% of sTAG	40°C	Han *et al.* (1999)
Carica papaya	Acidolysis of chicken fat with caprylic acid	23.4 mol% incorporated 8:0	65°C	Lee and Foglia (2000)
Rhizopus delemar	Acidolysis of tripalmitin with arachidonic acid	59.0 mol% incorporated 20:4	40°C	Shimada *et al.* (2000)
Rhizopus javanicus	Transesterification between LCT and MCT	74 mol% of sTAG	50°C	Mogi *et al.* (2000)
Rhizomucor miehei	Transesterification of tri-EPA and ethyl caprylate	64.8 mol% of sTAG	40°C	Irimescu *et al.* (2000)

LCT, long-chain triacylglycerol. MCT, medium-chain triacylglycerol. EPA, eicosapentaenoic acid. tri-EPA, trieicosapentaenoylglycerol.

al., 1999), but this is not suitable on a large scale due to the use of organic solvents and the high costs.

To shift the reaction equilibrium towards the synthesis of sTAG, vacuum is frequently applied to remove water formed during the reaction. Using this

Table 6. Composition of various fractions of interesterified palm olein having properties similar to cocoa butter[a]

	Triacylglycerols (%)							
Product	**OOL**	**POL**	**POO**	**POP**	**SOO**	**POS**	**SOS**	**SSS**
Cocoa butter	0.2	2.4	2.2	18.9	2.4	41.3	29.7	1.9
Fraction 1	0.4	1.4	2.3	15.3	2.9	44.1	29.6	1.5
Fraction 2	0.7	2.1	6.0	16.0	6.1	38.7	23.1	0.7
Fraction 3	0.7	1.5	1.7	13.7	3.6	38.5	24.8	0.7

[a]Modified after Chong *et al.* (1992). Fractions 1, 2 and 3 represent disaturated–monounsaturated triacylglycerols obtained after fractionation using different solvents. The triacylglycerols listed above are regiospecific isomers. L, linoleic acid. O, oleic acid. P, palmitic acid. S, stearic acid.

approach, transesterification of tricaprylin with eicosapentaenoic acid (EPA) ethyl ester gave a molar yield of 87% after 24 h (Han and Yamane, 1999).

In addition to small-scale batch reactions, continuous systems have also been described. Xu *et al.* (1998a, 2000a, 2000c) developed a pilot-scale system based on a continuous enzyme bed reactor and a flat membrane reactor for acidolysis using Lipozyme RM IM. Under these conditions, the extent of acyl migration was minimized compared with that found in a pilot batch reactor (Xu *et al.*, 1998b).

The interesterification of two triacylglycerols has been studied using a natural vegetable oil (peanut oil) and triacylglycerols of medium-chain fatty acids (Soumanou *et al.*, 1997; Fomuso and Akoh, 1998). The highest amount of the desired sTAG formed was 35% after 24 h using *Rhizopus delemar* lipase or Lipozyme RM IM. *Candida rugosa* lipase was more active than in acidolysis reactions (Soumanou *et al.*, 1997). The low overall yield can be explained by the fact that the lipases must show activity towards both long-chain and medium-chain triacylglycerols. Using a surfactant-modified lipase from *Rhizopus javanicus* at a_w=0.33 gave a considerable increase in the yield (74%) of sTAG in the transesterification between long-chain and medium-chain triacylglycerols at 60°C (Mogi *et al.*, 2000) (Table 5).

2. *Two-step process*

In the two-step process, a pure triacylglycerol or a natural fat/oil is first subjected to alcoholysis using a 1,3-regiospecific lipase, yielding the *sn*-2-MAG. This can be separated from the fatty acid esters, for instance by crystallization. It is then esterified in the second step with two equivalents of a fatty acid, yielding the desired sTAG in high yield and purity (Figure 4). The key to success in this process is the use of alcoholysis instead of hydrolysis, because only then is the competing acyl migration of a fatty acid from the *sn*-2 position to the *sn*-1 or *sn*-3 positions avoided. The appropriate choice of

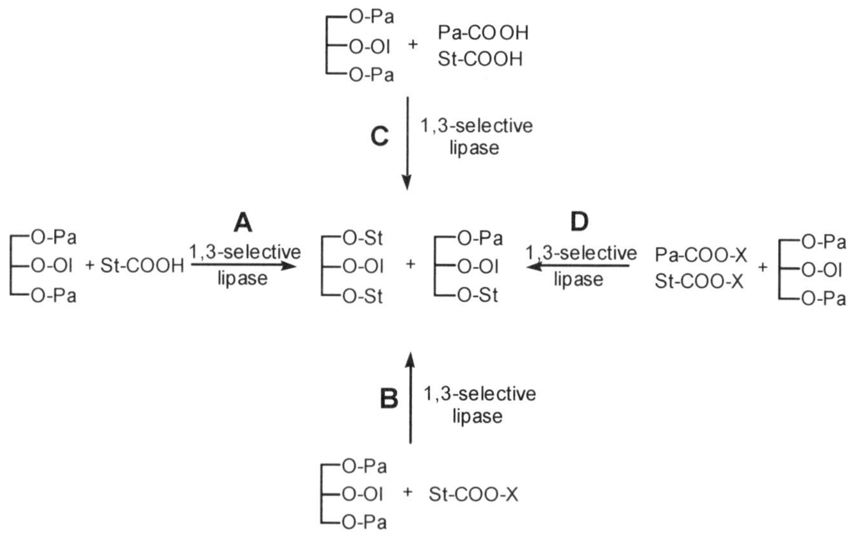

Figure 3. Possible approaches to the synthesis of cocoa butter equivalents: (A), acidolysis of palm oil mid-fractions with stearic acid; (B), interesterification between palm oil mid-fractions and stearic acid esters; (C), acidolysis of palm oil mid-fractions with mixtures of stearic and palmitic acid; (D), interesterification between palm oil mid-fractions and stearic and palmitic acid esters. Ol, oleic acid. Pa, palmitic acid. St, stearic acid. X, alcohol.

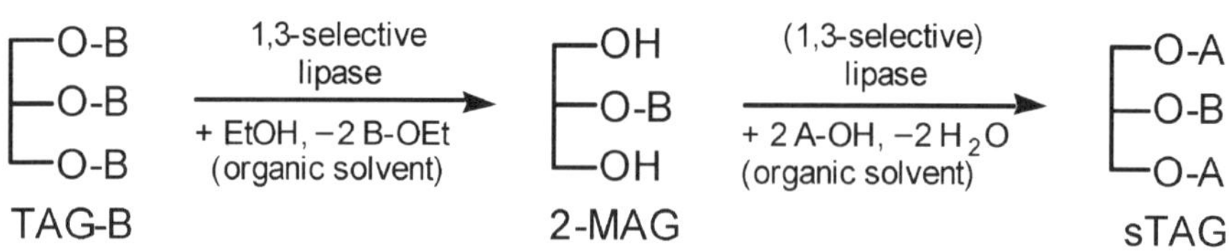

Figure 4. Principle of the two-step synthesis of sTAG by alcoholysis followed by esterification. A and B represent different acyl chains. EtOH, ethanol. 2-MAG, 2-monoacylglycerol.

material for lipase immobilization is also important, as the carrier can also promote acyl migration.

Although at first glance the two-step process may appear less suitable because of its greater complexity, this approach allows the synthesis of highly pure sTAG (Table 7). In principle, any 1,3-regioselective lipase should allow high yields of a desired sTAG by acidolysis in an one-step process, as the fatty acid bound to the *sn*-2 position should be unaffected. In practice, however, acyl migration occurs in an aqueous medium and leads to unwanted by-products.

Table 7. Comparison of conditions and yields of sTAG obtained by different enzymatic processes

Method	Enzyme	Solvent	Purity of sTAG (%, w/w)	Yield[a] (%, w/w)	References
Two-step synthesis	*Rhizopus delemar* lipase immobilized on 'EP 100'	MTBE/ *n*-hexane	91.2	71.8	Soumanou *et al.* (1998a)
Two-step synthesis	*Rhizopus delemar* lipase immobilized on 'Celite'	MTBE/ *n*-hexane	72	95	Schmid *et al.* (1998)
Acidolysis	Lipases immobilized on 'Celite'	*n*-hexane	0–36.2	NA	Soumanou *et al.* (1998b)
Interester-ification	Lipases immobilized on 'Celite'	*n*-hexane	46	NA	Soumanou *et al.* (1998b)
Interester-ification	Lipases immobilized on 'Hyflo Super Cel'	*n*-hexane	20–79	NA	Soumanou *et al.* (1997)

[a]Yields of 2-MAG obtained during the two-step process are expressed as the percentage of the theoretical yield of 33.33% (w/w). MTBE, methyl *tert*-butyl ether.

The key to the successful development of the two-step synthesis of sTAG was the discovery that high yields of 2-MAG can be obtained by alcoholysis of triacylglycerols in aprotic solvents such as methyl *tert*-butyl ether (MTBE) plus ethanol at controlled water activity (Millqvist Fureby *et al.*, 1994, 1996a, 1996b).

The successful transfer of these findings to the synthesis of sTAG was first demonstrated by Soumanou *et al.* (1998a). As MTBE is not allowed for food applications, alternative solvent systems had to be identified. In addition, methods for isolating 2-MAG, as well as suitable methods for the second reaction step (esterification of the 2-MAG with free fatty acids or fatty acid esters) had to be developed. Both reactions are rather fast (approximately 4–8 h for the first step; 1–3 h for the second step). The best results were obtained using *Rhizopus delemar* lipase as the catalyst at a_w=0.11 (Soumanou *et al.*, 1999).

This principle has also been used for the synthesis of 1,3-dioleoyl-2-palmitoyl-*sn*-glycerol (OPO), which can be used in infant nutrition products. Currently, this sTAG is manufactured by interesterification of tripalmitin with oleic acid using a lipase from *Rhizomucor miehei* (Lipozyme RM IM), and is marketed under the trade name 'Betapol'. Excess fatty acids are removed by evaporation, and remaining tripalmitin is removed by crystallization. However, the product contains only 65% palmitic acid in the *sn*-2 position. In contrast, alcoholysis of tripalmitin with ethanol for 24 h using *Rhizopus delemar* lipase immobilized on a polypropylene carrier ('EP-100') yielded 95% monopalmitin with purity >95% after crystallization. Subsequent esterification with oleic acid in *n*-hexane proceeded quantitatively within a few hours, and the final product (yield 70%) contained up to 92% palmitic acid in the *sn*-2 position (Schmid *et al.*, 1999). Alternatively, esterification in the absence of solvent under vacuum to remove water gave higher purity and approximately 100 fold higher productivity at comparable yields. More recent work has focused on the scale-up of this process. Using a mini-plant technique, it is now possible to synthesize OPO in a continuous manner for several weeks without significant decrease of productivity, product purity, or enzyme activity. The 2-MAG is separated from by-products using a membrane pervaporation module instead of crystallization.

sTAG containing medium-chain fatty acids in the *sn*-1 and *sn*-3 positions of glycerol and an unsaturated long-chain fatty acid in the sn-2 position have also been synthesized using this approach (Soumanou *et al.*, 1998a, 1998b).

Similarly, symmetric sTAG rich in docosahexaenoic acid (DHA) at the *sn*-2 position and caprylic acid at the *sn*-1,3 positions can be obtained (Irimescu *et al.*, 2001). Bonito oil was used as the substrate for ethanolysis, and the reaction was catalysed by *Candida antarctica* B lipase. After 2 h, a 92.5% yield of 2-MAG was obtained containing 43.5% DHA. After removal of the *Candida* lipase, Lipozyme RM IM was added directly without purification of the 2-MAG, and a final yield of 51% (w/w) sTAG was reported.

D. Vegetable oils for the synthesis of sTAG

The majority of vegetable oils and fats contain mono-, di- and tri-unsaturated fatty acids in the *sn*-2 position and saturated fatty acids in the *sn*-1,3 positions of glycerol (Table 8). They are frequently used as starting materials for use in shortenings, confectionery coatings, and margarines (Table 9). Cocoa butter is ideal for confectionery coatings as it melts between 25°C and 35°C. However, it is relatively expensive, and substitutes derived from enzymatic interesterification of vegetable oils have been developed (see Section C.1 above and Table 6).

Indeed, a wide range of sTAG have been produced with modified melting characteristics. Vegetable oils such as palm kernel or coconut oils are used in margarines after fractionation and hydrogenation (De Greyt *et al.*, 1998). Lipozyme TL IM has been successfully used in the interesterification of palm stearin and coconut oil in a pilot-scale batch stirred-tank reactor (1 kg or 300 kg scale) for the synthesis of margarine (Zhang *et al.*, 2001). Sunflower oil mixed with beef tallow and subjected to chemical interesterification produced a solid fat blend more appropriate for use in shortenings in the baking industry. Using the same approach, partially hydrogenated soybean oil mixed with several vegetable oils led to structured fats that are potentially suitable for the confectionery industry (Zeitoun *et al.*, 1993; Hurtova *et al.*, 1996). In these studies, the level of *trans* fatty acids, which are undesirable from a nutritional

Table 8. Major triacylglycerols of some vegetable oils used for the synthesis of sTAG[a]

Fat or oil	Major TAG		
Butter fat	PPB	PPC	POP
Cocoa butter	POS	SOS	POP
Coconut oil	DDD	CDD	CDM
Palm kernel oil	DDD	MOD	ODO
Corn (maize) oil	LLL	LOL	LLP
Cottonseed oil	PLL	POL	LLL
Rapeseed oil	LLL	OLL	POL
Olive oil	OOO	OOP	OLO
Palm oil	POP	POO	OLO
Peanut oil	OOL	POL	OLL
Canola oil	OOL	OOO	OOLn
Safflower oil	LLL	LLO	LLP
Soybean oil	LLL	LLO	LLP
Sunflower oil	LLL	LLO	LLP
Rapeseed oil	OOO	LOO	OOLn
Linseed oil	LnLnLn	LnLnL	LnLnO
Mustard seed oil	ErOEr	ErLEr	OOEr

[a]Modified after Small (1991). The three-letter symbols include all positional isomers of the triacylglycerols. B, butyric acid (4:0). C, capric acid (10:0). D, lauric acid (12:0). M, myristic acid (14:0). P, palmitic acid (18:0). S, stearic acid (18:0). O, oleic acid (18:1). L, linoleic acid 18:2). Ln, linolenic acid (18:3). Er, erucic acid (22:1).

Table 9. Important modifications of vegetable oils and fats based on their physical properties/applications

Oil/fat	Modification process	Applications	References
Coconut and palm kernel oils	Fractionation and hydrogenation	Margarines	De Greyt *et al.* (1998)
Palm oil	Enzymatic interesterification	Confectionery coatings	Coleman and Macrae (1980)
Olive oil	Enzymatic interesterification	Confectionery coatings	Chang *et al.* (1990)
Sunflower and rapeseed oils	Enzymatic interesterification	Confectionery coatings	Adlercreutz (1994)
Sunflower oil	Chemical interesterification	Shortenings	Rodriguez *et al.* (2001)
Soybean oil	Chemical interesterification	Confectionery	Zeitoun *et al.* (1993); Hurtova *et al.* (1996)
Soybean oil	Hydrogenation	Margarines	King *et al.* (2001)

Table 10. Examples of plant species rich in γ-linolenic acid (GLA)[a]

Family	Genus and species	Oil content of seed (%)	GLA content in oil (%)
Boraginaceae	*Borago officinalis*	28–31	17–25
	Anchusa azurea	21	11–13
	Anchusa capensis	29	10
	Buglosoides arvensis	11	6
	Echium asperrimum	19	10
	Echium sabulicola	20	11
	Echium vulgare	15–22	12–14
	Myosotis nemorosa	20	20
	Myosotis secunda	23	12
	Nonea vesicaria	10	9
	Symphytum officinale	21	27
Cannabaceae	*Cannabis sativa*	38	3–6
Onagraceae	*Oenothera biennis*	17–25	7–10
	Oenothera grandiflora	4	9
Saxifragaceae	*Ribes alpinum*	19	9
	Ribes nigrum	30	15–19
	Ribes rubrum	25	4–6
Scrophulariaceae	*Scrophularia marilandica*	38	10
	Scrophularia sciophila	22	10

[a]Modified after Guil-Guerrero *et al.* (2001); Clough (2001).

point of view, was reduced. More extensive reductions were possible in a mixture of supercritical carbon dioxide and hydrogen (King *et al.*, 2001).

Vegetable oils or fats have also been used as starting materials for the enzymatic synthesis of structured lipids with valuable dietetic or nutritional properties. Milk fat substitutes for use in baby foods have been produced enzymatically from palm oil mid-fractions (King and Padley, 1990). Using the same enzymatic process, long-chain saturated fatty acids (which have been implicated in atherogenesis) at the *sn*-2 position of peanut oil triacylglycerols have been replaced with oleic acid (Sridhar *et al.*, 1991).

An important group of seed oils are those containing GLA. GLA is used in the treatment of a wide variety of human diseases and is an important component of cosmetics. It is an intermediate in the biosynthesis of AA. The main sources of GLA (see Table 10) are seed oils from *Oenothera biennis* (evening primrose) (Hudson, 1984), *Borago officinalis* (borage) (Uzzan, 1986), and *Ribes nigrum* (blackcurrant) (Traitler *et al.*, 1984). Recently, Guil-Guerrero *et al.* (2001) presented the results of screening of 49 plant species as new sources of GLA (see also Chapter 9).

Specific plant oils have been used as starting materials to obtain sTAG rich in *n*-3 PUFA and/or GLA. Acidolysis with medium-chain fatty acids and evening primrose oil (Akoh *et al.*, 1996; Namal Senanayake and Shahidi, 1999b) or borage oil (Akoh and Moussata, 1998; Ju *et al.*, 1998; Namal Senanayake and Shahidi, 1999a; Shimada *et al.*, 1999) have been reported. Shimada *et al.* (1997c, 1998a) used *Rhizopus delemar* lipase to enrich the GLA content in the free fatty acid fraction to about 97%. *Candida rugosa* lipase, immobilized on different ion-exchange resins, discriminated 3 fold against GLA in the hydrolysis of borage oil (Schmitt-Rozieres *et al.*, 1999). Syed Rahmatullah *et al.* (1994a, 1994b) reported that Lipozyme RM IM was suitable for enrichment of GLA in esterification, but that *Candida rugosa* lipase was better for the hydrolysis of borage oil or evening primrose oil, resulting in GLA enrichment in the free fatty acid fraction (about 70% GLA) or acylglycerol fraction (about 50% GLA).

E. Fish oils for the synthesis of sTAG

Fish oils are the main source of EPA, DHA and some other long-chain *n*-3 PUFA (see also Chapter 8). They generally contain approximately 50 different fatty acids. The most prominent are listed in Table 11 for a range of common fish oils.

Fish oils can be used in the synthesis of sTAG containing PUFA or as a starting material for the production of fatty acid concentrates enriched in DHA

Table 11. Fatty acid compositions of selected fish oils used for the synthesis of sTAG and PUFA concentrates[a]

	Fatty acid content (mol%)				
Fatty acid	**Anchovy**	**Menhaden**	**Tuna**	**Sardine/pilchard**	**Cod liver oil**
14:0	9	9	3	8	4
15:0	1	1	1	1	ND
16:0	17	19	22	18	10
16:1	13	12	3	10	8
17:0	1	1	1	1	ND
18:0	3	3	6	3	2
18:1	10	11	21	13	25
18:2	1	1	1	1	2
18:3*n*-3	1	1	1	1	ND
18:4*n*-3	2	3	1	3	3
20:1	1	1	1	4	10
22:1	1	ND	3	3	7
20:5*n*-3	22	14	6	16	10
22:5*n*-3	2	2	2	2	1
22:6*n*-3	9	8	22	9	10

[a]Modified after Haraldsson and Hjaltason (2001). Note that variations within individual species related to season, area caught, age of fish, and diet are well known. ND, not detected.

or EPA. The sTAG usually contain medium-chain fatty acids at the *sn*-1 and *sn*-3 positions and PUFA at the *sn*-2 position. Enzymatic methods in this field have been reviewed recently (Shimada *et al.*, 1996; Jennings and Akoh, 1999; Xu *et al.*, 2000b; Zhou *et al.*, 2000). Table 12 lists the fatty acid compositions of tuna oil and of sTAG obtained from it after acidolysis.

The variety of fatty acids in fish oils usually makes it difficult to obtain pure concentrates of DHA or EPA, as no lipase shows perfect selectivity for these fatty acids. Homogeneous triacylglycerols containing DHA and EPA were synthesized by Haraldsson *et al.* (1989, 1995) The concentration of DHA and EPA (free fatty acids or esters) obtained by lipase-catalysed kinetic resolution and/or different physicochemical techniques, such as high-performance liquid chromatography (HPLC) or silver-ion chromatography, can exceed 90% (Haraldsson and Hjaltason, 2001). Figure 5 shows possible process schemes to obtain PUFA concentrates from fish oils using lipase catalysis.

Lipases useful for the concentration and purification of EPA and DHA can be divided into two categories:

(1) those that act preferentially on *n*-3 fatty acids other than EPA and DHA;
(2) those distinguishing between EPA and DHA.

The first category includes lipases from *Pseudomonas* sp. (Breivik *et al.*, 1997), *Pseudomonas fluorescens* (Rakshit *et al.*, 2000), *Candida antarctica* (lipase B) (Esteban Cerdan *et al.*, 1998), and *Geotrichum candidum* (Shimada

Table 12. Fatty acid composition (mol%) at *sn*-1(3) and *sn*-2 positions in tuna oil and sTAG obtained by acidolysis[a,b]

	Tuna oil		sTAG	
Fatty acid	***sn*-1(3)**	***sn*-2**	***sn*-1(3)**	***sn*-2**
8:0	ND	ND	41.9	0.5
14:0	1.9	1.9	0.9	1.8
16:0	14.7	14.7	3.7	7.1
16:1	4.9	4.9	1.1	1.1
17:0	0.8	0.8	0.3	0.5
17:1	1.1	1.1	0.3	0.3
18:0	5.0	5.0	0.7	0.4
18:1	15.4	15.4	2.5	3.2
18:2	1.0	1.0	0.3	0.3
20:4	2.0	2.0	0.7	0.6
20:5	3.5	3.5	1.6	2.1
22:6	8.8	8.8	7.8	12.4

[a]Modified after Shimada *et al.* (1996). [b]Acidolysis between tuna oil and caprylic acid, catalysed by *Rhizopus delemar* lipase. ND, not detectable.

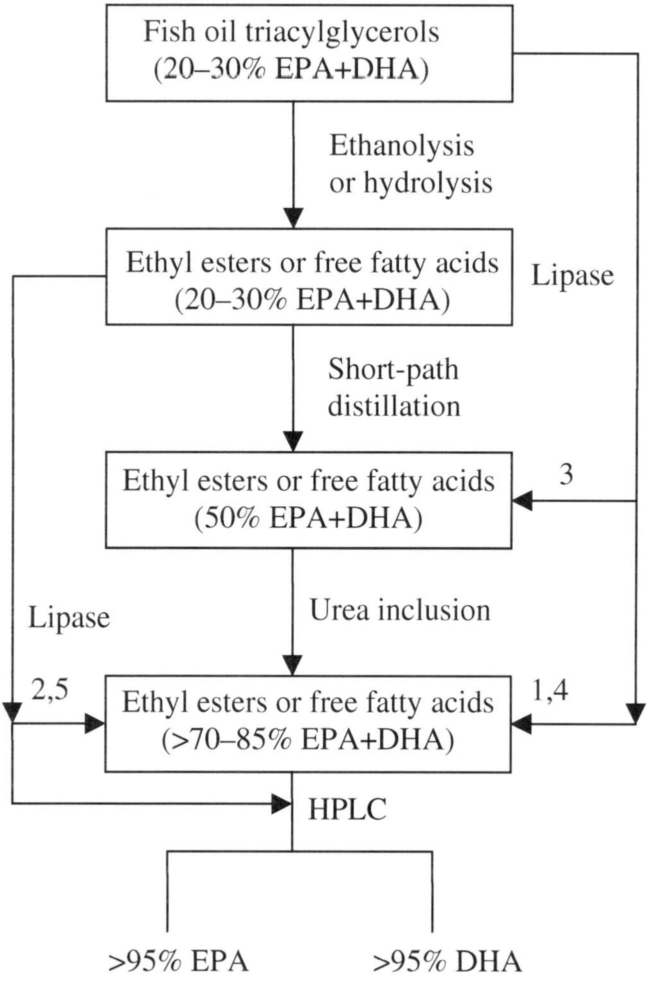

Figure 5. Possible processes for the synthesis of DHA and DPA concentrates from fish oils (modified after Haraldsson and Hjaltason, 2001). (1), *Pseudomonas fluorescens* lipase hydrolysis of tuna oil, resulting in 80% DHA+EPA in free fatty acids (Rakshit *et al.*, 2000). (2), *Rhizomucor miehei* lipase ethanolysis of free fatty acids of tuna oil, resulting in 74% DHA + 3% EPA in free fatty acids (Haraldsson and Kristinsson, 1998). (3), *Pseudomonas* sp. lipase ethanolysis of fish oil, resulting in 50% EPA+DHA in acylglycerols (Haraldsson *et al.*, 1997). (4), Two-step process: *Pseudomonas* sp. lipase ethanolysis, resulting in 46% EPA+DHA in acylglycerols; followed by urea fractionation, resulting in 85% EPA+DHA (Breivik *et al.*, 1997). (5), 'Novozymes 435' lipase esterification of PUFA from cod liver oil with glycerol, resulting in >70% EPA+DHA in triacylglycerols (Esteban Cerdan *et al.*, 1998).

et al., 1994a). The second includes lipases from *Candida rugosa* (Schmitt-Rozieres *et al.*, 2000), *Rhizopus delemar* (Shimada *et al.*, 1997a), and *Rhizomucor miehei* (Haraldsson and Kristinsson, 1998), which act preferentially on EPA. However, for *Candida rugosa* lipase this selectivity is under discussion (Wanasundara and Shahidi, 1998).

F. Microbial and algal fat modification

Oleaginous microorganisms accumulate lipids at more than 20% of their biomass. Oil-forming organisms are found among bacteria, molds, and algae. Lipids synthesized by bacteria are not commercially attractive because of the slow production rate and the difficulties of extraction. They may also be associated with toxic or allergenic factors. The most useful products are single-cell oils produced by molds, some of which are valuable sources of PUFA of the *n*-3 and *n*-6 series such as GLA (18:3*n*-6), dihomo-GLA (20:3*n*-6), AA (20:4 *n*-6), EPA (20:5*n*-3) and DHA (22:6*n*-3) (Table 13). Microbial oils can also be used for the production of cocoa butter equivalents, since mutants and hybrids of *Cryptococcus curvatus* with inactivated Δ^9-desaturase are able to synthesize lipids with a fatty acid composition close to that of cocoa butter (Leman, 1997). The fatty acid compositions of triacylglycerols from microalgae vary in type and total concentration and are affected by a range of factors, including species, stage of cell growth, light intensity, temperature, and composition of the growth medium (Kay, 1991).

High-value lipids can be obtained from true algae and some cyanobacteria i.e. *Botryococcus braunii, Dunaliella salina, Nannochloris* sp., *Manalanthus salina,* and *Chlorella pyrenoidosa* (Leman, 1997). Generally, these microalgae contain 3–6% fatty acid per dry weight (20–40% lipid in dry biomass). Approximately 25–65% of the fatty acids in microalgae currently produced by the major manufacturers are PUFA (Table 14). Sun Chlorella, a Japanese manufacturer, claims that their microalgae contain 81.8% unsaturated and 18.2% saturated fatty acids.

Lipases have been used for the enrichment of GLA (from 10.4% to 68.8%) (Mukherjee and Kiewitt, 1991) and AA (from 25% to 71%) (Shimada *et al.*, 1998b) starting from single-cell oils.

Robles Medina *et al.* (1999) described the use of *Candida antarctica* lipase for the high-yield synthesis of triacylglycerols rich in PUFA by esterification of glycerol and a PUFA concentrate obtained from microalgae. Using fatty acids from *Phaeodactylum tricornutum*, triacylglycerols containing 42.5% EPA were obtained. Starting with fatty acids derived from *Porphyridium cruentum*, triacylglycerols containing 43.4% AA and 45.6% EPA were produced.

Iwasaki *et al.* (1999) synthesized a sTAG of the MLM type using a single-cell oil rich in DHA and DPA (docosapentaenoic acid) produced by *Schizochytrium* sp. SR21. The targeted products were sTAG containing caprylic

Table 13. Major sources of PUFA[a]

PUFA	Source	PUFA content (% of total fatty acids)	Productivity (mg PUFA/g raw material)
n-6:			
GLA	Plants:		
	Evening primrose	6–10	NA
	Borage	19–25	NA
	Blackcurrant	17	NA
	Fungi:		
	Mortierella isabelina	10	55
	Mucor circinelloide	15–18	NA
DHGLA	Fungi:		
	Mortierella alpina 1S-4	7–23	28–123
AA	Algae:		
	Porphyridium cruentum	5–36	20–80
	Fungi:		
	Mortierella alpina 1S-4	58	275
n-3:			
EPA	Animals:		
	Fish oil	10–15	20
	Algae:		
	Chlorella minutissim	35–40	NA
	Fungi:		
	Mortierella alpina 1S-4	20	64
	Pythium irregulare	25	25
DHA	Animals:		
	Fish oil	7–14	NA
	Fungi:		
	Thraustochytrium aureum	49	70

[a]Modified after Leman (1997). NA, data not available. GLA, γ-linolenic acid. DHGLA, dihomo-γ-linolenic acid. AA, arachidonic acid. EPA, eicosapentaenoic acid. DHA, docosahexaenoic acid.

acid at the *sn*-1 and *sn*-3 positions and DHA or DPA at the *sn*-2 position. When *Pseudomonas* sp. KWI-56 lipase was used, 35% of the triacylglycerols obtained had the desired fatty acid composition, and 77–78% of these had the right configuration; whereas with *Rhizomucor miehei* lipase, 22% of the triacylglycerols obtained had the desired fatty acid composition, and all of these had PUFA in the *sn*-2 position.

G. Milk fat modification

Decreased consumption of high-fat milk and dairy products has promoted research into developing new processes for producing structured or modified milk fats with improved physical properties, improved digestibility, reduced

Table 14. Fatty acid composition of selected microalgae[a]

	Fatty acid content (mol%)		
Fatty acid	***Spirulina* sp.**	***Scenedesmus* spp.**	***Aphanizomenon flos-aquae***
12:0	0.4	0.8	NA
14:0	1.0	Trace	9.2
16:0	37.0	14.9	39.0
16:1	4.0	NA	9.4
16:2	4.5	Trace	NA
17:0	0.2	NA	1.5
18:0	0.6	0.6	0.9
18:1	5.0	8.0	11.7
18:2	24.0	19.7	6.5
18:3 (α)	21.0	0.4	19.3
18:3 (γ)	0.7	29.6	NA

[a]Modified after Kay (1991). NA, not analysed.

Table 15. Lipase-catalysed interesterification of milk fat

Enzyme	Reaction system	References
Mucor miehei ('Lipozyme 10.000L')	Microemulsion[a]	Safari *et al.* (1994b)
Mucor miehei ('Lipozyme 10.000L')	Organic solvents[b]	Safari *et al.* (1993)
Rhizopus niveus ('Lipase N')	Reverse micelle[c]	Safari *et al.* (1994a)
Pseudomonas fluorescens	Variable amount of water	Kalo *et al.* (1989)
Pseudomonas fragi	Microemulsion[d]	Pabai *et al.* (1995)
Pseudomonas fluorescens	Iso-octane	Kalo *et al.* (1989)

[a]Microemulsion containing 'Brig 35' and 1-heptanol as co-surfactants at 1% (v/v) water. [b]Hexane, or hexane:chloroform (70:30, v/v), or hexane:ethyl acetate (70:30, v/v), at different water contents. [c]System containing hexane at different ratios of water per mol surfactant. [d]Microemulsion free co-surfactant system containing 'Span 60' and 'Tween 60' (48:52, v/v).

Table 16. Changes in the positional distribution of selected fatty acids in the lipase-catalysed interesterification of butter fat in a reverse micelle system[a,b]

	sn-2 position (%)		*sn*-1,3 position (%)	
Fatty acid	**Untreated**	**Interesterified**	**Untreated**	**Interesterified**
12:0	5.9	4.2	3.9	5.0
14:0	18.1	19.3	11.5	11.2
16:0	32.1	37.2	27.0	19.9
18:0	8.1	9.0	7.4	4.5
18:1	19.1	18.9	17.5	23.8

[a]Modified after Safari *et al.* (1994a). [b]Interesterification catalysed by *Rhizopus niveus* lipase in a reverse micelle system containing 4.4% water.

caloric value, and enhanced flavor (Balcao and Malcata, 1998b). Intramolecular interesterification is the method most often used for the enzymatic modification of milk fat (Table 15). Although regioselective exchange by lipase catalysis yields products with the same overall fatty acid composition, it also leads to changes in physical properties (Table 16).

Balcao *et al.* (1998b) modified butter fat by combined hydrolysis and interesterification, catalysed by *Mucor javanicus* (now renamed *Mucor circinelloides*) lipase immobilized on hydrophobic hollow fibers. The product showed improved nutritional properties due to a decrease in the contents of lauric, myristic, and palmitic acids and a concomitant increase in the content of monounsaturated fatty acids in the triacylglycerol fraction.

Acidolysis of anhydrous butter fat with oleic acid has been described in several publications by Balcao *et al.* (Balcao and Malcata, 1997, 1998a; Balcao *et al.*, 1998a, 1998c). They found that *Mucor circinelloides* (previously named *Mucor javanicus*) lipase immobilized on hollow fibers exhibited limited selectivity towards short-chain fatty acids and mainly catalysed the exchange of myristic, palmitic, and stearic acids in triacylglycerols with oleic acid. The modified butter fat had a broadened melting temperature, and contained 27% more oleic acid, and less lauric (–7.5%), myristic (–5.7%), and palmitic (–6.1%) acids (Balcao *et al.*, 1998a). Christensen and Hølmer (1993) subjected butter oil to acidolysis using a concentrate of polyenoic fatty acids (18:1, 18:2, 20:5, 22:6). Using Lipozyme RM IM, they obtained a product similar to human milk fat in total fatty acid composition and triacylglycerol structure.

Milk fat in its original form is liquid at higher temperatures and thus not suitable for baking applications, industrial shortenings, pastries, margarines, etc. However, a change in melting characteristics and solid fat content can be achieved by enzymatic interesterification with rapeseed oil (Kalo *et al.*, 1986), canola oil (Rousseau and Marangoni, 1998b, 1998c), or palm stearin (Lai *et al.*, 2000; Nor Hayati *et al.*, 2000).

Garcia *et al.* (2001) efficiently incorporated conjugated linoleic acid (CLA) into butter oil by acidolysis in a packed-bed reactor containing Chirazyme L2 (immobilized *Candida antarctica* lipase B); approximately 80–90% of the free CLA was esterified (Table 17).

H. Methods for sTAG analysis

Independently of whether chemical or enzymatic methods are used for synthesis, the sTAG produced have to be analysed carefully. Besides the total fatty acid composition and the content of partial glycerol esters and free fatty acids, a regiospecific analysis of the fatty acid distribution is of utmost importance. In addition, sophisticated analytical methods are required to optimize a specific reaction and to measure changes in the sTAG composition during synthesis.

Table 17. Synthesis of modified lipids enriched in CLA

Enzyme	Substrate	Incorporation of CLA[a]	References
'Lipozyme IM'	Palm oil + CLA lauryl esters	30 wt%	McNeill *et al.* (1999)
'Lipozyme IM60'; 'Chirazyme L2'	Corn oil + CLA	11 wt%	Martinez *et al.* (1999)
Candida rugosa lipase	Menhaden oil + CLA	20 mol%	Garcia *et al.* (2000)
C. antarctica lipase; *Mucor miehei* lipase	Milk fat + CLA	6 wt%	Garcia *et al.* (2001)

[a]Content of CLA in triacylglycerols. CLA, conjugated linoleic acid.

Recently, Yamane (2000) presented a review of methods for the determination of reaction mixture composition and the analysis of sTAG.

One of the oldest methods used for triacylglycerol structure analysis is based on enzymatic hydrolysis by 1,3-regioselective porcine pancreatic lipase (PPL) followed by determination of the contents of *sn*-2-MAG, *sn*-1(3),2-diacylglycerols, and free fatty acids (Brockerhoff, 1965). Fatty acids present in the various glycerol ester fractions are converted into the corresponding methyl esters and analysed by gas chromatography or HPLC. Unfortunately, PPL cannot be used for all sTAG as it does not reliably hydrolyse fatty acids shorter than C_{12}, PUFA longer than C_{20}, and fatty acids with oxygenated groups. The limited range of fatty acids which can be analysed makes this approach less attractive, as sTAG containing short- to medium-chain fatty acids and PUFA are of particular interest.

Alternatively, triacylglycerol structure can be determined by chemical reaction based on a Grignard reagent (e.g. ethyl magnesium bromide), which yields *sn*-1(3),2-diacylglycerols. This method is more reliable, but is tedious and time-consuming (Brockerhoff, 1967).

Myher *et al.* (1996) used a Grignard reagent and PPL digestion to obtain diacylglycerols which were then analysed by thin-layer chromatography, chiral HPLC, or chiral-phase liquid chromatography – mass spectroscopy with electrospray ionization. Arcos *et al.* (2000) presented a method for the regioselective analysis of sTAG composition at the *sn*-1,3 and *sn*-2 positions using a conventional HPLC technique and a *sn*-1,3-selective lipase from *Rhizopus arrhizus*. Williams *et al.* (1995) described an approach using lipase from *Rhizopus arrhizus* or phospholipase A_2 for the analysis of the positional distribution of fatty acids in phospholipids and triacylglycerols. This method takes advantage of the fact that methanolic NaOH methylates only esters (linked to glycerol), while methanolic HCl methylates both free and esterified fatty acids.

One of the most powerful techniques for the analysis of sTAG is ^{13}C-NMR spectroscopy or improved ultra-high resolution NMR. Recently, Zamora *et al.* (2001) used this technique to screen virgin olive oils. Chandler *et al.* (2001) determined the selectivity of lipase-catalysed acidolysis reactions based on ^{13}C-NMR spectroscopy.

Silver-ion HPLC (Han *et al.*, 1999; Iwasaki *et al.*, 1999) and chiral HPLC (Iwasaki *et al.*, 2001) have also been used for the analysis of sTAG. Silver-ion HPLC enables the separation of unsaturated species according to the number, geometric configuration, and position of the double bond(s), whereas chiral HPLC enables the separation of stereoisomers after converting partial glycerols into diastereomeric derivatives.

Evaporative light-scattering detectors (ELSD) are frequently used for HPLC analysis (Christie, 1991). A solvent is evaporated in a stream of carrier gas in a heating chamber, and the solution containing the analyte is nebulized and passes in the form of droplets through a light beam, which is reflected and refracted. The amount of scattered light is measured and corresponds to the concentration of analyte. With this detector it is possible to observe separated lipid classes and to determine the positional distribution of fatty acids in triacylglycerols such as butter fat (Kermasha *et al.*, 1993; Homan and Andersen, 1998; Nordback *et al.*, 1998; Mu *et al.*, 2000). This applies even if the fatty acids are saturated and therefore not detectable using conventional HPLC detector systems.

For analysis of the reaction composition during the synthesis of sTAG, thin-layer chromatography in combination with a flame ionization detector (TLC–FID; 'Iatroscan') is a useful technique. Unfortunately, only partial glycerols, triacylglycerols, and free fatty acids can be quantified. There is no separation of different triacylglycerol species by this technique (Tatara *et al.*, 1983).

Gas chromatography is only useful for the determination of fatty acid methyl esters and some partial glycerols, as triacylglycerols are not generally volatile enough for detection. However, special techniques for sample injection have made it possible to analyse triacylglycerol species by high-temperature gas chromatography (Han *et al.*, 1999; Iwasaki *et al.*, 1999).

Modified triacylglycerols can occur in different polymorphic forms, and these influence the melting properties of the fat, the softening point, the solids content, etc. Analysis of such properties is commonly performed using differential scanning calorimetry (DSC). DSC has been used to determine the physical properties of sTAG formed from interesterification products of triolein and tristearin (Seriburi and Akoh, 1998) and from acidolysis reactions between stearic acid and coconut oil (Rao *et al.*, 2001).

I. Future outlook

The examples given in this chapter demonstrate that enzymatic synthesis methods are superior to chemical interesterifications for the production of sTAG. Strong research efforts in the past decade have led to the development of numerous processes, using a broad range of natural starting materials, to obtain value-added products with important and highly desirable nutritional and medical properties. It is to be expected that these achievements will lead to further commercialization of lipase-catalysed reactions for the production of sTAG.

References

Adlercreutz, P (1994) Enzyme-catalyzed lipid modification. *Biotechnol. Genet. Eng. Rev.*, 12, 231–254.

Akoh, CC (1998) Structured lipids. In: *Food Lipids: Chemistry, Nutrition, and Biotechnology* (CC Akoh and DB Min, eds), Marcel Dekker, Inc, New York, pp.699–727.

Akoh, CC and Moussata, CO (1998) Lipase-catalyzed modification of borage oil: incorporation of capric and eicosapentaenoic acids to form structured lipids. *J. Am. Oil Chem. Soc.*, 75, 697–701.

Akoh, CC and Yee, LN (1997) Enzymatic synthesis of position-specific low-calorie structured lipids. *J. Am. Oil Chem. Soc.*, 74, 1409–1413.

Akoh, CC, Jennings, BH and Lillard, DA (1996) Enzymatic modification of evening primrose oil: incorporation of *n*-3 polyunsaturated fatty acids. *J. Am. Oil Chem. Soc.*, 73, 1059–1062.

Anthonsen, HW, Baptista, A, Drablos, F, Martel, P, Petersen, SB, Sebastiao, M and Vaz, L (1995) Lipases and esterases: a review of their sequences, structure and evolution. *Biotechnol. Annu. Rev.*, 1, 315–371.

Arcos, JA, Garcia, HS and Hill CG, Jr (2000) Regioselective analysis of the fatty acid composition of triacylglycerols with conventional high-performance liquid chromatography. *J. Am. Oil Chem. Soc.*, 77, 507–512.

Auerbach, MH, Chang, PW, Kosmark, R, O'Neill, JJ and Philips, JC (1998) Salatrim: a family of reduced-calorie structured lipids. In: *Structural Modified Food Fats: Synthesis, Biochemistry, and Use* (AB Christophe, ed.), AOCS Press, Champaign, pp.89–120.

Babayan, VK (1987) Medium chain triglycerides and structured lipids. *Lipids*, 22, 417–420.

Babayan, VK and Rosenau, JR (1991) Medium-chain-triglyceride cheese. *Food Technol.*, 45(2), 111–114.

Balcao, VM and Malcata, FX (1997) Lipase-catalyzed modification of butterfat *via* acidolysis with oleic acid. *J. Mol. Catal. B: Enzymatic*, 3, 161–169.

Balcao, VM and Malcata, FX (1998a) Interesterification and acidolysis of butterfat with oleic acid by *Mucor javanicus* lipase: changes in the pool of fatty acid residues. *Enzym. Microb. Technol.*, 22, 511–519.

Balcao, VM and Malcata, FX (1998b) Lipase catalyzed modification of milkfat. *Biotechnol. Adv.*, 16, 309–341.

Balcao, VM and Malcata, FX (1998c) On the performance of a hollow-fiber bioreactor for acidolysis catalyzed by immobilized lipase. *Biotechnol. Bioeng.*, 60, 114–123.

Balcao, VM, Kemppinen, A, Malcata, FX and Kalo, PJ (1998a) Lipase-catalyzed acidolysis of butterfat with oleic acid: characterization of process and product. *Enzym. Microb. Technol.*, 23, 118–128.

Balcao, VM, Kemppinen, A, Malcata, FX and Kalo, PJ (1998b) Modification of butterfat by selective hydrolysis and interesterification by lipase: process and product characterization. *J. Am. Oil Chem. Soc.*, 75, 1347–1358.

Bednarski, W and Adamczak, M (2002) Modified lipids and fat mimetics. In: *Chemical and*

Functional Properties of Food Lipids (ZE Sikorski and A Kolakowska, eds), CRC Press, Boca Raton, pp.309–324.

Benjamin, S and Pandey, A (1998) *Candida rugosa* lipase: molecular biology and versatility in biotechnology. *Yeast*, 14, 1069–1087.

Berger, M and Schneider, MP (1991) Lipases in organic solvents: the fatty acid chain length profile. *Biotechnol. Lett.*, 13, 641–645.

Bornscheuer, UT (1999) Recent advances in the lipase-catalyzed biotransformation of fats and oils. *Recent Res. Dev. Oil Chem.*, 3, 93–106.

Bornscheuer, UT (2000) *Enzymes in Lipid Modification*, Wiley-VCH, Weinheim.

Bornscheuer, UT (2001) Lipase-catalyzed synthesis of structured triacylglycerols. *Lipid Technol. Newsl.*, 7, 104–107.

Bornscheuer, UT and Kazlauskas, RJ (1999) *Hydrolases in Organic Synthesis: Regio- and Stereoselective Biotransformation*, Wiley-VCH, Weinheim.

Breivik, H, Haraldsson, GG and Kristinsson, B (1997) Preparation of highly purified concentrates of eicosapentaenoic acid and docosahexaenoic acid. *J. Am. Oil Chem. Soc.*, 74, 1425–1429.

Brockerhoff, H (1965) Stereospecific analysis of triglycerides. *J. Lipid Res.*, 6, 10–15.

Brockerhoff, H (1967) Stereospecific analysis of triglycerides: an alternative method. *J. Lipid Res.*, 8, 167–169.

Catoni, E, Cernia, E and Palocci, C (1996) Different aspects of "solvent engineering" in lipase biocatalysed esterification. *J. Mol. Catal. B: Enzymatic*, 105, 79–86.

Chandler, IC, Quinlan, PT and McNeill, GP (1998) Lipase-catalyzed synthesis of chiral triglycerides. *J. Am. Oil Chem. Soc.*, 75, 1513–1518.

Chandler, IC, Howarth, OW and Crout, DHG (2001) Measuring stereoselectivity in lipase-catalyzed acidolysis reactions by ultra-high resolution ^{13}C nuclear magnetic resonance. *J. Am. Oil Chem. Soc.*, 78, 953–958.

Chang, MK, Abraham, G and John, VT (1990) Production of cocoa butter-like fat from interesterification of vegetable oils. *J. Am. Oil Chem. Soc.*, 67, 832–834.

Chong, CN, Hoh, YM and Wang, CW (1992) Fractionation procedures for obtaining cocoa butter-like fat from enzymatically interesterified palm olein. *J. Am. Oil Chem. Soc.*, 69(2), 137–140.

Christensen, TC and Hølmer, G (1993) Lipase catalyzed acyl-exchange reactions of butter oil. Synthesis of a human milk fat substitute for infant formulas. *Milchwissenschaft*, 48, 543–548.

Christie, WW (1991) Recent developments in high-performance liquid and gas chromatography of lipids. *Corps Gras*, 38, 155–160.

Clough, PM (2001) Specialty vegetable oils containing γ-linolenic acid and stearidonic acid. In: *Structured and Modified Lipids* (FD Gunstone, ed.), Marcel Dekker, New York, pp.75–117.

Coleman, MH and Macrae, AR (1977) Rearrangement of fatty acid esters in fat reaction reactants. German Patent Application DE 2 705 608.

Coleman, MH and Macrae, AR (1980) Fat process and composition. UK Patent Application 1 577 933.

De Greyt, W, Huyghebaert, A and Kellens, M (1998) Chemical and physicochemical modification of lipids. In: *Structural Modified Food Fats: Synthesis, Biochemistry, and Use* (AB Christophe, ed.), AOCS Press, Champaign, pp.1–19.

Diks, RMM and Bosley, JA (2000) The exploitation of lipase selectivities for the production of acylglycerols. In: *Enzymes in Lipid Modification* (UT Bornscheuer, ed.), Wiley-VCH, Weinheim, pp.3–22.

Esteban Cerdan, L, Robles Medina, A, Gimenez Gimenez, A, Ibanez Gonzalez, MJ and Molina Grima, E (1998) Synthesis of polyunsaturated fatty acid-enriched triglycerides by lipase-catalyzed esterification. *J. Am. Oil Chem. Soc.*, 75, 1329–1337.

Fitch Haumann, B (1997) Structured lipids allow fat tailoring. *Inform*, 8, 1004–1011.

Foglia, TA and Valivety, R (1997) Carica papaya latex-catalyzed synthesis of structured triacylglycerols. *J. Am. Oil Chem. Soc.*, 74, 1447–1450.

Fomuso, LB and Akoh, CC (1997) Enzymatic modification of triolein: incorporation of caproic and butyric acids to produce reduced-calorie structured lipids. *J. Am. Oil Chem. Soc.*, 74, 269–272.

Fomuso, LB and Akoh, CC (1998) Structured lipids: lipase-catalyzed interesterification of tricaproin and trilinolein. *J. Am. Oil Chem. Soc.*, 75, 405–410.

Gandhi, NN and Mukherjee, KD (2001a) Reactivity of medium-chain substrates in the interesterification of tripalmitin catalyzed by papaya lipase. *J. Am. Oil Chem. Soc.*, 78, 965–968.

Gandhi, NN and Mukherjee, KD (2001b) Synthesis of designer lipids using papaya (*Carica papaya*) latex lipase. *J. Mol. Catal. B: Enzymatic*, 11, 271–277.

Gill, I and Valivety, R (1997a) Polyunsaturated fatty acids, Part 1: occurrence, biological activities and applications. *Trends Biotechnol.*, 15, 401–409.

Gill, I and Valivety, R (1997b) Polyunsaturated fatty acids, Part 2: biotransformation and biotechnological applications. *Trends Biotechnol.*, 15, 470–478.

Garcia, HS, Arcos, JA, Ward, DJ and Hill, CG, Jr (2000) Synthesis of glycerides containing *n*-3 fatty acids and conjugated linoleic acid by solvent-free acidolysis of fish oil. *Biotechnol. Bioeng.*, 70(5), 587–591.

Garcia, HS, Arcos, JA, Keough, KJ and Hill, CG, Jr (2001) Immobilized lipase-mediated acidolysis of butteroil with conjugated linoleic acid: batch reactor and packed bed reactor studies. *J. Mol. Catal. B: Enzymatic*, 11(4–6), 623–632.

Guil-Guerrero, JL, Gracia Maroto, FF and Gimenez Gimenez, A (2001) Fatty acid profiles from forty-nine plant species that are potential new source of γ-linolenic acid. *J. Am. Oil Chem. Soc.*, 78, 677–684.

Gunstone, FD (2001) *Structured and Modified Lipids*, Marcel Dekker, New York.

Haas, MJ, Bailey, DG, Baker, W, Berka, TR, Cichowicz, DJ, Derewenda, ZS, Genuario, RR, Joerger, RD, Klein, RR, Scott, K and Woolf, DJ (1999) Biochemical and molecular biological characterization of a lipase produced by the fungus *Rhizopus delemar*. *Fett/Lipid*, 101, 364–370.

Halling, PJ, Peilow, AD, Macrae, AR and Valivety, RH (1994) Relationship between water activity and catalytic activity of lipases in organic media. *Eur. J. Biochem.*, 222, 461–466.

Han, JJ, Iwasaki, Y and Yamane, T (1999) Monitoring of lipase-catalyzed transesterification between eicosapentaenoic acid ethyl ester and tricaprylin by silver ion high-performance liquid chromatography and high-temperature gas chromatography. *J. Am. Oil Chem. Soc.*, 76, 31–39.

Han, JJ and Yamane, T (1999) Enhancement of both reaction yield and rate of synthesis of structured triacylglycerol containing eicosapentaenoic acid under vacuum with water activity control. *Lipids*, 34, 989–995.

Haraldsson, GG and Hjaltason, B (2001) Fish oils as sources of important polyunstaturated fatty acids. In: *Structured and Modified Lipids* (FD Gunstone, ed.), Marcel Dekker, New York, pp.313–350.

Haraldsson, GG and Kristinsson, B (1998) Separation of eicosapentaenoic acid and docosahexaenoic acid in fish oil by kinetic resolution using lipase. *J. Am. Oil Chem. Soc.*, 75, 1551–1556.

Haraldsson, GG, Hoskuldsson, PA, Sigurdsson, ST, Thorsteinsson, F and Gudbjarnason, S (1989) The preparation of triglycerides highly enriched with ω-3 polyunsaturated fatty acids *via* lipase catalyzed interesterification. *Tetrahedron Lett.*, 30, 1671–1674.

Haraldsson, GG, Gudmundsson, BO and Almarsson, O (1995) The synthesis of homogeneous triglycerides of eicosapentaenoic acid and docosahexaenoic acid by lipase. *Tetrahedron*, 51, 941–952.

Haraldsson, GG, Kristinsson, B, Sigurdardottir, R, Gudmundsson, GG and Breivik, H (1997) The preparation of concentrates of eicosapentaenoic acid and docosahexaenoic acid by lipase-catalyzed transesterification of fish oil with ethanol. *J. Am. Oil Chem. Soc.*, 74(11), 1419–1424.

Hellyer, SA, Chandler, IC and Bosley, JA (1999) Can the fatty acid selectivity of plant lipases be predicted from the composition of the seed triglyceride? *Biochim. Biophys. Acta*, 1440, 215–224.

Hills, MJ, Kiewitt, I and Mukherjee, KD (1990) Enzymatic fractionation of fatty acids: enrichment of γ-linolenic acid and docosahexaenoic acid by selective esterification catalyzed by lipases. *J. Am. Oil Chem. Soc.*, 67, 561–564.

Homan, R and Andersen, MK (1998) Rapid separation and quantitation of combined neutral and polar lipid classes by high-performance liquid chromatography and evaporative light-scattering mass detection. *J. Chromatogr. B*, 708, 21–26.

Huang, Y-S and Akoh, CC (1996) Enzymatic synthesis of structured lipids: transesterification of triolein and caprylic acid ethyl ester. *J. Am. Oil Chem. Soc.*, 73, 245–250.

Hudson, BJF (1984) Evening primrose (*Oenothera* spp.) oil and seed. *J. Am. Oil Chem. Soc.*, 61, 540–543.

Hurtova, S, Schmidt, S, Zemanovic, J, Simon, P and Sekretar, S (1996) Random interesterification of fats blends with alkali catalyst. *Fett/Lipid*, 98, 60–65.

Irimescu, R, Yasui, M, Iwasaki, Y, Shimidzu, N and Yamane, T (2000) Enzymatic synthesis of 1,3-dicapryloyl-2-eicosapentaenoylglycerol. *J. Am. Oil Chem. Soc.*, 77, 501–506.

Irimescu, R, Furihata, K, Hata, K, Iwasaki, Y and Yamane, T (2001) Two-step enzymatic synthesis of docosahexaenoic acid-rich symmetrically structured triacylglycerols via 2-monoacylglycerols. *J. Am. Oil Chem. Soc.*, 78, 743–748.

Iwasaki, Y and Yamane, T (2000) Enzymatic synthesis of structured lipids. *J. Mol. Catal. B: Enzymatic*, 10, 129–140.

Iwasaki, Y, Han, JJ, Narita, M, Rosu, R and Yamane, T (1999) Enzymatic synthesis of structured lipids from single cell oil of high docosahexaenoic acid content. *J. Am. Oil Chem. Soc.*, 76, 563–569.

Iwasaki, Y, Yasui, M, Ishikawa, T, Irimescu, R, Hata, K and Yamane, T (2001) Optical resolution of asymmetric triacylglycerols by chiral-phase high-performance liquid chromatography. *J. Chromatogr.*, 905, 111–118.

Jaeger, K-E, Ransac, S, Dijkstra, BW, Colson, C, van Heuvel, M, Misset, O, Eggert, T, Eipper, A and Reetz, MT (1994) Bacterial lipase. *FEMS Microb. Rev.*, 15, 29–63.

Jaeger, K-E, Dijkstra, BW and Reetz, MT (1999) Bacterial biocatalysts: molecular biology, three-dimensional structures, and biotechnological applications of lipases. *Annu. Rev. Microbiol.*, 53, 315–351.

Jennings, BH and Akoh, CC (1999) Enzymatic modification of triacylglycerols of high eicosapentaenoic and docosahexaenoic acids content to produce structured lipids. *J. Am. Oil Chem. Soc.*, 76, 1133–1137.

Ju, Y-H, Huang, F-H and Fang, C-H (1998) The incorporation of *n*-3 polyunsaturated fatty acids into acylglycerols of borage oil *via* lipase-catalyzed reactions. *J. Am. Oil Chem. Soc.*, 75, 961–965.

Kalo, P, Vaara, K and Antila, M (1986) Changes in triglyceride composition and melting properties of butter fat solid fraction/rapeseed oil mixtures induced by lipase catalysed inter-esterification. *Fette Seifen Anstrichmittel.*, 9, 362–365.

Kalo, P, Huotari, H and Antila, M (1989) Chemical composition of butter fat interesterified with *Pseudomonas fluorescens* lipase at various temperatures. *Finn. J. Dairy Sci.*, 47, 29–38.

Kay, RA (1991) Microalgae as food and supplement. *Crit. Rev. Food Sci. Nutr.*, 30, 555–573.

Kazlauskas, RJ and Bornscheuer, UT (1998) Biotransformations with lipases. In: *Biotechnol-*

ogy-Series 8a: Biotransformations I (H-J Rehm, G Reed, A Puhler and P Stadler, eds), Wiley-VCH, Weinheim, pp.37–191.

Kennedy, JP (1991) Structured lipids: fats of the future. *Food Technol.*, 45(11), 76–83.

Kermasha, S, Kubow, S, Safari, M and Reid, A (1993) Determination of the positional distribution of fatty acids in butterfat triacylglycerols. *J. Am. Oil Chem. Soc.*, 70, 169–173.

King, DM and Padley, FB (1990) Milk fat substitutes. European Patent Application EP 0 209 327.

King, JW, Holliday, RL, List, GR and Snyder, JM (2001) Hydrogenation of vegetable oils using mixtures of supercritical carbon dioxide and hydrogen. *J. Am. Oil Chem. Soc.*, 78, 107–113.

Klemann, LP, Aji, K, Chrysam, MM, D'Amelia, RP, Henderson, JM, Huang, AS, Otterburn, MS, Yarger, RG, Boldt, G and Roden, A (1994) Random nature of triacylglycerols produced by the catalysed interesterification of short- and long-chain fatty acid triglyceride. *J. Agric. Food Chem.*, 42, 443–446.

Kluesener, BW, Stipp, GK and Yang, DK (Procter & Gamble Co) (1992) Selective esterification of long chain fatty acid monoglycerides with medium chain fatty acids. US Patent 5,142,071.

Knauf, VC and Del Vecchio, AJ (1998) Genetic engineering of crops that produce vegetable oil. In: *Food Lipids: Chemistry, Nutrition, and Biotechnology* (CC Akoh and DB Min, eds), Marcel Dekker, New York, pp.779–805.

Kritchevsky, D (1998) Fats and oils in human health. In: *Food Lipids: Chemistry, Nutrition, and Biotechnology* (CC Akoh and DB Min, eds), Marcel Dekker, New York, pp.449–461.

Laane, C, Boeren, S, Vos, K and Verger, C (1987) Rules for optimization of biocatalysis in organic solvents. *Biotechnol. Bioeng.*, 30, 81–87.

Lai, OM, Ghanali, HM, Cho, F and Chong, CL (2000) Physical properties of lipase-catalyzed transesterified blends of palm stearin and anhydrous milk fat. *Food Chem.*, 70, 215–219.

Lee, K-T and Akoh, CC (1998) Structured lipids: synthesis and applications. *Food Rev. Int.*, 14, 17–34.

Lee, K-T and Foglia, TA (2000) Synthesis, purification, and characterization of structured lipids produced from chicken fat. *J. Am. Oil Chem. Soc.*, 77, 1027–1034.

Leman, J (1997) Oleaginous microorganisms: an assessment of the potential. *Adv. Appl. Microbiol.*, 43, 195–243.

Mangos, TJ, Jones, KC and Foglia, TA (1999) Lipase-catalyzed synthesis of structured low-calorie triacylglycerols. *J. Am. Oil Chem. Soc.*, 76, 1127–1132.

Marangoni, AG and Rousseau, D (1995) Engineering triacylglycerols: the role of interesterification. *Trends Food Sci. Technol.*, 6, 329–335.

Martinez, CE, Vinay, JC, Brieva, R, Hill, CG, Jr and Garcia, HS (1999) Lipase-catalyzed interesterification (acidolysis) of corn oil and conjugated linoleic acid in organic solvents. *Food Biotechnol.*, 13, 183–193.

Matsuo, T, Sawamura, N, Hashimoto, Y and Hashida, W (Fuji Oil Co) (1981) Method for enzymatic transesterification of lipid and enzyme used therein. European Patent 0 035 883.

McNeill, GP and Sonnet, PE (1995) Low-calorie triglyceride synthesis by lipase-catalyzed esterification of monoglyceride. *J. Am. Oil Chem. Soc.*, 72, 1301–1307.

McNeill, GP, Rawlins, C and Peilow, AC (1999) Enzymatic enrichment of conjugated linoleic acid isomers and incorporation into triglycerides. *J. Am. Oil Chem. Soc.*, 76, 1265–1268.

Megremis, CJ (1991) Medium-chain triglycerides: a nonconventional fat. *Food Technol.*, 45(2), 108–114.

Millqvist Fureby, A, Adlercreutz, P and Mattiasson, B (1994) Lipase-catalyzed alcoholysis of triglycerides for the preparation of 2-monoglycerides. *Enzyme Microb. Technol.*, 16, 1042–1047.

Millqvist Fureby, A, Adlercreutz, P and Mattiasson, B (1996a) Glyceride synthesis in a solvent-free system. *J. Am. Oil Chem. Soc.*, 73, 1489–1495.

Millqvist Fureby, A, Virto, C, Adlercreutz, P and Mattiasson, B (1996b) Acyl group migrations in 2-monoolein. *Biocatal. Biotrans.*, 14, 89–111.

Miura, S, Ogawa, A and Konishi, H (1999) A rapid method for enzymatic synthesis and purification of the structured triacylglycerol, 1,3-dilauroyl-2-oleoyl glycerol. *J. Am. Oil Chem. Soc.*, 76, 927–931.

Mogi, K-I, Nakajima, M and Mukataka, S (2000) Transesterification reaction between medium- and long-chain fatty acid triglycerides using surfactant-modified lipase. *Biotechnol. Bioeng.*, 67, 513–519.

Mu, H, Kalo, P, Xu, X and Hoy, C-E (2000) Chromatographic methods in the monitoring of lipase-catalyzed interesterification. *Eur. J. Lipid. Sci. Technol.*, 102, 202–211.

Mukherjee, KD (1994) Plant lipases and their application in lipid biotransformation. *Prog. Lipid Res.*, 33, 165–174.

Mukherjee, KD (1998) Lipid biotechnology. In: *Food Chemistry: Chemistry, Nutrition, and Biotechnology* (CC Akoh and DB Min, eds), Marcel Dekker, New York, pp.589–640.

Mukherjee, KD and Hills, MJ (1994) Lipases from plants. In: *Lipases: Their Structure, Biochemisty and Application* (P Woolley and SB Petersen, eds), Cambridge University Press, Cambridge, pp.49–75.

Mukherjee, KD and Kiewitt, I (1991) Enrichment of γ-linolenic acid from fungal oil by lipase-catalysed reactions. *Appl. Microbiol. Biotechnol.*, 35, 579–584.

Myher, JJ, Kuksis, A, Geher, K, Park, PW and Diersen-Schade, DA (1996) Stereospecific analysis of triacylglycerols rich in long-chain polyunsaturated fatty acids. *Lipids*, 31, 207–215.

Namal Senanayake, SPJ and Shahidi, F (1999a) Enzymatic incorporation of docosahexaenoic acid into borage oil. *J. Am. Oil Chem. Soc.*, 76, 1009–1015.

Namal Senanayake, SPJ and Shahidi, F (1999b) Enzyme-assisted acidolysis of borage (*Borago officinalis* L.) and evening primrose (*Oenothera biennis* L.) oils: incorporation of omega-3 polyunsaturated fatty acids. *J. Agric. Food Chem.*, 47, 3105–3112.

Nor Hayati, IN, Aminah, A, Mamot, S, Nor Aini, IN and Noor Lida, HM (2000) Melting characteristic and solid fat content of milk fat and palm stearin blends before and after enzymatic interesterification. *J. Food Lipids*, 7, 175–193.

Nordback, J, Lundberg, E and Christie, WW (1998) Separation of lipid classes from marine particulate material by HPLC on a polyvinyl alcohol-bonded stationary phase using dual-channel evaporative light-scattering detection. *Marine Chem.*, 60, 165–175.

Ollis, DL, Cheah, E, Cygler, M, Dijkstra, B, Frolow, F, Franken, S, Harel, M, Remington, SJ and Silman, I (1992) The α/β hydrolase fold. *Protein Eng.*, 5, 197–211.

Pabai, F, Kermasha, S and Morin, A (1995) Lipase from *Pseudomonas fragi* CRDA 323: partial purification, characterization and interesterification of butter fat. *Appl. Microbiol. Biotechnol.*, 43, 42–51.

Parkin, KL (1998) Biosynthesis of fatty acids and storage lipids in oil-bearing seed and fruit tissues of plants. In: *Food Lipids: Chemistry, Nutrition, and Biotechnology* (CC Akoh and DB Min, eds), Marcel Dekker, New York, pp.729–778.

Pinsirodom, P and Parkin, KL (1999) Fatty acid and product selectivities of potato tuber lipid acyl hydrolase in esterification reactions with glycerol in organic media. *J. Am. Oil Chem. Soc.*, 76, 1119–1125.

Quinlan, P and Moore, S (1993) Modification of triglycerides by lipases: Process technology and its application to the production of nutritionally improved fats. *Inform*, 4, 580–585.

Rakshit, SK, Vasuhi, R and Kosugi, Y (2000) Enrichment of polyunsaturated fatty acids from tuna oil using immobilized *Pseudomonas fluorescens* lipase. *Bioprocess Eng.*, 23, 251–255.

Rao, R, Sankar, KY, Sambaiah, K and Lokesh, BR (2001) Differential scanning calorimetric studies on structured lipids from coconut oil triglycerides containing stearic acid. *Eur. Food Res. Technol.*, 212, 334–343.

Robles Medina, A, Esteban Cerdan, L, Gimenez Gimenez, A, Camacho Paez, B, Ibanez Gonzalez, MJ and Molina Grima, E (1999) Lipase-catalyzed esterification of glycerol and polyunstarurated fatty acids from fish and microalgae oils. *J. Biotechnol.*, 70, 379–391.

Rodriguez, A, Castro, E, Salinas, MC, Lopez, R and Miranda, M (2001) Interesterification of tallow and sunflower oil. *J. Am. Oil Chem. Soc.*, 78, 431–436.

Rousseau, D and Marangoni, AG (1998a) Chemical interesterification of food lipids: theory and practice. In: *Food Lipids: Chemistry, Nutrition, and Biotechnology* (CC Akoh and DB Min, eds), Marcel Dekker, New York, pp.251–281.

Rousseau, D and Marangoni, AG (1998b) Tailoring the textural attributes of butter fat/canola oil blends via *Rhizopus arrhizus* lipase-catalyzed interesterification. 1. Compositional modifications. *J. Agric. Food Chem.*, 46, 2368–2374.

Rousseau, D and Marangoni, AG (1998c) Tailoring the textural attributes of butter fat/canola oil blends via *Rhizopus arrhizus* lipase-catalyzed interesterification. 2. Modification of physical properties. *J. Agric. Food Chem.*, 46, 2375–2381.

Safari, M, Kermasha, S and Pabai, F (1993) Interesterification of butter fat by lipase from *Mucor miehei* in organic solvent media. *Food Biotechnol.*, 7, 265–273.

Safari, M, Kermasha, S, Lamboursain, L and Sheppard, JD (1994a) Interesterification of butterfat by lipase from *Rhizopus niveus* in reverse micellar systems. *Biosci. Biotech. Biochem.*, 58, 1553–1557.

Safari, M, Kermasha, S, Pabai, F and Sheppard, JD (1994b) Interesterification of butter fat by lipase from *Mucor miehei* in microemulsion system. *J. Food Lipids*, 1, 247–263.

Schmid, RD and Verger, R (1998) Lipases: interfacial enzymes with attractive applications. *Angew. Chem. Int. Ed.*, 37, 1608–1633.

Schmid, U, Bornscheuer, UT, Soumanou, MM, McNeill, GP and Schmid, RD (1998) Optimization of the reaction conditions in the lipase-catalyzed synthesis of structured triglycerides. *J. Am. Oil Chem. Soc.*, 75, 1527–1531.

Schmid, U, Bornscheuer, UT, Soumanou, MM, McNeill, GP and Schmid, RD (1999) Highly selective synthesis of 1,3-oleoyl-2-palmitoylglycerol by lipase catalysis. *Biotechnol. Bioeng.*, 64, 678–684.

Schmidt-Dannert, C (1999) Recombinant microbial lipases for biotechnological applications. *Bioorg. Med. Chem.*, 7, 2123–2130.

Schmitt-Rozieres, M, Vanot, G, Deyris, V and Comeau, L-C (1999) *Borago officinalis* oil: fatty acid fractionation by immobilized *Candida rugosa* lipase. *J. Am. Oil Chem. Soc.*, 76, 557–562.

Schmitt-Rozieres, M, Deyris, V and Comeau, L-C (2000) Enrichment of polyunsaturated fatty acids from sardine cannery effluents by enzymatic selective esterification. *J. Am. Oil Chem. Soc.*, 77, 329–332.

Seriburi, V and Akoh, CC (1998) Enzymatic interesterification of triolein and tristearin: chemical structure and differential scanning calorimetric analysis of the products. *J. Am. Oil Chem. Soc.*, 75, 711–716.

Shahidi, F and Wanasundara, UN (1998) Omega-3 fatty acid concentrates: nutritional aspects and production technologies. *Trends Food Sci. Technol.*, 9, 230–240.

Shimada, Y, Maruyama, K, Okazaki, S, Nakamura, M, Sugihara, A and Tominaga, Y (1994a) Enrichment of polyunsaturated fatty acids with *Geotrichum candidum* lipase. *J. Am. Oil Chem. Soc.*, 71, 951–954.

Shimada, Y, Sugihara, A and Tominaga, Y (1994b) Microbial lipase: structure and production. *Bioprocess Eng.*, 19, 359–371.

Shimada, Y, Sugihara, A, Maruyama, K, Nagao, T, Nakayama, S, Nakano, H and Tominaga,

Y (1996) Production of structured lipid containing docosahexaenoic and caprylic acids using immobilized *Rhizopus delemar* lipase. *J. Ferment. Bioeng.*, 81, 299–303.

Shimada, Y, Maruyama, K, Sugihara, A, Moriyama, S and Tominaga, Y (1997a) Purification of docosahexaenoic acid from tuna oil by two-step enzymatic method: hydrolysis and selective esterification. *J. Am. Oil Chem. Soc.*, 74, 1441–1446.

Shimada, Y, Sugihara, A, Nakano, H, Nagao, T, Suenaga, M, Nakai, S and Tominaga, Y (1997b) Fatty acid specificity of *Rhizopus delemar* lipase in acidolysis. *J. Ferment. Bioeng.*, 83, 321–327.

Shimada, Y, Sugihara, A, Shibahiraki, M, Fujita, H, Nakano, H, Nagao, A, Terai, T and Tominaga, Y (1997c) Purification of γ-linolenic acid from borage oil by two-step enzymatic method. *J. Am. Oil Chem. Soc.*, 74, 1465–1470.

Shimada, Y, Sakai, N, Sugihara, A, Fujita, H, Honda, Y and Tominaga, Y (1998a) Large-scale purification of γ-linolenic acid by selective esterification using *Rhizopus delemar* lipase. *J. Am. Oil Chem. Soc.*, 75, 1539–1543.

Shimada, Y, Sugihara, A, Minamigawa, Y, Higashiyama, K, Akimoto, K, Fujikawa, S, Komemushi, S and Tominaga, Y (1998b) Enzymatic enrichment of arachidonic acid from *Mortierella* single-cell oil. *J. Am. Oil Chem. Soc.*, 75, 1213–1217.

Shimada, Y, Suenaga, M, Sugihara, A, Nakai, S and Tominaga, Y (1999) Continuous production of structured lipid containing γ-linolenic and caprylic acids by immobilized *Rhizopus delemar* lipase. *J. Am. Oil Chem. Soc.*, 76, 189–193.

Shimada, Y, Nagao, T, Hamasaki, Y, Akimoto, K, Sugihara, A, Fujikawa, S, Komemushi, S and Tominaga, Y (2000) Enzymatic synthesis of structured lipid containing arachidonic and palmitic acids. *J. Am. Oil Chem. Soc.*, 77, 89–93.

Small, DM (1991) The effect of glyceride structure on absorption and metabolism. *Annu. Rev. Nutr.*, 11, 413–434.

Smith, RE, Finley, JW and Leveille, GA (1994) Overview of Salatrim™, a family of low-calorie fats. *J. Agric. Food Chem.*, 42, 432–434.

Soumanou, MM, Bornscheuer, UT, Menge, U and Schmid, RD (1997) Synthesis of structured triglycerides from peanut oil with immobilized lipase. *J. Am. Oil Chem. Soc.*, 74, 427–433.

Soumanou, MM, Bornscheuer, UT and Schmid, RD (1998a) Two-step enzymatic reaction for the synthesis of pure structured triacylglycerides. *J. Am. Oil Chem. Soc.*, 75, 703–710.

Soumanou, MM, Bornscheuer, UT, Schmid, U and Schmid, RD (1998b) Synthesis of structured triglycerides by lipase catalysis. *Fett/Lipid*, 100, 156–160.

Soumanou, MM, Bornscheuer, UT, Schmid, U and Schmid, RD (1999) Crucial role of support and water activity on the lipase-catalyzed synthesis of structured triglycerides. *Biocatal. Biotrans.*, 16, 443–459.

Sridhar, R, Lakshminarayana, G and Kaimal, TNB (1991) Modification of selected edible vegetable oils to high oleic oils by lipase-catalyzed ester interchange. *J. Agric. Food Chem.*, 39, 2069–2071.

Stadler, P, Kovac, A and Paltauf, F (1995) Understanding lipase action and selectivity. *Croatica Chemica Acta*, 68, 649–674.

Stipp, GK and Kluesener, BW (Procter & Gamble Co) (1992) Selective esterification of long chain fatty acid monoglycerides with medium chain fatty acid anhydrides. US Patent 5,142,072.

Syed Rahmatullah, MSKS, Shukla, VKS and Mukherjee, KD (1994a) Enrichment of γ-linolenic acid from evening primrose oil and borage oil via lipase-catalyzed hydrolysis. *J. Am. Oil Chem. Soc.*, 71, 569–573.

Syed Rahmatullah, MSKS, Shukla, VKS and Mukherjee, KD (1994b) γ-Linolenic acid concentrates from borage and evening primrose oil fatty acids via lipase-catalyzed esterification. *J. Am. Oil Chem. Soc.*, 71, 563–567.

Tatara, T, Fuji, T, Kawase, T and Minagawa, M (1983) Quantitative determination of tri-, di-, monooleins and free oleic acid by thin layer chromatography-flame ionization detector system using internal standards and boric acid impregnated chromarod. *Lipids*, 18, 732–736.

Traitler, H, Winter, H, Richlin, U and Ingenbleek, Y (1984) Characterization of gamma-linolenic acid in *Ribes* seed. *Lipids*, 19, 923–928.

Uzzan, A (1986) Borage oil: an oil with a solid future. *Rev. Fr. Corps Gras*, 33, 385–389.

Verger, R (1997) "Interfacial activation" of lipases: facts and artifacts. *Trends Biotechnol.*, 15, 32–38.

Wanasundara, UN and Shahidi, F (1998) Lipase-assisted concentration of *n*-3 polyunsaturated fatty acids in acylglycerols from marine oils. *J. Am. Oil Chem. Soc.*, 75, 945–951.

Williams, JP, Khan, MU and Wong, D (1995) A simple technique for the analysis of positional distribution of fatty acids on di- and triacylglycerols using lipase and phospholipase A2. *J. Lipid Res.*, 36, 1407–1412.

Xu, X (2000) Production of specific-structured triacylglycerols by lipase-catalyzed reactions: a review. *Eur. J. Lipid. Sci. Technol.*, 287–303.

Xu, X, Balchen, S, Hoy, C-E and Adler-Nissen, J (1998a) Production of specific-structured lipids by enzymatic interesterification in a pilot continuous enzyme bed reactor. *J. Am. Oil Chem. Soc.*, 75, 1573–1579.

Xu, X, Skands, ARH, Hoy, C-E, Mu, H, Balchen, S and Adler-Nissen, J (1998b) Production of specific-structured lipids by enzymatic interesterification: elucidation of acyl migration by response surface design. *J. Am. Oil Chem. Soc.*, 75, 1179–1186.

Xu, X, Balchen, S, Jonsson, G and Adler-Nissen, J (2000a) Production of structured lipids by lipase-catalyzed interesterification in a flat membrane reactor. *J. Am. Oil Chem. Soc.*, 77, 1035–1041.

Xu, X, Fomuso, LB and Akoh, CC (2000b) Modification of menhaden oil by enzymatic acidolysis to produce structured lipids: optimization by response surface design in a packed bed reactor. *J. Am. Oil Chem. Soc.*, 77, 171–176.

Xu, X, Fomuso, LB and Akoh, CC (2000c) Synthesis of structured triacylglycerols by lipase-catalyzed acidolysis in a packed bed bioreactor. *J. Agric. Food Chem.*, 48, 3–10.

Yamane, T (2000) Lipase-catalyzed synthesis of structured triacylglycerols containing polyunsaturated fatty acids: monitoring the reaction and increasing the yield. In: *Enzymes in Lipid Modification* (UT Bornscheuer, ed.), Wiley-VCH, Weinheim, pp.148–169.

Yang, TH, Jang, Y, Han, JJ and Rhee, JS (2001) Enzymatic synthesis of low-calorie structured lipids in a solvent-free system. *J. Am. Oil Chem. Soc.*, 78, 291–296.

Yankah, VV and Akoh, CC (2000) Lipase-catalyzed acidolysis of tristearin with oleic or caprylic acids to produce structured lipids. *J. Am. Oil Chem. Soc.*, 77, 495–500.

Zamora, R, Alba, V and Hidalgo, FJ (2001) Use of high-resolution ^{13}C nuclear magnetic resonance spectroscopy for the screening of virgin olive oils. *J. Am. Oil Chem. Soc.*, 78, 89–94.

Zeitoun, MAM, Neff, WE, List, GR and Mounts, TL (1993) Physical properties of interesterified fats blends. *J. Am. Oil Chem. Soc.*, 70, 461–471.

Zhang, H, Xu, X, Nilsson, J, Mu, H, Adler-Nissen, J and Hoy, C-E (2001) Production of margarine fats by enzymatic interesterification with silica-granulated *Thermomyces lanuginosa* lipase in a large-scale study. *J. Am. Oil Chem. Soc.*, 78, 57–64.

Zhou, D, Xu, X, Mu, H, Hoy, C-E and Adler-Nissen, J (2000) Lipase-catalyzed production of structured lipids via acidolysis of fish oil with caprylic acid. *J. Food Lipids*, 7, 263–274.

CHAPTER 7

Phytosterols

PIA SALO, INGMAR WESTER and ANU HOPIA

Raisio Life Sciences Ltd, Raisio, Finland

A. Introduction

Tools to improve cardiovascular health are of great importance in countries where life expectancy is significantly increasing and where the prevalence of lifestyle-related diseases such as cardiovascular disease and type II diabetes is constantly growing. Atherosclerosis is still the leading cause of mortality and morbidity in affluent societies. By changing diet and lifestyle the risk factors for cardiovascular mortality and morbidity can be lowered. Among these risk factors, the importance of total cholesterol and low-density lipoprotein (LDL) cholesterol levels in blood serum is well established. An elevated serum concentration of cholesterol and especially of LDL cholesterol is a key risk factor for the development of atherosclerosis and its clinical manifestations, which include coronary heart disease and cerebrovascular and peripheral artery diseases.

The average diet provides a large number of components involved in improving and maintaining cardiovascular health such as polyunsaturated fatty acids and soluble fiber. Plant sterols (phytosterols), especially 4-desmethyl sterols, have recently gained much scientific and commercial interest due to the introduction of foods enriched in plant sterols and dietary supplements with enhanced serum cholesterol-lowering efficacy. Due to the structural similarity between cholesterol and 4-desmethyl plant sterols, the intake of optimal amounts of plant sterols and stanols decreases intestinal cholesterol absorption and leads to a 6–15% reduction in serum LDL cholesterol concentrations. Plant sterols are already present in the normal diet but generally at levels too low to provide significant cholesterol-lowering properties. However, functional foods enriched with plant sterols provide the most efficient dietary tool to lower serum cholesterol levels, compared with any other dietary component (Table 1).

Functional foods enriched in plant sterols are currently (as of August 2002) commercially available in Europe, the USA, South America, Australia, and Asia.

Table 1. The effects of selected ingredients on serum LDL cholesterol levels[a]

Ingredient	Dose (g/day)	Effect on serum LDL cholesterol
Soluble fibre	5–10	–3% to –5%
Plant stanols/sterols	2–3	–6% to –15%
Soy protein	High intake	Small reductions when it replaces animal protein
n-3 fatty acids	High intake	It is suggested that higher intakes reduce the risk of coronary events; mechanisms are unknown

[a]Data taken from Van Horn and Ernst (2001).

Research on the health effects and safety of plant sterols is extensive. New food formats are continually being introduced to the market. The purpose of this review is to summarize current scientific understanding of the role of plant sterols as cholesterol-lowering food ingredients, and to review the different types of plant sterol-enriched foods available on the market.

B. Chemistry of plant sterols

Plant sterols are analogues of cholesterol (a zoosterol) occurring in plant materials and, as in mammalian cells, they provide structure and function in cell membranes in addition to serving as a starting material for the biological production of hormones. Sterols have the tetracyclic ring system specific to steroids, a hydroxyl group at C-3, and a side chain of variable length at C-17. The IUPAC-IUB rules for numbering sterols and for ring lettering are based on the cholestane skeleton shown for cholesterol in Figure 1. The molecular structures of natural plant sterols have been reviewed by Akihisa *et al.* (1991), who reported that 266 different sterols and related compounds had been recognized in plant and marine materials by 1991.

Plant sterols can be divided, on both structural and biosynthetic grounds, into 4-desmethyl, 4-monomethyl, and 4,4-dimethyl sterols based on the number of methyl groups at C-4 in the tetracyclic ring system (Figure 2). Plant sterols are not endogenously synthesized in the human body and are therefore derived solely from dietary sources.

1. 4-Desmethyl sterols

4-Desmethyl sterols, with no methyl group at C-4, are the main class of sterols in plants and thus in the human diet. The proportion of 4-desmethyl sterols among total sterols in common crude vegetable oils is >85% (Piironen *et al.*, 2000a). Sitosterol, the principal sterol in most plants, is usually accompanied by its 22-dehydro analogue, stigmasterol. The molecular structures of the most important plant sterols and stanols (see Section B.5 below) are shown in

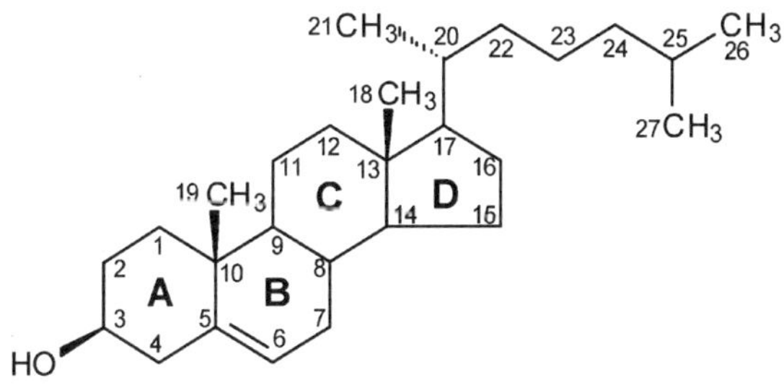

Cholesterol

Figure 1. The IUPAC–IUB rules for numbering sterols and ring lettering, shown here for cholesterol.

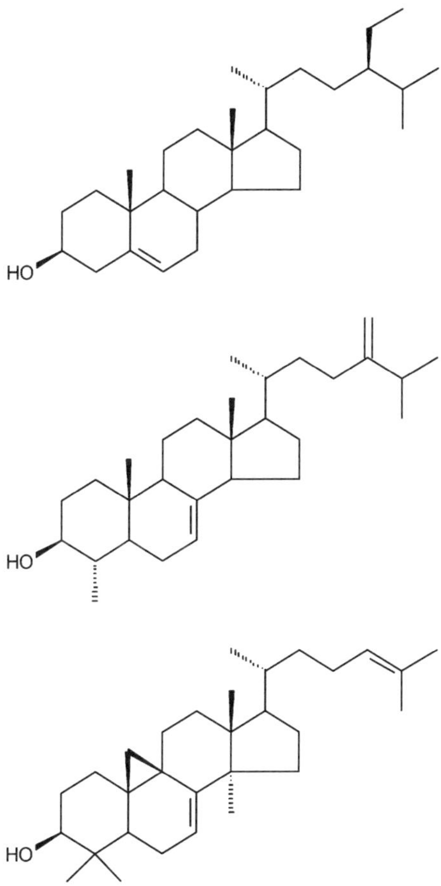

4-Desmethyl sterol

Sitosterol

4-Methyl sterol

Gramisterol

4,4-Dimethyl sterol

Cycloartenol

Figure 2. Molecular structures of examples from the three main sterol sub-classes: sitosterol, a 4-desmethyl sterol with no methyl group at C-4; gramisterol, a 4-monomethyl sterol; and cycloartenol, a 4,4-dimethyl sterol.

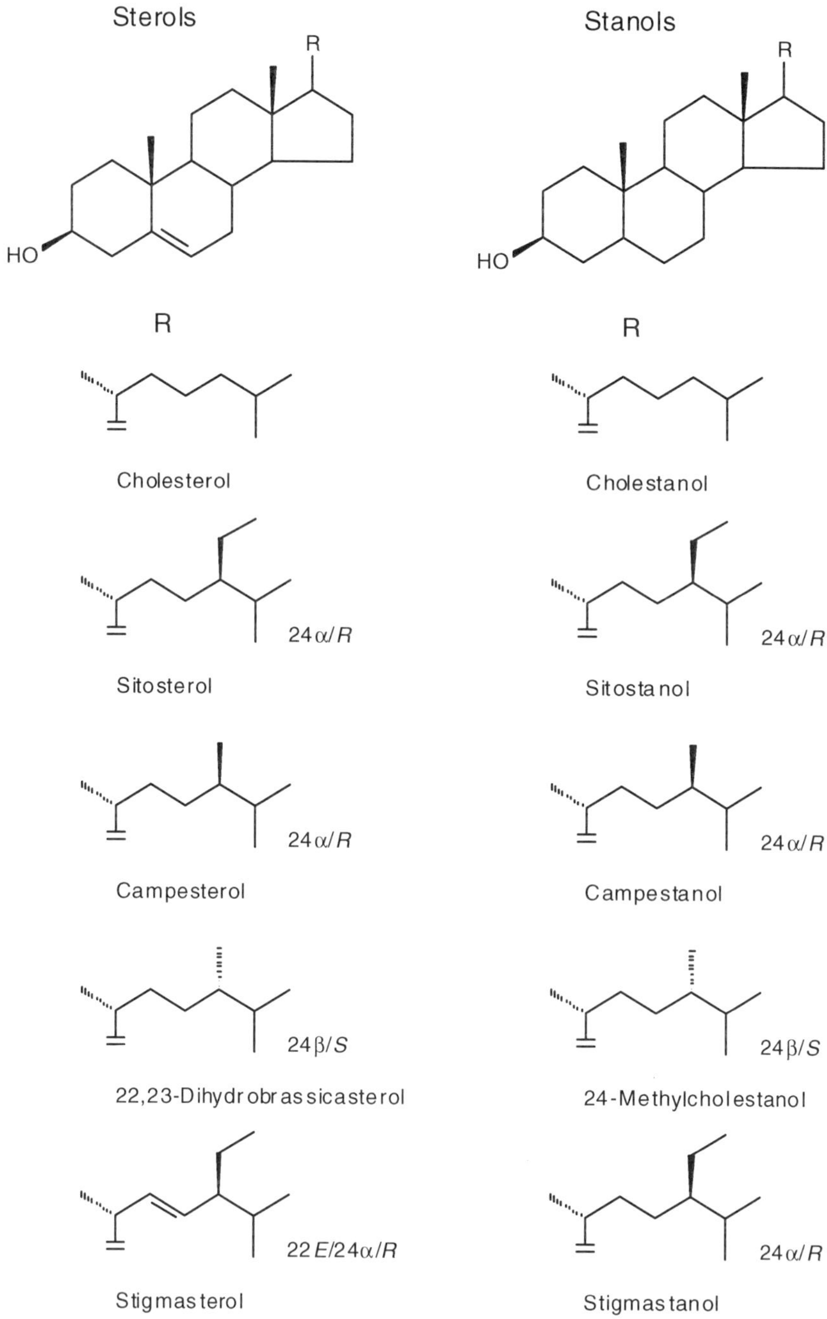

Figure 3. Molecular structures of the most important 4-desmethyl sterols and stanols.

Figure 3. Another abundant plant sterol is 24-methylcholesterol, frequently referred to as campesterol in the scientific literature. However, 24-methylcholesterol is usually a mixture two epimers: 24α-methylcholesterol (campesterol) and 24β-methylcholesterol (22,23-dihydrobrassicasterol). These epimers can be separated by high-performance liquid chromatography (HPLC) (Akihisa *et al.*, 1986; Chitwood and Patterson, 1991) and by capillary gas–liquid chromatography on long, very polar capillary columns (Thompson *et al.*, 1981). Furthermore, their ratio can be determined by ^{1}H or ^{13}C nuclear magnetic resonance (NMR) spectroscopy (Nes *et al.*, 1977; Nes and Parish, 1989). Thus 35% of "campesterol" in soybean sterols has been claimed to be 22,23-dihydrobrassicasterol (Wright *et al.*, 1978).

Cholesterol is also a 4-desmethyl sterol, and the similarity between its ring structure and the substitution at C-24 to those of some plant sterols (Figure 3) seem to be critical elements for the important effects of plant sterols on cholesterol absorption. Only 4-desmethyl sterols (not 4-methyl or 4,4-dimethyl sterols) exert this effect.

2. *4-Methyl sterols*

4-Methyl sterols occur naturally in only minor amounts as they are precursors of the desmethyl sterols. In commercial soybean sterols, 98.1% are 4-desmethyl sterols, 1.3% are 4-methyl sterols and 0.6% are 4,4-dimethyl sterols (Sierksma *et al.*, 1999). Compared with 4-desmethyl sterols, the number of different 4-methyl sterol compounds is small.

Kritchevsky *et al.* (1999) studied the effects of 4-methyl sterols from algae and of sitosterol on cholesterol metabolism in rats. He found that 4-methyl sterols (in an algal sterol blend containing 50% 4-methyl sterols and 40% desmethyl sterols) did not lower serum or liver lipid levels and confirmed the hypolipidemic properties of sitosterol.

3. *4,4-Dimethyl sterols*

4,4-Dimethyl sterols are also intermediates in plant sterol biosynthesis and can be found in only trace amounts in most plants. Some plant materials such as rice bran oil (see Chapter 4) have elevated levels of 4,4-dimethyl sterols such as cycloartenol and 24-methyl cycloartenol. The content of 4,4-dimethyl sterols in rice bran oil sterols varies, probably with the rice variety and processing conditions. According to de Deckere and Korver (1996), crude rice bran oil contains 1.8, 0.4 and 1.2 g/100 g oil of 4-desmethyl, 4-methyl and 4,4-dimethyl sterols, respectively, mainly as esters of ferulic and *p*-coumaric acids.

4. *Triterpenes*

The molecular structures of triterpene alcohols resemble those of sterols,

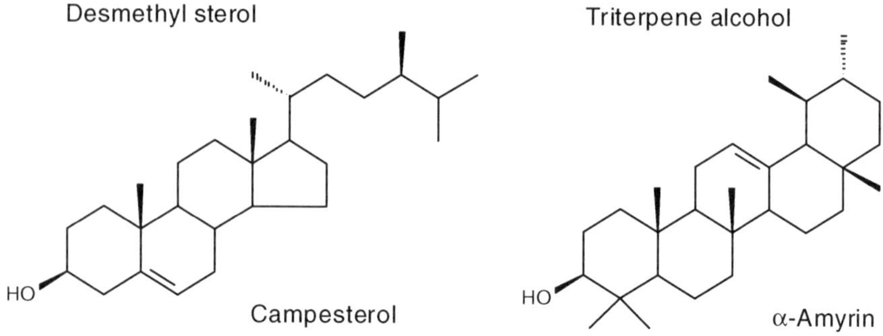

Figure 4. Molecular structures of campesterol (a 4-desmethyl sterol) and α-amyrin (a triterpene alcohol).

especially the 4,4-dimethyl sterols, so triterpene alcohols have also been studied for their possible effects on cholesterol absorption. In contrast to plant sterols with a tetracyclic structural unit, triterpene alcohols normally have a five-ring structure (Figure 4). Triterpene alcohols such as β-amyrin, butyrospermol, α-amyrin, and lupeol are present in high abundance in shea nut oil. Shea nut sterols contain approximately 80% triterpene alcohols, occurring mainly as cinnamates and acetates (Peers, 1977).

5. *Plant stanols – saturated sterols*

Plant stanols are saturated sterols lacking double bonds in the ring structure or the side chain (see Figure 3). The biosynthesis, biological function, and importance of plant sterols to human nutrition have been reviewed by Piironen *et al.* (2000a).

Small amounts of sitostanol and campestanol, which are 5,6-saturated analogues of the main 4-desmethyl sterols (sitosterol and campesterol), occur in plant materials such as coniferous trees (pine, spruce) and certain grains (rye, wheat and corn/maize).

6. *Effects of different sterol groups on cholesterol absorption*

Sitosterol and campesterol differ from cholesterol in the presence of an ethyl and methyl group respectively at C-24 in the side chain (Figure 3). These small differences have a considerable impact on the fate of these compounds in the digestive tract. The absorption of sitosterol, the most common plant sterol, is only 10% that of cholesterol, whereas campesterol (24-methylcholesterol) has a somewhat higher absorption. This was demonstrated in a study by Miettinen *et al.* (2000), in which the absorption of sterols was compared before and after

Table 2. Absorption of certain 4-desmethyl sterols of interest to functional foods before and during sitostanol ester margarine consumption[a]

Sterol	Absorption (%)	
	Before	During
Campesterol[b]	14.8	6.3
Sitosterol	4.5	2.1
Campestanol	2.3	0.2
Sitostanol	2.6	0.1

[a]Data taken from Miettinen *et al.* (2000). [b]Total absorption of campesterol and 22,23-dihydrobrassicasterol.

switching from a regular diet to a diet containing sitostanol ester margarine (Table 2). Hydrogenation of sitosterol to the corresponding sitostanol further reduces its absorption to very low levels.

Interestingly, only 4-desmethyl sterols seem to affect cholesterol absorption and to lower serum total and LDL cholesterol levels (Weststrate and Meijer, 1998; Sierksma *et al.*, 1999; Vissers *et al.*, 2000). This suggests that similarity of ring structure and of side chain length are crucial elements for the effects on cholesterol absorption. The absorption rate seems to depend both on saturation of the double bond in ring B, and the length of side chain substitution at C-24. Heinemann *et al.* (1993) found an inverse relation between the absorbability of plant sterols and their efficiency in inhibiting cholesterol absorption. As discussed by Vissers *et al.* (2000), the 4,4-dimethyl sterols tested have a side chain of equal length to that of cholesterol, whereas triterpene alcohols do not have a side chain. 4,4-Dimethyl sterols and triterpene alcohols have no cholesterol-lowering effect, although a triterpene alcohol blend surprisingly reduced serum carotenoid levels very efficiently (Weststrate *et al.*, 1998). This finding indicates that the effects of different plant sterol sub-classes on cholesterol absorption and on serum carotenoid levels are not linked, but most probably depend on different mechanisms of action. Similarly it was concluded that the cholesterol-lowering effect of a sterol blend obtained from rice bran oil, consisting of 46% 4,4-dimethyl and 48% 4-desmethyl sterols, was due to the 4-desmethyl component of the sterol blend only (Vissers *et al.*, 2000).

In nature, plant sterols occur in free form, as sterol fatty acid esters, as sterol glucosides, and as esterified sterol glucosides (Figure 5). However, most dietary plant sterols are in free or fatty acid ester forms. Additionally, minor amounts of plant sterols are present as esters of ferulic and *p*-coumaric acids in some plants like rice, corn (maize), and rye. Vegetable oils and fats are a major source of 4-desmethyl sterols, mainly in free and fatty acid ester forms, but other important dietary sterol sources, such as grains, also contain sterol glucosides and esterified sterol glycosides (Piironen *et al.*, 2000a).

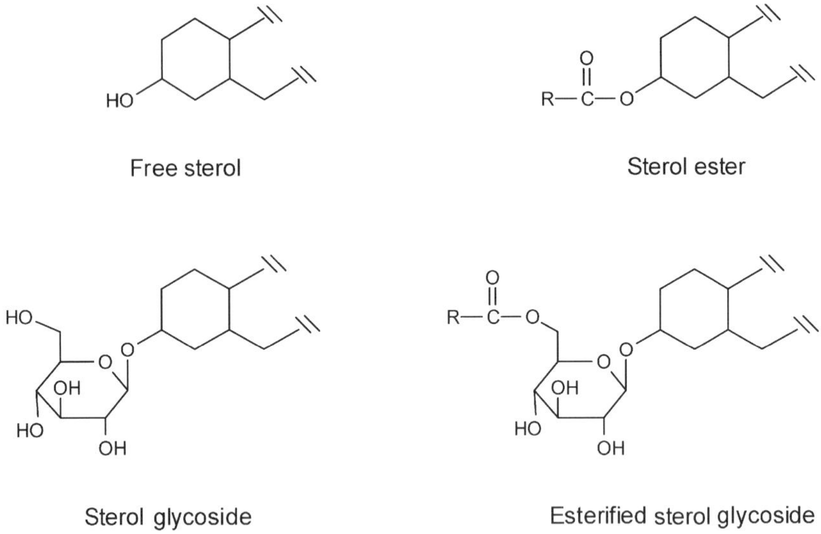

Figure 5. Partial molecular structures (showing ring A and its substituents) of a free sterol, sterol fatty acid ester, sterol glucoside, and esterified sterol glucoside.

C. Dietary sources of 4-desmethyl sterols

As a result of the commercial interest in plant sterols for food products, the sterol content of major food components has recently attracted attention (Piironen *et al.*, 2000a, 2000b; Normén *et al.*, 1999). Additionally the dietary intake of plant sterols has been estimated. Schothorts and Jekel (1999) estimated average sterol intakes in the Netherlands by analysing duplicate 24 h diet samples collected in 1994. The mean daily intakes of cholesterol and of total plant sterols were 202 mg and 146 mg respectively.

Phillips *et al.* (1999) analysed the plant sterol content of 9 experimental US diets and showed that dietary intakes of plant sterols increased with increasing intakes of polyunsaturated fatty acids and decreasing intakes of saturated fatty acids. In that study, average US diets were found to contain 223–273 mg cholesterol and 138–171 mg total plant sterols, whereas American Heart Association (AHA) National Cholesterol Education Program "Step 1" and "Step 2" diets contain 264 mg and 274 mg cholesterol, and 189 mg and 208 mg total phytosterols, respectively. Average Western intakes vary from study to study, and are summarized in Table 3.

The main sources of plant sterols in Western diets are vegetable oils, margarines prepared from vegetable oils, and grain products (Table 3) (Piironen *et al.*, 2000a, 2000b). The 5,6-saturated derivatives of the main plant sterols (i.e. campestanol and sitostanol) occur naturally in only trace amounts. The

Table 3. Average Western intake levels, sources, and metabolism of dietary sterols[a]

Parameter	Cholesterol	Plant sterols	Plant stanols
Dietary intake	300–500 mg/day	200–400 mg/day	10–60 mg/day
Dietary sources	Egg yolk, butter, dairy products, meat, crustaceans	Vegetable oils, corn (maize), beans, grains	Grains (wheat, rye, corn/maize)
Endogenous synthesis	800–1200 mg/day	Not synthesized	Not synthesized
Absorption	40–60%	5–15%	0.1–2%
Plasma concentration	140–320 mg/dl	0.3–1.7 mg/dl	0.01 mg/dl
Excretion	40–60%	85–95%	>98%

[a]Data taken from Piironen *et al.* (2000a, 2000b).

main dietary sources of stanols are wheat, rye, and corn (maize). The 4-desmethyl sterol fraction of wheat and rye typically contains 15–20% stanols. Tall oil, a commercial source of plant sterols (see next section), contains approximately 15–20% plant stanols in the 4-desmethyl sterol fraction.

Since the absorption of plant sterols and stanols is low, dietary intakes can be measured from feces. The fecal output – and thus the dietary intake – of the plant sterols sitosterol and campesterol during a control margarine diet was 375 mg/day in a Finnish study (Gylling and Miettinen, 1999b). The daily intake of sitostanol in Finnish diets amounts to 20–60 mg (Miettinen *et al.*, 1993), a major part being derived from grain products such as wheat and rye (Piironen *et al.*, 2000a, 2000b). The daily stanol (sitostanol plus campestanol) content of US diets is in the range of 12–23 mg (Phillips *et al.*, 1999).

D. Commercial sources of plant sterols

Currently plant sterols are derived on a commercial scale from vegetable oils and from wood. Sterol blends from both these sources have been extensively studied regarding their safety parameters and their cholesterol-lowering efficacy, both in animals and in humans. Free and esterified soy sterols, free tall oil sterols, and stanol esters from both sources have obtained GRAS ("generally recognized as safe") status in the USA.

Vegetable oil sterols are obtained as a by-product in the refining of soybean oil (Clark, 1996). In the final refining step of vegetable oils and fats (deodorization), volatile compounds are distilled from the vegetable oil under reduced pressure with steam injection, at high temperature. The purpose of this process is to obtain vegetable oil or fat that is bland in taste and has good oxidative stability. Depending on the conditions used, 15–30% of the tocopherols and free sterols are removed. Deodorizer distillates generally contain 10–30% unsaponifiables, containing 40% phytosterols and 15% tocopherols (Daguet, 2000). These distillates have long been used as a source of natural tocopherols (vitamin E). The sterol blend is a by-product in the isolation of natural tocopherols. It is purified, usually by crystallization from solvents, to fulfil the specifications outlined in GRAS documents and other critical safety evaluations. The potential supply of crude sterols from vegetable oil distillates is currently estimated to be 5000–6000 tonnes/year.

Another source of plant sterols is tall oil. This is a by-product of the wood pulp industry formed when coniferous woods are digested under alkaline conditions ("tall" means "pine" in Swedish, but "pine oil" has long been used as a commercial name for the lighter volatile terpene fractions of wood extraction). The fatty and resinous substances contained in the wood are brought into alkaline aqueous solution and are recovered as "skimmings". These are the sodium salts (soaps) of fatty and rosin acids, together with neutral substances such as plant sterols and long-chain fatty alcohols. After acidulation, the tall oil soap is processed by a sequence of distillations into different fatty acid and rosin acid products.

Sterols can either be extracted directly from the soap with different solvent mixtures or they can be recovered from the tall oil pitch remaining after the first distillation. The tall oil soap used has a high content of unsaponifiables, with the sterol content depending on the type of coniferous woods used. Typically, sterol levels in tall oil soaps are 3–5% (w/w). Most of the sterols are found in the pitch fraction remaining after distillation; the sterol content of this fraction can be as high as 10% (w/w) (Holmbom and Erä, 1978). Subsequent processing involves solvent extraction, short-path distillation, and selective recrystallization.

There remains an unused potential source of wood sterols in the side-streams of the wood pulping industry. At the end of 2001, the global production of wood-derived sterols was estimated to be 1500 tonnes/year. However, the maximum potential has been estimated to be 20 times higher. Constraining factors are the high capital investment, with sterols being, so far, the only product. In contrast, when sterols are recovered from vegetable oil distillates most of the capital investment has already been covered in the production of natural tocopherols.

Typical sterol compositions from commercial wood and from vegetable oils are shown in Table 4. The compositions of these two sterol blends differ although sitosterol is the principal sterol in both. Wood-derived sterols contain no 22,23-dihydrobrassicasterol and only low levels of stigmasterol. Wood-derived sterols contain relatively high levels of stanols, whereas the stanols from vegetable oils are produced mainly during partial or complete hydrogenation of the oils. The major vegetable oils, such as soybean, sunflower, and canola (low erucic acid rapeseed) oil do not contain stanols, while corn (maize) oil contains only minute amounts (Schuhmann and Scheller, 1996).

Table 4. Average sterol composition of commercial plant sterols[a]

	Relative content (%, w/w of total sterols)	
Sterol	**Wood-derived**	**Vegetable oil-derived**
Sitosterol	72	45
Campesterol	8.2	26.8[b]
Stigmasterol	0.3	19.3
Brassicasterol	0	1.6
Sitostanol	15.3	2.1
Campestanol	1.6	0.8
Other minor sterols	2.6	4.4

[a]Data based on sterol contents of 10 consecutive batches of commercial products delivered to Raisio Benecol Ltd (Raisio, Finland) in 1999. [b]Approximately 30% of which is estimated to be 22,23-dihydrobrassicasterol, based on ^{13}C-NMR.

Other potential sources of plant sterols are rapeseed sterols obtained from the production of fatty acid methyl esters for biodiesel fuel, and sterols contained in other food processing by-products such as maize fiber. However, these sterol blends do not currently have the required s ıfety and efficacy data for commercial use in functional foods.

E. Health effects of sterols

The cholesterol-lowering property of plant sterols was first recognized in the early 1950s when Pollak (1953) showed that dietary sitosterol significantly reduced serum cholesterol levels in humans. Since then, numerous studies on the cholesterol-lowering and other physiological effects of plant sterols and on the absorption of cholesterol and plant sterols have been published. Most of the studies are reported in reviews by Pollak and Kritchevsky (1981), Ling and Jones (1995), Cater (1999, 2000), Nguyen (1999), Miettinen and Gylling (1999), and Plat and Mensink (2001).

1. Cholesterol-lowering mechanism of plant sterols and stanols

During the process of absorption, cholesterol dissolved in the lipid core of micelles is transported from the lumen of the small intestine across the intestinal wall and into the lymph. Because the solubility of cholesterol in aqueous systems is low, its absorption depends on the formation of detergent structures (mixed micelles) in the small intestine. These are composed mainly of bile salts, phospholipids, digestion products of fats such as fatty acids and monoacylglycerols, cholesterol (of which 90% is in free form), and fat-soluble micronutrients (Figure 6).

The precise mechanism of cholesterol transport from the micelles into the

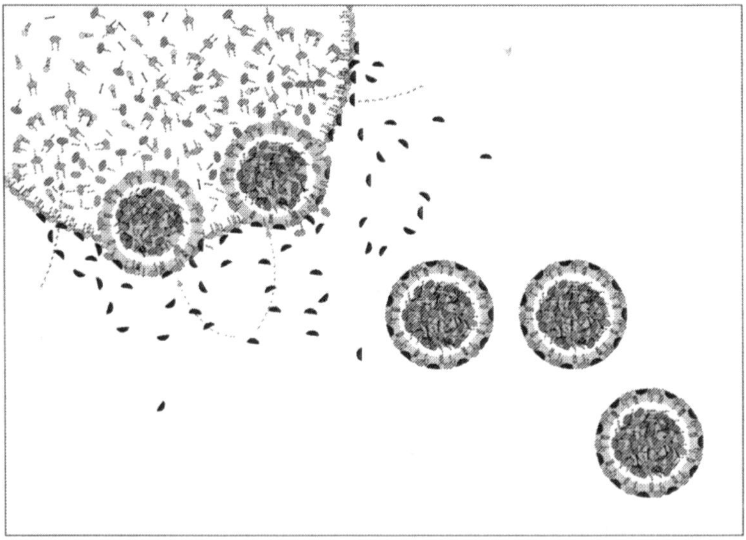

Figure 6. Because the solubility of cholesterol in aqueous systems is low, its absorption depends on the formation of detergent structures (mixed micelles) in the small intestine. Initially, when dietary fat enters the stomach and passes into the small intestine, it takes the form of relatively large lipid droplets (shown in grey). Bile acids (shown in black) reduce the surface tension in the lipid droplets, leading to the formation of smaller structures (mixed micelles). Mixed micelles consist of an outer "shell" of bile acids, monoacylglycerols, phospholipids and lysolecithin, and an inner core of digestion products of fats such as fatty acids, monoacylglycerols, cholesterol (of which 90% in free form), and fat-soluble micronutrients.

intestinal mucosal cells (enterocytes) is yet not fully understood. However, it is known that micelles are not absorbed intact, but that the various components are taken up by the enterocytes at different rates, probably involving both active (involving binding proteins) and passive transport mechanisms.

Of the total amount of cholesterol entering the digestive tract (1.0–1.5 g/day) approximately one-third is derived from the diet and two-thirds from bile. Plant sterols and stanols decrease serum cholesterol concentrations by interfering with cholesterol absorption (Figure 7). Dual-isotope studies have shown that, depending on the experimental set-up, the ingestion of plant stanols reduces cholesterol absorption efficiency by 50–80% (Heinemann *et al.*, 1991; Miettinen and Vanhanen, 1994; Gylling and Miettinen, 1994; Jones *et al.*, 2000). Plant sterols reduce cholesterol absorption by up to 50% when measured by the dual-isotope method (Jones *et al.*, 2000), isotope kinetic technique (Lees *et al.*, 1977), or intestinal perfusion (Heinemann *et al.*, 1991).

This decrease in absorption applies to both dietary and endogenously synthesized biliary cholesterol (Gylling *et al.*, 1997), which explains why plant sterols and stanols have an effect even in those consuming low-cholesterol

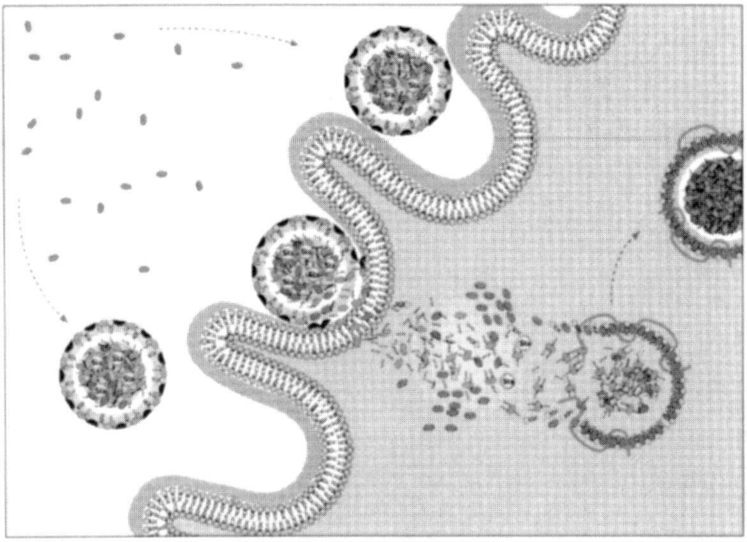

Figure 7. Mixed micelles carry the solubilized hydrophobic lipids through the aqueous digest medium into close proximity with the intestinal mucosal cells (enterocytes). Micelles are not absorbed intact, but the various components are taken up by the enterocytes at independent rates. Intracellular esterification of cholesterol by acyl-coenzyme A prevents its transport back to the intestinal lumen. After esterification, cholesterol esters are assembled into chylomicrons and secreted into the lymphatic system and, finally, to the blood circulation.

diets (Gylling *et al.*, 1997; Heinemann, *et al.*, 1991; Becker *et al.*, 1993). Indeed, Gylling *et al.* (1997) showed that the ingestion of 2.6 g/day of plant stanols in fatty acid ester form reduced total cholesterol absorption by 45%, from 7.37 mg/kg body weight when subjects consumed their normal home diet to 4.03 mg/kg body weight when using a stanol ester margarine. Dietary cholesterol absorption was reduced by 50% (from 1.53 to 0.76 mg/kg body weight) and biliary cholesterol absorption was reduced by 44% (from 5.84 to 3.27 mg/body weight).

Several mechanisms by which plant sterols might affect cholesterol absorption have been proposed (Pollak and Kritchevsky, 1981; Ling and Jones, 1995). The interference with intestinal cholesterol absorption by plant sterols is probably related to the close resemblance in the chemical structures of these molecules, although the precise mechanisms are not yet fully understood. At present, reduced micellar solubility of cholesterol is thought to be the main mechanism by which plant sterols reduce cholesterol absorption. Ikeda *et al.* (1989) elegantly proved that cholesterol solubility in mixed micelles is reduced by plant sterols and plant stanols, in studies on rats. For 10 days rats were fed a diet enriched with cholesterol (0.5%, w/w of diet) alone, or together with added sitosterol (0.5%) or sitostanol (0.5%). Sitosterol and sitostanol reduced the micellar solubility of cholesterol in the intestinal lumenal contents by 24% and 53%, respectively.

Another possible mechanism by which plant sterols and stanols may interfere with cholesterol absorption is that they reduce the amount of free cholesterol in the micellar phase by inhibiting the hydrolysis of cholesterol esters. Miettinen *et al.* (2001) showed that plant stanol esters reduced the amount of free cholesterol in the micellar phase by 34%, but increased the amount of esterified cholesterol in the micellar and oil phases from 10% to 35%.

Cholesterol esterase (EC 3.1.1.13) catalyses both the hydrolysis and esterification of sterols, and is the only pancreatic enzyme with high hydrolytic activity for cholesterol esters (Swahn *et al.*, 1997). The hydrolysis reaction equilibrium for cholesterol fatty acid esters under normal physiological conditions in the small intestine is driven towards complete hydrolysis (Rudd and Brockman, 1984). Miettinen *et al.* (2000) found that the degree of esterification of both cholesterol and plant sterols was about 10% in the intestinal excreta of patients with colectomies. The degree of esterification of cholesterol was similar whether patients consumed a control diet or a diet supplemented with a high-sterol spread, indicating that the presence of elevated amounts of plant sterols or stanols does not affect the hydrolysis equilibrium averaged over the whole passage of the small intestine. However, based on the data of Miettinen (2001), elevated intakes of esterified plant sterols do change the degree of esterification of cholesterol and other plant sterols in the upper part of the small intestine.

Plat *et al.* (2000) recently showed that 2.5 g/day of plant stanols as fatty acid esters consumed with a meal in one daily dose was as effective at reducing serum total and LDL cholesterol levels as the same amount divided into three doses taken throughout the day (Figure 8). This indicates that additional mechanisms at the absorptive sites of the enterocytes or within the enterocytes must play a role in the overall effect of plant sterols on cholesterol absorption. In an *in vitro* cell model, Field *et al.* (1997) showed that sitosterol was taken up by Caco-2 cells less efficiently than cholesterol. Furthermore, plant sterols and stanols are absorbed into the cell walls of the digestive tract in animals (Bhattacharyya and Lopez, 1979; Sanders *et al.*, 2000).

Irrespective of the actual mechanism behind the reduction of cholesterol absorption by plant sterols, solubilization in the emulsified fat phase of the food digest is a prerequisite for plant sterols to be incorporated into the micelles. Plant sterols need to be incorporated into mixed micelles or other liposome-type entities with micellar-type structure before they can enter into the enterocytes or their cell walls. Esterified sterols and stanols are fat-soluble and can directly enter and mix homogeneously into the fat phase of a food digest. On the other hand, the ability of free sterols and stanols to reduce cholesterol absorption is highly dependent on the extent to which they are first solubilized

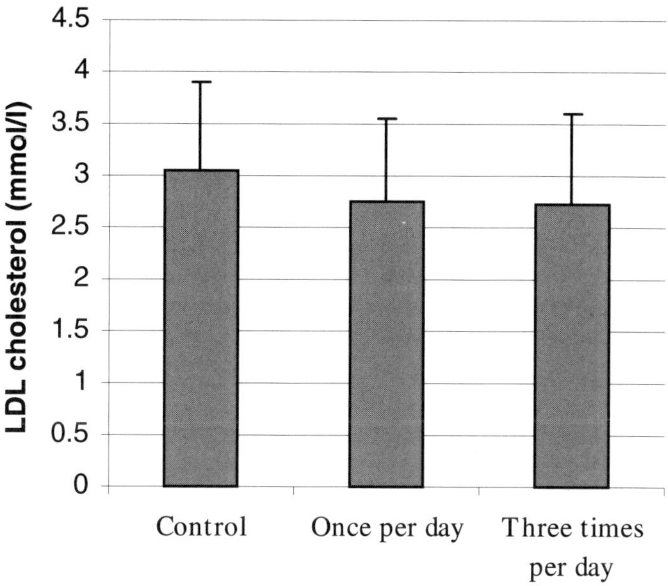

Figure 8. Plat *et al.* (2000) showed in a randomized, placebo-controlled trial that 2.5 g of plant stanol in ester form reduced serum total cholesterol and LDL cholesterol levels to the same extent irrespective of whether it was consumed as one daily dose or divided into three doses taken throughout the day.

in the emulsified fat phase of the food digest in the stomach. As a consequence, the physical form of the free sterols and stanols poses restrictions on the food matrix used to transport them into the digestive system. Indeed, data from clinical studies with free sterols shows that their efficacy is reduced when they are dispersed in protein-rich, low-fat food matrices such as low-fat yoghurts rather than, for example, fat-based spreads (Jones *et al.*, 1999; Volpe *et al.*, 2001). Thus the presence of fat in the food matrix and the way in which the free sterol or stanol is dispersed within that fat phase are of key importance. This was shown in a study comparing the impacts of crystalline and lecithin-micellized sitostanol on cholesterol absorption when ingested as part of a standardized breakfast (Ostlund *et al.*, 1999). In that study, 1000 mg of crystalline sitostanol reduced cholesterol absorption by only 11%, whereas 300 mg of sitostanol administered as lecithin micelles reduced cholesterol absorption significantly, by 34% (Figures 9 and 10). However, a diet high in fat might promote the efficacy of even crystalline free sterols.

The problem of availability for incorporation into the fat phase of the food digest does not arise with esterified plant sterols and stanols, since such materials, by definition, constitute fatty acid-containing lipids and will, therefore,

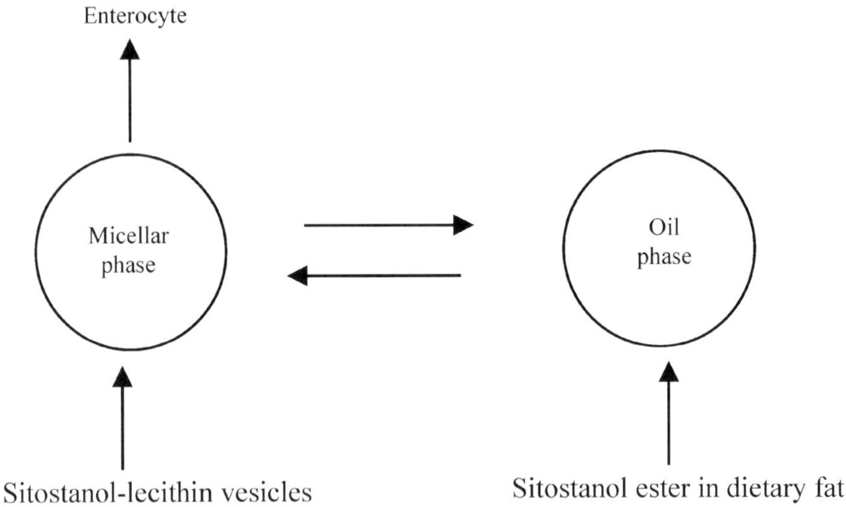

Figure 9. Interaction of sitostanol with intestinal phases. Sitostanol may be introduced directly into the intestinal micellar phase as a sitostanol–lecithin complex, or sitostanol ester may be dissolved in dietary fat for entry into the oil phase. Unesterified or non-complexed sitostanol does not readily enter either the oil or the micellar phase. Taken from Ostlund *et al.* (1999). Reproduced with permission by the *American Journal of Clinical Nutrition*. © Am. J. Clin. Nutr. American Society for Clinical Nutrition.

be effectively incorporated into the fat phase of the food digest. This also explains why fatty acid esters of plant sterols and stanols are effective independent of the food matrix. The use of fatty acid esters of plant sterols and stanols depends on their efficient hydrolysis by pancreatic cholesterol esterase (EC 3.1.1.13) to release the free sterols and stanols, the form that actively reduces cholesterol absorption at the site of the forming micelles.

The optimal effect is achieved when plant sterols and stanols are ingested as part of a meal, preferably consumed as a food. To date there is no scientific evidence that plant sterols, either in unesterified or esterified form, would be effective if consumed between meals or under fasting conditions. Additional studies are needed to find out to what extent digestible macrocomponents such as fats and proteins are needed for the effective action of both esterified and unesterified sterols. Optimal effects are probably obtained when sterol-enriched foods are consumed as part of the main meal. More research, however, is needed on this issue.

With decreased cholesterol absorption from the intestine, the liver increases both the synthesis of cholesterol from its precursors (Becker *et al.*, 1993; Gylling *et al.*, 1997; Jones *et al.*, 2000) and the uptake of cholesterol by increasing the expression of LDL receptors (Plat, 2001). This further reduces the amount of cholesterol circulating in the bloodstream.

Compared with cholesterol, plant sterols are poorly absorbed, which suggests

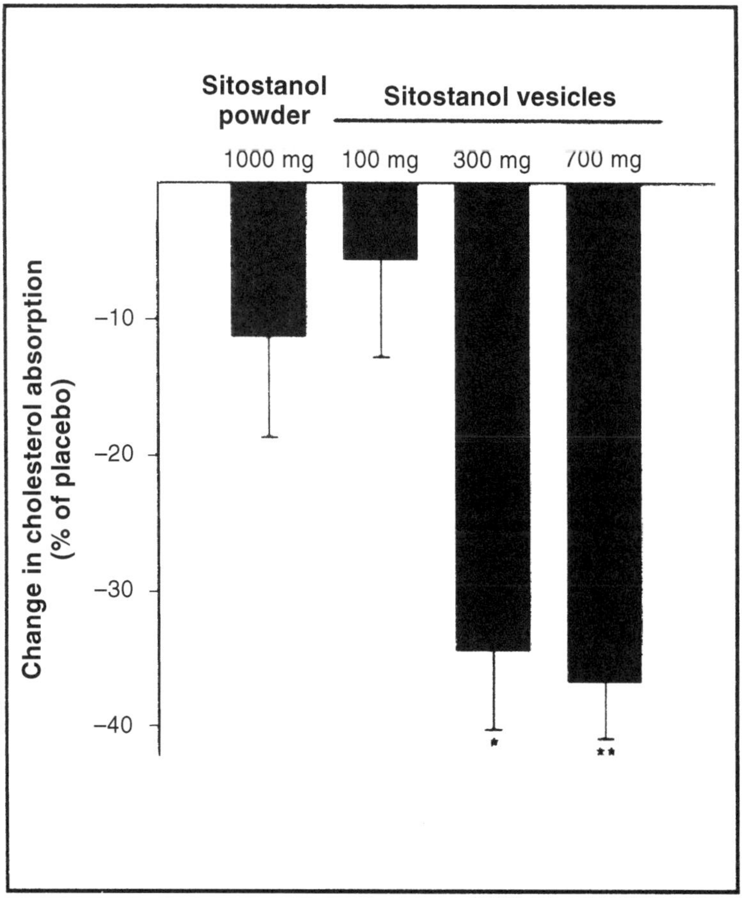

Figure 10. Sitostanol administered in lecithin micelles (vesicles) is more effective than crystalline (powdered) lecithin at reducing cholesterol absorption. *,**Significantly different from placebo: *P=0.01, **P=0.003. Taken from Ostlund *et al.* (1999). Reproduced with permission by the *American Journal of Clinical Nutrition*. © Am. J. Clin. Nutr. American Society for Clinical Nutrition.

that the enterocytes differentiate between their molecular structures. The absorption of plant stanols is even lower than the absorption of sterols. The reason for this difference is still not clear. However, it is known that plant sterols and stanols are poorly esterified by the intestinal enzyme acyl-coenzyme A (CoA) cholesterol acyltransferase (ACAT) (Field and Mathur, 1983). Since esterification is needed for effective incorporation into chylomicrons, the absence of an efficient esterification mechanism may explain the poor absorption of plant sterols from enterocytes into the lymph. The higher esterification rate of campesterol compared with sitosterol and stigmasterol might explain the higher absorption rate of campesterol (Bhattacharyya, 1981).

2. *Early studies of phytosterols*

The cholesterol-lowering properties of phytosterols were first demonstrated in the early 1950s by Peterson (1951), who fed cholesterol and plant sterols to chicks and found that soy sterols inhibited the increase in cholesterol concentration otherwise induced by cholesterol feeding. Shortly afterwards, Pollak (1953) showed the same effect in humans, to whom he administered 5–10 g of crude sitosterol daily for up to 8 months. In similar studies in rabbits, he observed that sitosterol was poorly absorbed and that, if present in excess, it blocked the absorption of cholesterol and prevented atherosclerosis. Poor absorption and lack of any endogenous synthesis of sitosterol in humans were subsequently demonstrated by Salen *et al.* (1970).

In the first human studies, free non-esterified sitosterol was administered as powder, granules, or liquid suspension with daily intakes of 12–18 g. These studies showed 6–17% decreases in serum cholesterol concentrations with wide individual variations (Farquhar *et al.*, 1956; Kalliomäki *et al.*, 1957; Lees *et al.*, 1977; Schlierf *et al.*, 1978), but resulted in marked increases in plant sterol levels in plasma – especially of campesterol, which is absorbed better than sitosterol. This raised safety concerns in light of the description of sitosterolemia two years previously. Sitosterolemia is a recessively inherited disorder characterized by excessive absorption and high plasma levels of plant sterols, resulting in the premature development of atherosclerosis.

In 1976 and 1977 Sugano *et al.* reported that β-sitostanol had a greater hypocholesterolemic effect than β-sitosterol in cholesterol-fed rats. Similar results were obtained with rabbits (Ikeda *et al.*, 1981). In addition, sitostanol decreased the formation of aortic atheromas caused by cholesterol feeding more than sitosterol. Similar results have recently been reported in animal models, where the cholesterol-lowering effect of plant sterol blends with different compositions in cholesterol-challenged animals were shown to be dependent on the sitostanol content (Ntanios and Jones, 1998; Ntanios *et al.*, 1998a, 1998b).

The first description of the use of plant stanols to lower plasma cholesterol levels in humans was by Heinemann *et al.* (1986) in a small uncontrolled study. They showed that the administration of capsules of sitostanol dispersed in monoacylglycerol and sunflower oil at a dose of 1.5 g/day lowered LDL cholesterol levels by 15%. Similar concentrations of sitosterol and sitostanol infused directly into the small intestine decreased cholesterol absorption by 50% and 85%, respectively (Heinemann *et al.*, 1991). Becker *et al.* (1993) obtained impressive results with low-dose sitostanol in an uncontrolled study with 9 children suffering from familial hypercholesterolemia. LDL cholesterol levels decreased by 33% when the children consumed 1.5 g sitostanol daily, and by 20% when they consumed 6 g sitosterol daily, suggesting that sitostanol is more effective than sitosterol at lowering LDL cholesterol levels.

In contrast to these studies, an uncontrolled study by Denke (1995) showed no significant lowering of LDL cholesterol levels in response to free sitostanol incorporated in capsules and taken at a dose of 3 g/day. This negative outcome was useful in drawing attention to the importance of the physical state of plant sterols and stanols in determining their efficacy.

Since free plant sterols can be used in margarines at a concentration of up to 3% (w/w) without causing quality problems (Sierksma *et al.*, 1999), an average margarine consumption of 25 g/day would provide only 0.75 g of plant sterol (and even less stanol). This limitation has been the main stimulus for the development of the more lipid-soluble fatty acid esters of phytosterols and phytostanols.

3. *Clinical studies with unesterified and esterified sterols and stanols*

Only a few clinical studies have been published on the effects of low-dose unesterified sterols or stanols. Miettinen and Vanhanen (1994) reported no effect (versus control) with free sterols. A study by Pelletier *et al.* (1995), in which free sterols were incorporated into butter, was a crossover study with no wash-out period. It found that plasma LDL cholesterol levels were 15% lower in those with a daily intake of 740 mg soy sterols in 50 g butter than in those consuming a "control" butter. Unfortunately, however, the "control" butter contained 18% more cholesterol than the sterol-enriched butter, and the paper reported no baseline values that would have allowed independent calculation of the relative serum cholesterol changes induced by the butter alone and by the butter plus free sterols.

Sierksma *et al.* (1999) reported that plasma total and LDL cholesterol levels were 3.8% and 6% lower, respectively, in those consuming a spread containing 0.8 g free sterols daily than in those consuming a control spread. Volpe *et al.* (2001) incorporated soy sterols into a low-fat yoghurt-based drink. With 1 g of sterols, serum total and LDL cholesterol levels were reduced (but apparently not significantly) by 4.4% and 6.2%, respectively, compared with a control. In an uncontrolled follow-up study with 2 g of sterols, serum total and LDL cholesterol levels decreased by 11.2% and 15.6%, respectively, from baseline levels. Tikkanen *et al.* (2001) recently studied the effects of free sterols incorporated into various low-fat food products, such as yoghurt, meat, and jam. Serum LDL cholesterol levels were reduced by 8% compared with a control group when sterol doses of up to 5 g/day were used.

Esterified soy sterols and free stanols were compared by Nestel *et al.* (2001) in a study in which the sterols (2.5 g/day) were added to breakfast cereal, bread, and margarine. Plasma median total cholesterol levels decreased by 8.5% with sterol esters and by 3.5% with the free stanol, and plasma median LDL cholesterol levels decreased by 13.6% and 8.5% with sterol esters and free stanol, respectively.

During the past decade, major interest has focused on plant stanol and sterol esters and their efficacy in reducing serum LDL cholesterol levels. Hypocholesterolemic effects have been reported:

- in normocholesterolemic individuals (Niinikoski *et al.*, 1997; Weststrate and Meijer, 1998; Hendriks *et al.*, 1999; Plat *et al.*, 2000; Plat and Mensink, 2000);
- in individuals with mild to moderate hypercholesterolemia (Vanhanen *et al.*, 1993, 1994; Miettinen *et al.*, 1995; Weststrate and Meijer, 1998; Andersson *et al.*, 1999; Hendriks *et al.*, 1999; Hallikainen and Uusitupa, 1999; Hallikainen *et al.*, 2000b) or with familial hypercholesterolemia (Vuorio *et al.*, 2000);
- in women with coronary heart disease (Gylling *et al.*, 1997);
- in men with non-insulin-dependent diabetes (Gylling and Miettinen, 1994) and in conjunction with cholesterol-lowering statin therapy (Blair *et al.*, 2000; Neil *et al.*, 2001);
- in normocholesterolemic children (Tammi *et al.*, 2000) and in children with slightly elevated cholesterol levels or with familial hypercholesterolemia (Gylling *et al.*, 1995; Vuorio *et al.*, 2000).

Controlled studies have been carried out using crossover or parallel study designs. Their duration has mainly been short, typically 4–8 weeks, except for the North Karelia Stanol Ester Study (Miettinen *et al.*, 1995) and the TNO Study on Sterol Ester Spreads (Brink and Hendriks, 2000), each of which lasted for 52 weeks.

Clinical studies have shown that the minimum amount of sterols and stanols required to produce a significant serum cholesterol-lowering effect is about 0.8–1.0 g/day. However, recent studies have focused on higher intake levels of 2–3 g/day of plant sterols and stanols in their esterified, fat-soluble forms. In many studies these amounts have been shown to provide the close-to-maximum cholesterol-lowering effect, but they are much higher than the typical daily intake from average diets. In populations consuming typical Western diets, the intake of sitosterol is about 150–350 mg/day and that of stanols is about 50 mg/day (Miettinen *et al.*, 1990). Vegetarians consume about 500 mg/day of sitosterol and about 60 mg/day of campesterol. Thus, the usual daily intake of plant sterols and stanols is generally thought to be so low that it does not have a significant impact on serum cholesterol levels, although a recent study has shown that plant sterol intake from regular dietary sources such as edible oils may also have an effect on cholesterol absorption (Ostlund *et al.*, 2002).

Table 5 summarizes data from published randomized double-blind controlled trials testing foods containing stanols or sterols. The stanols and sterols were esterified, added to margarine (or occasionally mayonnaise, olive oil, butter or yoghurt), and used alongside a placebo margarine or other foods to make the

Table 5. Effects of esterified plant sterols and stanols on serum total cholesterol and LDL cholesterol levels in adults

Reference	*n*	Sterol/ stanol ester	Dose (g/day)	Duration of study (weeks)	% Change in cholesterol: Total	% Change in cholesterol: LDL
Vanhanen *et al.* (1993)	67	Stanol	3.1	6	–7	–9
Miettinen and Vanhanen (1994)	31	Stanol	0.7–0.8	9	NS	–8
Vanhanen *et al.* (1994)	15	Stanol	0.7–0.8	9	NS	NS
	15	Stanol	1.8–1.9	6	–9	–15
Gylling and Miettinen (1994)	11	Stanol	3	6	–6	–9
Miettinen *et al.* (1995)	153	Stanol	1.8	24	–9	–9
	153	Stanol	2.6	52	–10	–13
Gylling and Miettinen (1996)	8	Stanol	3	7	–11	–14
Niinikoski *et al.* (1997)	24	Stanol	2.2	5	–10	–13
Gylling *et al.* (1997)	22	Stanol	3	7	–8	–15
Weststrate and Meijer	95	Stanol	2.5	3.5	–7	–13
(1998)	95	Sterol	3.3	3.5	–8	–13
Ayesh *et al.* (1999)	21	Sterol	8.6	3 or 4	–18	–23
Gylling and Miettinen	23	Stanol	3.18	6	–4	–8
(1999a)	23	Stanol	3.16	6	–6	–10
	21	Stanol	2.43	5	–8	–12
Hallikainen and Uusitupa	55	Stanol	2.31	8	–10.6	–13.7
(1999)	55	Stanol	2.16	8	–8.1	–8.6
Hendriks *et al.* (1999)	80	Sterol	0.83	3.5	–5	–7
	80	Sterol	1.61	3.5	–6	–9
	80	Sterol	3.24	3.5	–7	–10
Andersson *et al.* (1999)	61	Stanol	2	8	–7	–7
	61	Stanol	2	8	–9	–12
Nguyen *et al.* (1999)	318	Stanol	2.2	8	–5	–5
	318	Stanol	2.7	8	–7	–10
Hallikainen *et al.* (2000b)	22	Stanol	0.8	4	–2.8	–1.7
	22	Stanol	1.6	4	–6.8	–5.6
	22	Stanol	2.4	4	–10.3	–9.7
	22	Stanol	3.2	4	–11.3	–10.4
Hallikainen *et al.*	34	Stanol	2.0	4	–9.2	–12.7
(2000a)	34	Sterol	2.0	4	–7.3	–10.4
Jones *et al.* (2000)	15	Sterol	1.96	3	–9	–13
	15	Stanol	1.57	3	NS	–6
Miettinen *et al.* (2000)	11	Stanol	2	1	–16	–14
Plat and Mensink (2000)	112	Stanol	3.8	8	–8	–13
	112	Stanol	4.0	8	–8	–13
Plat *et al.* (2000)	39	Stanol	2.47	4	–6	–9
	39	Stanol	2.46	4	–7	–10
Brink and Hendriks (2000)	185	Sterol	1.6	52	–5	–7
Maki *et al.* (2001)	219	Sterol	1.1	5	–5.2	–7.6
	219	Sterol	2.2	5	–6.6	–8.6
Davidson *et al.* (2001)	84	Sterol	3	8	NS	NS
	84	Sterol	6	8	NS	NS
	84	Sterol	9	8	NS	NS
Mensink *et al.* (2002)	60	Stanol	3	4	–9	–14

n, Number of study participants. NS, Not statistically significant.

Table 6. Effects of plant sterols and stanols on serum LDL cholesterol levels in children

Reference	*n*	Mean age (range)	Sterol/stanol	Dose (g/day)	Duration of study (weeks)	% Reduction in serum LDL cholesterol
Becker *et al.* (1992)[a]	7	8 (5–11)	Free sitosterol	6.0	12	18*
Becker *et al.* (1993)[a]	9	12 (10–15)	Free sitosterol	6.0	12	19*
			Free sitostanol	1.5	12	33*
			Free sitostanol	1.5	28	29*
Gylling *et al.* (1995)	14	9 (2–15)	Sitostanol ester	3.0	6	15
Williams *et al.* (1999)	19	4 (2–5)	Sitostanol ester	3.0	4	16
Vuorio *et al.* (2000)[a]	24	9 (3–13)	Sitostanol ester	2.2	12	18
Tammi *et al.* (2000)	81	6	Sitostanol ester	1.5	12	7.5**

[a]Uncontrolled studies. *n*, Number of study participants. *,**Statistically significant: *$p<0.05$, **$p<0.01$.

trials double-blind. The results of clinical studies carried out with children are summarized in Table 6.

Three clinical studies have attempted to compare the cholesterol-lowing efficacy of esterified soy sterols and stanols (Table 7). Weststrate and Meijer (1998) compared a soy sterol ester margarine and a stanol ester margarine in a setting where the two spreads differed in their amount of total sterols, the degree of esterification of the sterols, and in their fatty acid composition. In that study, similar reductions were found in serum total and LDL cholesterol concentrations between the groups using the two spreads. In a small crossover study, Jones *et al.* (2000) showed that a plant sterol ester spread (delivering 1.87 g sterols per day) reduced plasma LDL cholesterol concentrations more efficiently than a stanol ester spread (1.56 g stanols per day) did. Subjects were randomized to one of 6 predetermined Latin squares, each of which included 3 sequenced test periods and 3 subjects. Interestingly, the effects of plant sterol and stanol esters as a function of time differed from the responses seen in other studies: the cholesterol-lowering effect was only observed after 2–3 weeks, whereas previous studies had shown evidence of such an effect after 1–2 weeks. Finally, Hallikainen *et al.* (2000a) carried out a comparative study that is the only one to date to use study margarines of equal fatty acid composition and containing almost equal amounts of sterols and stanols. The effects of a stanol ester spread (containing 2.02 g/20 g total sterols) and a sterol ester spread (containing 2.06 g/20 g total sterols) were compared in 34 mildly hypercholesterolemic subjects. Total serum cholesterol concentrations reduced by 9.2% and 7.3%, and LDL cholesterol concentrations by 12.7% and 10.4% with the stanol and sterol spreads, respectively.

Table 7. Studies comparing the serum cholesterol-lowering effects of esterified sterols and stanols

Reference	Study design	*n*	Duration of study (weeks)	Spread dose	Sterol/stanol content of spread [S:M:P][a]	Change in serum total cholesterol[b]	Change in serum LDL cholesterol[b]
Weststrate and Meijer (1998)	Randomized, double-blind	95	3.5	30 g/day	Sterol spread [1:2:3] 3.2 g/100 g sterols 0.1 g/100 g stanols	−8.3%	−13%
	Incomplete Latin square				Stanol spread [1:2:1] 0.5 g/100 g sterols 2.2 g/100 g stanols	−7.3%	−13%
	5 treatments 4 periods				Control spread [1:2:2] 0.1 g sterols		
Jones *et al.* (2000)	Randomized double-blind crossover design	15	3	23 g/day	Sterol spread [1:2:2] 1.87 g/100 g sterols 0.05 g/100 g stanols	−9.1%	−13.2%
	Latin square				Stanol spread [1:2:2] 1.56 g/100 g stanols 0.02 g/100 g sterols	−5.5%	−6.4%
	3 test margarines 3 diets				Control spread [1:2:2] <0.1 g/100 g sterols <0.1 g/100 g stanols		
Hallikainen *et al.* (2000a)	Randomized double-blind repeated-measures design	34	4	20 g/day	Sterol spread [1:2:1] 1.98 g/100 g sterols 0.09 g/100 g stanols	−7.3%	−10.4%
	Latin square				Stanol spread [1:2:1] 1.92 g/100 g stanols 0.10 g/100 g sterols	−9.2%	−12.7%
	3 test margarines 3 periods				Control spread [1:2:1] <0.1 g/100 g sterols 0 g/100 g stanols		

[a][S:M:P] indicates the saturated:monounsaturated:polyunsaturated fatty acid ratio in the spreads. [b]Change in cholesterol concentrations calculated in comparison with control spreads. *n*, Number of study participants.

4. *Effects of background diet*

Reductions in serum LDL cholesterol levels attributable to a given dose of stanol esters appear to be of similar magnitude regardless of whether the fat content of the background diet is high (Vanhanen *et al.*, 1993; Miettinen *et al.*, 1995) or low (Hallikainen and Uusitupa 1999; Hallikainen *et al.*, 2000a), whether the ratio of polyunsaturated to saturated fatty acids is low (Miettinen *et al.*, 1995) or high (Vanhanen *et al.*, 1993; Hallikainen and Uusitupa 1999), and whether the intake of dietary cholesterol is greater (Miettinen *et al.*, 1995) or less than 200 mg/day (Hallikainen and Uusitupa, 1999).

A similar effect is expected to be true of sterol esters. The combined effect of adding stanol esters and increasing the ratio of polyunsaturated to saturated fatty acids in the diet was a 19% net reduction in serum LDL cholesterol levels, while adding stanol esters to the usual diet yielded a 12% reduction (Andersson *et al.*, 1999; Hallikainen *et al.*, 2000a). Thus, consumption of a spread with added sterol or stanol esters probably further enhances the cholesterol-lowering efficacy of a diet with an improved fatty acid profile.

5. *Additive effect with cholesterol-lowering medication*

Of the cholesterol-lowering medications available to date, statins are currently the first choice for treating patients with hypercholesterolemia. They lower serum LDL cholesterol concentrations by 24–50% (Knopp, 1999) by blocking the synthesis of cholesterol through inhibition of the action of 3-hydroxy-3-methyl-glutaryl coenzyme A (HMG-CoA) reductase in the liver.

In a recent study by Blair *et al.* (2000), 167 hypercholesterolemic subjects already being treated with statins added 3 g of plant stanols as esters to their daily diet. A regular canola oil-based spread was used as a placebo treatment. The placebo reduced serum total and LDL cholesterol concentrations by 5% and 7%, respectively. Adding stanol esters to the spread further reduced total and LDL cholesterol concentrations by an additional 7% and 10%, respectively. Thus, in patients already on statin medication, adding a stanol ester-enriched spread to the diet resulted in a 12% reduction in total cholesterol and a 17% reduction in LDL cholesterol concentrations.

In patients with non-insulin-dependent diabetes, 40 mg/day of pravastatin decreased endogenous cholesterol synthesis and, consequently, serum LDL cholesterol concentrations by 38%. Adding 3 g plant stanols as esters to the diet resulted in a further decrease (up to 44%) in LDL cholesterol concentrations (Gylling and Miettinen, 1996). Similarly, Neil *et al.* (2001) showed that the combination of a sterol-containing margarine with statin treatment leads to additional improvements in serum lipid levels.

These studies show that the combination of sterols or stanols with statins leads to a greater reduction in serum total and LDL cholesterol concentrations than statins alone – or, alternatively, allows a lowering of the statin dose.

6. *Dose–response of esterified plant sterols and stanols*

Most of the clinical studies on sterol esters and stanol esters have been carried out with typical doses of about 1–3 g/day. In combination, these data support the notion that an optimal serum LDL cholesterol-lowering effect is achieved at an intake in the region of 2 g/day, and that increasing the dose beyond this does not yield significant improvements. It is important to note that the LDL cholesterol-lowering effect obtainable with a small daily dose of <1 g sterols or stanols per day (4–6%) can be statistically significant in well-controlled clinical studies with large numbers of subjects. But this order of reduction is relatively small and easily confounded by the normal day-to-day variation (±10%) in serum cholesterol levels. Thus, it has been argued that there is no scientific justification for using daily doses of sterol or stanol esters that are either <1 g or >3 g.

The cholesterol-lowering efficacy of sterols (Hendriks *et al.*, 1999; Maki *et al.*, 2001) and stanols (Vanhanen *et al.*, 1994; Wester, 1999; Hallikainen *et al.*, 2000b) appears to be dose-dependent.

Two published studies have examined the effects of a stanol ester intake of <1 g stanol/day, but neither showed any significant decrease in serum LDL cholesterol levels compared with placebo (Vanhanen *et al.*, 1994; Hallikainen *et al.*, 2000b). Hallikainen *et al.*'s study is the only dose–response study on stanol esters published so far. Significant decreases in LDL cholesterol concentration compared with placebo were observed for doses of 1.6 g stanol/day or above, but no further meaningful reduction in LDL cholesterol concentration was seen when the dose was increased from 2.3 g/day to 3.2 g/day. At a stanol dose of 0.8 g/day a significant reduction was seen in the serum apolipoprotein B concentration, a sensitive marker of reduced levels of atherogenic lipoproteins.

Maki *et al.* (2001) evaluated the effects of two doses (1.1 g and 2.2 g sterols) of sterol ester incorporated into a spread. These led to reductions in total serum cholesterol concentrations of 5.2% and 6.6%, and in LDL cholesterol concentrations of 7.6% and 8.4%, respectively.

Davidson *et al.* (2001) analysed daily plant sterol ester doses of 0 g, 3 g, 6 g, and 9 g sterol for safety measures. During the 8-week intervention, no adverse effects were noted. Surprisingly, serum lipid responses were modest. With the dose of 9 g/day, only non-significant reductions of 5.8% and 9.0% were seen for serum total and LDL cholesterol concentrations, respectively.

7. *Short-term and long-term efficacy*

Most of the published studies have been short-term and have involved the consumption of sterol or stanol esters for 1–2 months (Table 5). Miettinen *et al.* (2000) showed that the effect on serum LDL cholesterol levels is exerted within 1 week of starting the daily consumption of stanol esters. Nguyen *et al.* (1999) and Hallikainen and Uusitupa (1999) concluded that most of the reduction was already apparent after 2 weeks with stanol ester spreads. In addition, Plat and Mensink (2000) evaluated the effects of high- and low-campestanol ester spreads in an 8-week study and showed that reductions in total and LDL cholesterol concentrations after 4 weeks and 8 weeks were virtually identical and already maximal at 4 weeks. However, in the North Karelia Stanol Ester Study (Miettinen *et al.*, 1995), the effect of stanol esters (2.6 g/day) on LDL cholesterol levels was greater at 12 months than at 6 months. This difference may have been due to a selection bias, since only half the subjects remained on that dose for the second 6 months. The lack of significant waning of the effect on LDL cholesterol levels in any of the studies is noteworthy, implying the absence of any long-term counteracting influence resulting from increased endogenous cholesterol synthesis.

As with the North Karelia Stanol Ester Study, a 12-month study on sterol esters incorporated into margarine was conducted recently in the Netherlands (Brink and Hendriks, 2000; Hendriks *et al.*, 2001). Adults (185) consumed 1.6 g/day of soy sterols as esters incorporated into margarine. Serum total and LDL cholesterol concentrations reduced rapidly (within 3 weeks) by 5% and 7%, respectively, and remained at the same level for the duration of the experiment. However, surprisingly, the reductions in cholesterol levels were only seen in the male and not in the female participants (Figure 11).

8. *Frequency of intake*

Most studies have investigated the effects of plant stanols ingested in two or, more usually, three divided doses each day together with meals. This has been based on the notion that stanols need to be present in the intestinal lumen postprandially in order to displace cholesterol from mixed micelles and thereby inhibit its absorption.

Plat *et al.* (2000) recently compared, over 4 weeks, the effects of a margarine-based daily dose of stanol ester (2.5 g stanol) given in one single or three divided doses. As discussed above (see Figure 8), the decrease in LDL cholesterol concentration on the single-dose regimen (9.4%) did not differ significantly from that on the three divided doses (10.4%). The persistence of the cholesterol-lowering effect from the single dose suggests that stanols may not only compete with cholesterol for micellar solubilization, but may also have an additional, longer-lasting effect, possibly involving binding to a cholesterol transporter at the surface of intestinal mucosal cells.

9. *Effects on plasma plant sterols*

The ingestion of plant sterols and stanols also interferes with the absorption of sterols other than cholesterol by displacing them from the mixed micelles. The more hydrophobic sterol molecules tend to displace the more hydrophilic molecules from the micelles. Sitostanol decreases the absorption and thus also the serum concentrations of sterols (Vanhanen and Miettinen 1992; Gylling *et al.*, 1997, 1999; Miettinen *et al.*, 2000). Whether these changes have any clinical significance probably depends on whether plasma plant sterol levels

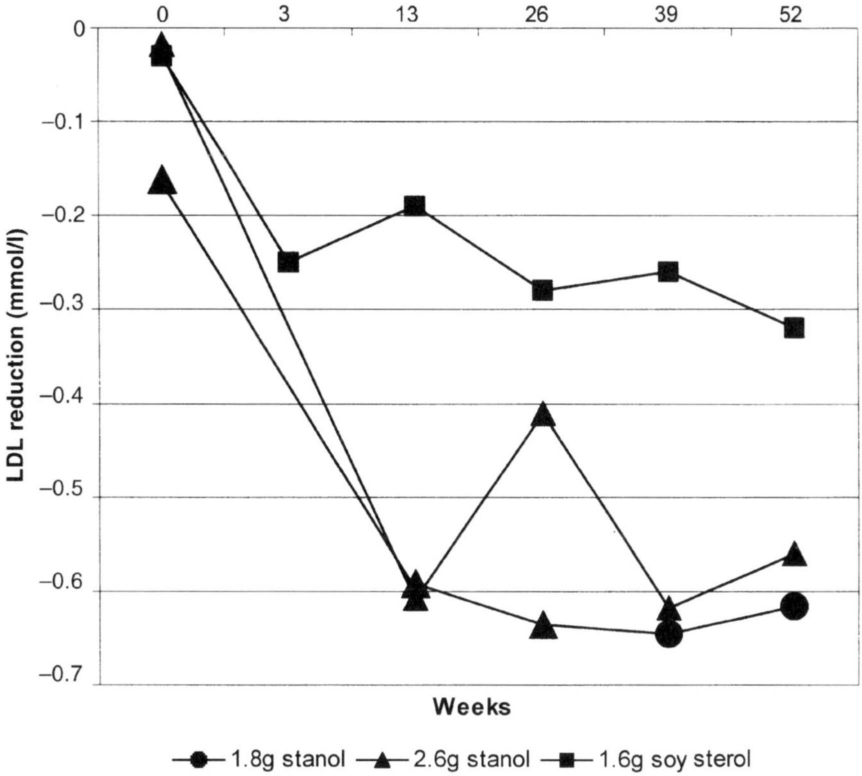

Figure 11. The long-term efficacy and safety of stanol ester and sterol ester spreads have been evaluated in two separate trials: the North Karelia Stanol Ester Study (Miettinen *et al.*, 1995) and the TNO Study on Sterol Ester Spreads (Brink and Hendriks, 2000). In the stanol ester study, 2.6 g stanol ester per day (half of participants were switched to 1.8 g stanol ester after the first 26 weeks) reduced serum total and LDL cholesterol concentrations by 10% and 14%, respectively. In the sterol ester study, 1.6 g soy sterol per day reduced serum total and LDL cholesterol concentrations by 5% and 7%, respectively.

were already raised above normal levels in the basal state, as has been reported in children with familial hypercholesterolemia and in heterozygous carriers of the gene for phytosterolemia.

10. The benefits of sterol and stanol ester consumption

The risk of mortality due to coronary heart disease rises exponentially with increasing serum cholesterol concentration and the benefits of its reduction are well established. From prospective cohort studies and controlled intervention trials, Law *et al.* (1994) deduced that a 10% reduction in serum cholesterol concentration produces a 50% reduction in ischemic heart diseases at age 40, falling to a 20% reduction at age 70. On average, approximately 3% (AHA "Step 1" diet) or 5.6% (AHA "Step 2" diet) reductions in serum cholesterol concentrations can be achieved by dietary counselling of free-living subjects (Tang *et al.*, 1998). In addition, in mildly to moderately hypercholesterolemic subjects, serum LDL cholesterol concentrations are typically reduced by 10–15% with the use of plant sterol or plant stanol based food products. This reduction would be expected to reduce the risk of heart diseases by approximately 25% (Law, 2000) and thus have an important potential role in the prevention of coronary heart disease.

F. The development of sterol-enriched foods

1. *Foods or capsules?*

Plant sterol-enriched foods have been on the market since 1995, when a plant stanol ester-enriched margarine was first introduced onto the Finnish market. Before that, plant sterols had been available only as dietary supplements. However, the popularity of the capsules was limited by the fact that the efficacy of plant sterols in capsule form had not been clinically proven. A new era of interest in plant sterols started with the paper (Miettinen *et al.*, 1995) indicating the potential of plant stanol ester-enriched foods from the points of view of both clinical efficacy and commercial potential. As already discussed, the efficacy of plant sterols is dependent on their effective solubilization in the fat phase of the food digest. This is better achieved if the plant sterol is already homogeneously incorporated into a fatty food matrix. Thus, for optimal efficacy, plant sterols should be digested with meals, underlining the fact that foods are the optimal carriers for these ingredients.

2. *What type of food is optimal as a plant sterol carrier?*

Early studies on plant sterols, based on animal studies (Mattson *et al.*, 1977), had already suggested that it would be better to use plant sterols in esterified form because of their better solubility in fat. However, a few years later Mattson *et al.* (1982) showed in one human study that esterified sterols were not as potent as sterols in free form. Pollak (1985) suggested that the incorporation of β-sitosterol into butter and margarines would be a feasible dietary prophylactic tool for lowering serum cholesterol levels.

Today several aspects need to be considered when developing foods enriched in plant sterols and stanols. The food itself should already have a "healthy" profile in its category, with low salt, fat, cholesterol, and saturated fatty acid contents. Furthermore, local eating habits need to be considered, as sterol-enriched foods should be staple foods, making them easy to incorporate into every meal. A nutritionally balanced fatty acid composition, and a lower fat or energy content than other typical foods of the same category would help consumers to recognize sterol-enriched foods as "heart-healthy" choices. Data from several clinical studies have convincingly shown that plant sterols and plant stanols, as part of recommended diets, enhance the overall reduction of serum total and LDL cholesterol levels (Hallikainen *et al.*, 1999, 2000a; Andersson *et al.*, 1999).

It is important to find a food form in which an adequately high level of active ingredient can be incorporated and for which it is easy for consumers to estimate the adequate daily intake of the active ingredient. Concepts such as achieving half or all the recommended daily amount from one portion, such as a cup of yoghurt or a glass of milk, have proven to be easy for consumers to accept. As of mid-2002, several products which include one recommended daily intake of active ingredient are available, for example in Argentina where a daily recommended amount of plant sterol ester is provided in one glass of milk or milk–juice drink. In Finland, consumers have a choice, where the recommended daily optimal intake of plant stanol ester is provided in one cup of yoghurt or 2–3 portions of other foods such as margarine.

3. Physical and technological properties of sterols and their fatty acid esters

Sterols may be added to foods in the free form or as fatty acid esters. However, the physical properties of crystalline plant sterol and stanol preparations, such as the high melting point and very low solubility of the free sterols in both fat and water phases, limit their applicability in foods. Without further development, these ingredients would have remained as dietary supplements mainly.

The esterification of plant stanols with unsaturated fatty acids changes their physical properties from sticky powders with a high melting point (142°C) to fat-like substances with a melting point of 43°C (Figure 12). Use of these esterified sterols and stanols increases the technological possibilities of incorporating them into foods at levels high enough to obtain an optimal serum cholesterol-lowering effect without having any undesired impact on the quality and palatability of the products.

In addition to esterification there are some other techniques by which the technological limitations of free sterols have been reduced. Liposomal sterol aggregates can be produced by dissolving free sterols or stanols in a melt of food emulsifiers with a sterol:emulsifier ratio of <1:1 (preferably 1:2), blending the mixture into a protein-containing product such as milk, and carrying out homogenization (Dahlsten *et al.*, 2000). This can then be used to produce various milk-based products such as chocolate milk, sour milk, yoghurt, soft unripened cheese products, and hard cheeses.

Ostlund (1999) studied the solubility of free sterols in artificial bile in the

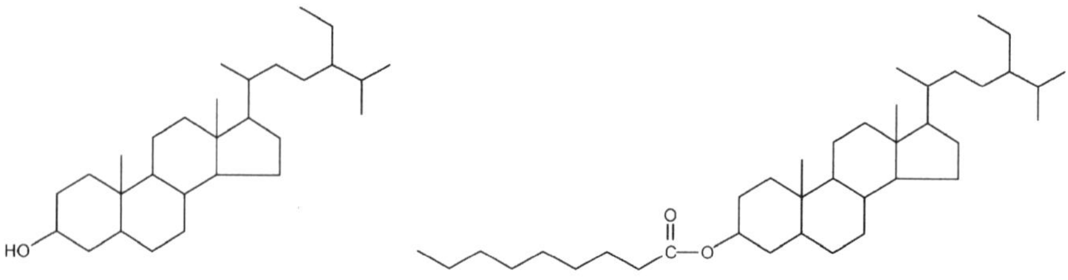

Plant stanol
- Crystalline powder
- Restricted solubility
- Dropping point 142°C

Plant stanol ester
- Fat like
- Fat soluble
- Dropping point 43°C

Figure 12. Esterification-induced changes to the physical properties of plant stanols.

presence of compounds found in the gut such as taurocholate, monoolein and diolein, and lecithin and lysolecithin. A 1:1:0.5 mixture of sitostanol, lecithin and lysolecithin had the greatest solubility (97.9%), whereas a mixture of sitostanol and lecithin alone had a solubility of only 38.2%. Furthermore, a micellar or liposomal aggregate of sitostanol and lecithin (1:1 molar ratio), produced by sonication in water and lyophilization, had enhanced solubility (89.7%) compared with the sitostanol–lecithin mixture dried from chloroform (38.2%). Micelles based on sodium stearoyl lactylate, lecithin, or lysolecithin were equally effective at increasing the solubility of free sitostanol. The emulsifier–sitostanol micelles obtained were water-soluble. Sitostanol–lecithin micelles were more effective at reducing cholesterol absorption than sitostanol in powder form (Ostlund *et al.*, 1999). Akashe and Miller (2001) suggested the use of mesophase-stabilized compositions for the delivery of cholesterol-reducing sterols and stanols in food products.

Diacylglycerols increase the solubility of free sterols in vegetable oils. An oil containing >15% (w/w) diacylglycerols will solubilize 4.7% (w/w) of free sterols (Goto *et al.*, 1999). A sterol-enriched diacylglycerol oil has a greater cholesterol-lowering effect than a regular vegetable oil containing the same amount of sterols (Meguro *et al.*, 2001) (see also Chapter 5).

Early research focused on grinding or milling free sterols to enhance their solubility (Ong, 1978). More recently, efficient milling, together with the addition of ingredients such as sugar to decrease the size of the sterol crystals, has been suggested as a way of incorporating free sterols into foods (Tiainen *et al.*, 1998). So far, however, there are no efficacy data available for such preparations.

Fat-based preparations of free sterols can also be produced by dissolving free sterols in hot oils and fats, and emulsifying the blend by mixing before cooling

(Yliruusi *et al.*, 1999). It is too soon to predict how well such techniques will effectively improve the ability to incorporate sterols into various food forms.

G. Commercial applications

Most sterol-enriched foods on the market today are based on esterified sterols or stanols. Currently, free sterols lack regulatory approval in many countries, limiting the introduction of free sterol-enriched foods onto the market. Examples of commercially available sterol-enriched foods are listed in Table 8.

1. Oils and fats

Spreads were the first commercial application of foods enriched with plant stanol esters, and they remain by far the most widely known applications among consumers and food companies. The first products were launched by Raisio Group Ltd as 'Benecol' margarine in Finland in 1995. A few years later, in 1999, Benecol products were launched in the USA, the UK, Belgium, the

Table 8. Foods enriched with sterols and countries where these products are or have been commercially available

Food type	Active ingredients	Countries where available
Spreads	Sterol esters and stanol esters	Australia, Belgium, Brazil, Canada, UK and Ireland, Finland, France, Germany, Holland, Japan, Portugal, South Africa, Spain, Switzerland, Sweden, USA
Cream cheese-type vegetable oil-based spread	Stanol esters	Belgium, Finland, Holland, UK and Ireland
Yoghurt	Stanol esters	Belgium, Finland, Holland, UK and Ireland
Ready-to-eat-cereals	Free sterols	USA (test marketing)
Snack/cereal bars	Stanol esters and free sterols	UK and Ireland, USA
Milk	Sterol esters and stanol esters	Argentina, UK and Ireland, United Arab Emirates
Milk–juice drink	Sterol esters	Argentina
Orange juice	Free sterols	USA (test marketing)
Salad dressings, oil, mayonnaise	Stanol esters and free sterols	Argentina, Israel, Japan, USA
Drinks	Free sterols	Korea
Mayonnaise-based salads	Stanol esters	Finland
Meat products (frankfurters, chicken meat balls)	Stanol esters	Finland

Table 9. Nutritional profiles of two low-fat margarines available in Finland

Nutritional parameter	'Keiju' low-fat margarine	'Light Benecol'	Difference
Absorbable fat	40%	32%	–20%
Energy content[a]	360 kcal	300 kcal	–17%
SAFA[a]	11 g	5 g	–55%
MUFA[a]	19 g	18 g	–5%
PUFA[a]	10 g	9 g	–10%
Plant stanol[a]		8 g	

[a]Figures are given per 100 g product. SAFA, saturated fatty acids. MUFA, monounsaturated fatty acids. PUFA, polyunsaturated fatty acids.

Netherlands, and Ireland. Thus, the longest experience of consumption of plant stanol ester-enriched products is a 6-year period in Finland. Sterol ester-enriched spreads were first launched in 1999 in the USA by Lipton Ltd as 'Take Control' margarine. Today sterol ester-enriched margarines are available worldwide under brand names such as 'Take Control', 'Flora ProActive', 'Becel ProActive'', and 'Rama Proactive' from Unilever/Lipton, or 'Logicol' from Goodman Fielder Ltd. Other spreads include vegetable oil-based cream cheeses available in Finnish, UK, and other European markets.

An additional benefit that low-fat sterol-enriched spreads may provide over other sterol-enriched foods is that the diet can simultaneously be changed towards accepted dietary guidelines, as discussed above (see Section C.4). The combined nutritional benefit when changing from a regular reduced-fat spread to a sterol-enriched spread is demonstrated by the case of 'Light Benecol' (Table 9). The absorbable fat, and consequently the energy content, can be decreased by 20%, as part of the fat in the product is replaced by non-absorbed stanol. Furthermore, due to its relatively high melting range (30–40°C), plant stanol esterified with canola oil fatty acids can be used to replace the hard stock of the spread. This has an interesting effect on the overall fatty acid composition of the spread. The content of saturated fatty acids is reduced by 55% while monounsaturated and polyunsaturated fatty acids are reduced by only 5% and 10% respectively (Table 9). The structure-providing effect of sterol and stanol esters can be utilized to fully replace hard fats, which contain nutritionally less desirable saturated and *trans* fatty acids, in fat blends for spreads (Wester, 1998).

Another important benefit of spreads over some other food applications is the "self-limiting" nature of daily consumption. Spreads are not normally consumed at doses above the recommended 25 g/day so the risk of over-consuming plant sterols from this source is low. However, spreads are not a convenient food form for consumers "on-the-go", and for such people it might be difficult to obtain a near-optimal intake of plant sterols every day. Providing only one food type as a carrier also limits the number of people who can utilize plant sterols as a versatile dietary tool for lowering elevated serum cholesterol levels. Six years of experience in Finland indicate that there is a disadvantage to using only one type of food (spreads) as a carrier. It is difficult for consumers interested in benefiting from the cholesterol-lowering effect of stanol esters to obtain the optimal daily intake of stanols (2 g/day) recommended by the AHA (NCEP Expert Panel, 2001).

Another limitation of spreads is that in many cultures spreads are not important foods. Furthermore, many consumers tend to apply recommendations to reduce fat intakes by reducing their consumption of visible fats, including spreads.

Salad dressings enriched with plant sterols have been on the US market but are no longer available there. In Finland, plant stanol esters are available incorporated into mayonnaise in ready-meal salads. As fat-based emulsions, salad dressings and mayonnaise are good applications for sterol-enriched products.

Vegetable oils enriched with plant sterols are available on the Japanese market as 'Econa' brand sterol-enriched diacylglycerol oils produced by Kao Corp (see also Chapter 5).

2. *Dairy products*

Dairy foods have recently aroused interest as possible candidate foods for sterol enrichment. The first sterol-enriched dairy food applications were Benecol yoghurt enriched with plant stanol ester , which appeared in the UK in 1999, and milk and milk–juice drinks enriched with plant sterol ester, which were launched in Argentina in 2000 under the 'SereCol' brand. Milk containing free sterol was launched in Australia under the Logicol brand, but was withdrawn after the Australia New Zealand Food Authority (ANZFA; now replaced by Food Standards Australia New Zealand) decided to allow only spreads as carriers for plant sterols.

Dairy foods have several benefits as plant sterol carriers. Several low-fat dairy foods are considered to be a natural part of a healthy diet by many consumers in various food cultures. Furthermore, many dairy foods are staple foods, and are available as single servings into which it is easy to incorporate either one-half or all of the recommended daily amount of plant sterols, thus making it easier for consumers to estimate their daily intake of plant sterols. As many dairy products are oil-in-water emulsions, sterol and stanol esters can be added into the base milk during the homogenization step, as when replacing milk fat with vegetable oils in dairy products. Sterol and stanol esters are fatty substances and so provide dairy products with a creamy texture, yet a very low fat content. Mensink *et al.* (2002) have shown that stanol esters in yoghurt are as effective at lowering serum total and LDL cholesterol levels as they are in spreads.

3. Snacks

Snack bars or energy bars have been introduced to the market as providing a "take-with" food form with easy sterol dosing. The only products currently available on the market are Benecol energy bars in the USA, the UK and Ireland. Cereal bars enriched with free sterols have also been introduced for test marketing in the USA by Altus Foods under the 'Take Heart' brand.

4. Drinks

There are only a few sterol-enriched drinks, other than dairy products, currently on the market. However, drinks are a growing segment of the functional food market and it is likely that there will be more of these types of products in the future. The few examples include a juice–milk drink enriched with plant sterol esters in Argentina ('SereCol'), and a free sterol-enriched drink in Korea ('Ucole').

5. Others

Other food types enriched by plant sterols include meat products enriched with plant stanol ester (Atria Ltd, Finland). As sterol esters are lipids, they can be incorporated into practically any food form that contains at least some fat in it. However, the level of fat can be very low, as has been shown in low-fat yoghurts and in milk or milk–juice drinks, where the plant stanol or sterol ester provides virtually the only fat in the product.

H. Summary

Plant sterols (phytosterols), especially 4-desmethyl sterols, have recently gained much scientific and commercial interest due to the introduction of plant sterol-enriched foods and dietary supplements with marked serum cholesterol-lowering efficacy. Due to the structural similarity between cholesterol and 4-desmethyl plant sterols, the intake of optimal amounts of plant sterols and stanols (saturated sterols) lowers intestinal cholesterol absorption and results in a 6–15% reduction in serum LDL cholesterol concentrations.

Plant sterols and stanols can be classified on the basis of structure or biosynthesis. Only the 4-desmethyl sterols and stanols affect cholesterol absorption and decrease serum total and LDL cholesterol levels, suggesting that similarity of ring structure and side-chain length are crucial requirements for their effects on cholesterol absorption.

Knowledge of the serum cholesterol-lowering ability of plant sterols began in the early 1950s but interest was revived with the paper published in *The New England Journal of Medicine* by Miettinen *et al.* in 1995 and the simultaneous launch of the first plant sterol-enriched food, Benecol margarine, in Finland. These events indicated the potential of foods enriched with plant stanol esters from the point of view of both clinical efficacy and commercial potential.

4-Desmethyl sterols decrease serum cholesterol concentrations by interfering with cholesterol absorption. Plant sterols and stanols reduce the solubility of cholesterol in the micellar phase, thus reducing the amount of cholesterol available for absorption. Furthermore, some still-unknown mechanisms at the absorptive sites of the intestinal enterocytes or within the enterocytes also play a role in the overall effect of plant sterols on cholesterol absorption.

Irrespective of the actual mechanisms behind the reduction of cholesterol absorption by plant sterols, solubilization into the emulsified fat phase of the food digest is a prerequisite for plant sterols to be incorporated into the micelles. The physical properties of free, crystalline plant sterols and stanols limit their applicability in foods and their cholesterol-lowering effect in many food matrices, but these limitations can be partly solved by producing fat dispersions of plant sterols or emulsifier–sterol aggregates. Currently, however, plant sterols are mainly used as fatty acid esters in functional foods.

The serum cholesterol-lowering effect of plant sterols and stanols has been proven in several clinical studies. The hypocholesterolemic effects have been verified in normocholesterolemic individuals, in individuals with mild to moderate hypercholesterolemia or with familial hypercholesterolemia, in women with coronary heart disease, and in men with non-insulin-dependent diabetes – in conjunction with cholesterol-lowering statin therapy and irrespective of the background diet. In addition, studies have been conducted with normocholesterolemic children and with children with slightly elevated cholesterol levels, or with familial hypercholesterolemia.

The duration of the clinical studies has mainly been short, typically 4–8 weeks, except for two trials that each lasted 52 weeks: the North Karelia Stanol Ester Study (Miettinen *et al.*, 1995) and the TNO Study on Sterol Ester Spreads (Brink and Hendriks, 2000). The trials have shown that the cholesterol-lowering effect is gained within the first week of consuming these products and is permanent with continuing intake of the products.

According to the clinical studies, the minimum amount of sterols and stanols required to produce a significant serum cholesterol-lowering effect is about 0.8–1.0 g/day. However, a larger amount (2–3 g/day) of plant sterols and stanols in their esterified, fat-soluble form is required to provide the close-to-maximal and clinically significant cholesterol-lowering effect. In contrast, the efficacy of plant sterols and stanols in their unesterified form has been more modest and appears to be highly dependent on the food matrix or, alternatively, to require a higher dietary fat content.

The optimal effect is achieved when plant sterols and stanols are ingested as part of a meal, and preferably consumed as a food. As yet there is no evidence that plant sterols in unesterified or esterified form would be effective when consumed between meals or under fasting conditions. Additional studies are needed to find out what quantity of digestible macrocomponents such as fats and proteins are needed for the effective action of both esterified and unesterified sterols.

After being available on the market for 6 years, a wide range of different sterol-based foods have proven effective and safe in decreasing elevated serum total and LDL cholesterol levels. Currently they are widely available on the global food market. Spreads are still by far the most common food type, but dairy foods and drinks are increasing in this segment.

The regulatory status of plant sterol-enriched foods is still developing. An interim US Food and Drug Administration (FDA) rule allowing health claims on plant sterol and stanol ester products was recently approved. In many countries, lack of regulatory legislation is currently an important barrier to sterol-enriched foods entering the market.

References

Akashe, A and Miller, M (2001) Use of mesophase-stabilized compositions for delivery of cholesterol-reducing sterols and stanols in food products. US Patent 6,274,574 B1.

Akihisa, T, Swapnadip, T, Rosenstein, FU and Matsumoto, T (1986) Sterols of Cucurbitaceae: the configurations at C-24 of 24-Alkyl-Δ^5-, Δ^7- and Δ^8-sterols. *Lipids*, 21, 39–47.

Akihisa, T, Kokke, WCMC and Tamura, T (1991) Naturally occurring sterols and related compounds from plants. In: *Physiology and Biochemistry of Sterols* (GW Patterson and WD Nes, eds), American Oil Chemists' Society, Champaign, pp.172–228.

Andersson, A, Karlström, B, Mohsen, R and Vessby, B (1999) Cholesterol-lowering effects of a stanol ester-containing low-fat margarine used in conjunction with a strict lipid-lowering diet. *Eur. Heart J.*, Suppl. 1, S80–S90.

Ayesh, R, Weststrate, JA, Drewitt, PN and Hepburn, PA (1999) Safety evaluation of phytosterol esters. Part 5. Faecal short-chain fatty acid and microflora content, faecal bacterial enzyme activity and serum female sex hormones in healthy normolipidaemic volunteers consuming a controlled diet either with or without a phytosterol ester-enriched margarine. *Food Chem. Toxicol.*, 37, 1127–1138.

Becker, M, Staab, D and Von Bergman, K (1992) Long-term treatment of severe familial hypercholesterolemia in children: effect of sitosterol and benzafibrate. *Pediatrics*, 89, 138–142.

Becker, M, Staab, D and von Bergmann, K (1993) Treatment of severe familial hypercholesterolemia in childhood with sitosterol and sitostanol. *J. Pediatr.*, 122, 292–296.

Bhattacharyya, AK (1981) Uptake and esterification of plant sterols by rat small intestine. *Am. J. Physiol.*, 240, G50–G55.

Bhattacharyya, AK and Lopez, LA (1979) Absorbability of plant sterols and their distribution in rabbit tissues. *Biochim. Biophys. Acta*, 574, 146–153.

Blair, SN, Capuzzi, DM, Gottlieb, SO, Nguyen, T, Morgan, JM and Cater, NB (2000) Incremental reduction of serum total cholesterol and low-density lipoprotein cholesterol with the addition of plant stanol ester-containing spread to statin therapy. *Am. J. Cardiol.*, 86, 46–52.

Brink, E and Hendriks, H (2000) *Long-term Follow-up Study on the Use of a Spread Enriched with Plant Sterols (TNO Report No. V 99.869)*, TNO Nutrition and Food Research Institute.

Cater, NB (1999) Historical and scientific basis for the development of plant stanol ester foods as cholesterol-lowering agents. *Eur. Heart J.*, Suppl. 1, S36–S44.

Cater, NB (2000) Plant stanol ester: Review of cholesterol-lowering efficacy and implications for coronary heart disease risk reduction. *Prevent. Cardiol.*, 3, 121–130.

Chitwood, DJ and Patterson, GW (1991) Separation of epimeric pairs of C-24 alkylsterols by reversed-phase high performance liquid chromatography of the free sterols at subambient temperature. *J. Liquid Chromatogr.*, 14, 151–163.

Clark, JP (1996). Tocopherols and sterols from soybeans. *Lipid Technol.*, 8, 111–117.

Daguet, D (2000) Phytosterols: highly promising compounds. *Lipid Technol.*, 12, 77–80.

Dahlsten, C-J, Burling, H and Strinning, O (2000) Preparation of cholesterol reducing edible products by mixing plant sterols/stanols in a melt of a food emulsifier. World Patent Application WO 00/33669.

Davidson, MH, Maki, KC, Umporowicz, DM, Ingram, KA, Dicklin, MR, Schaefer, E, Lane, RW, McNamara, JR, Ribaya-Mercado, JD, Perrone, G, Robins, SJ and Franke, WC (2001) Safety and tolerability of esterified phytosterols administered in reduced-fat spread and salad dressing to healthy adult men and women. *J. Am. Coll. Nutr.*, 20, 307–319.

de Deckere, EAM and Korver, O (1996) Minor constituents of rice bran oil as functional foods. *Nutr. Rev.*, 54, 11, S120–S126.

Denke, MA (1995) Lack of efficacy of low-dose sitostanol therapy as an adjunct to a cholesterol-lowering diet in men with moderate hypercholesterolemia. *Am. J. Clin. Nutr.*, 61, 392–396.

Farquhar, JW, Smith, RE and Dempsey, ME (1956) The effect of beta-sitosterol on the serum lipids of young men with arteriosclerotic heart disease. *Circulation*, 14, 77–82.

Field, FJ and Mathur, SN (1983) Beta-sitosterol: esterification by intestinal acylcoenzyme A: cholesterol acyltransferase (ACAT) and its effect on cholesterol esterification. *J. Lipid Res.*, 24, 409–417.

Field, FJ, Born, E and Mathur, SN (1997) Effect of micellar beta-sitosterol on cholesterol metabolism in CaCo-2 cells. *J. Lipid Res.*, 38, 348–360.

Goto, N, Nishide, T, Tanaka, Y and Yasukawa, T (1999) Oil and fat composition containing phytosterol. World Patent Application WO 99/59423.

Gylling, H and Miettinen, TA (1994) Serum cholesterol and cholesterol and lipoprotein metabolism in hypercholesterolaemic NIDDM patients before and during sitostanol ester-margarine treatment. *Diabetologia*, 37, 773–780.

Gylling, H and Miettinen, TA (1996) Effects of inhibiting cholesterol absorption and synthesis on cholesterol and lipoprotein metabolism in hypercholesterolemic non-insulin-dependent diabetic men. *J. Lipid Res.*, 37, 1776–1785.

Gylling, H and Miettinen, TA (1999a) Cholesterol reduction by different plant stanol mixtures and with variable fat intake. *Metabolism*, 48, 575–580.

Gylling, H and Miettinen, TA (1999b) Phytosterols, analytical and nutritional aspects. In: *Functional Foods – a New Challenge for Food Chemists, Vol. 1* (R Lasztity, W Phannhauser, L Simon-Sarkadi and S Tömösközi, eds), Publishing Company of TUB, Budapest, p.109.

Gylling, H, Siimes, MA and Miettinen, TA (1995) Sitostanol ester margarine in dietary treatment of children with familial hypercholesterolemia. *J. Lipid Res.*, 36, 1807–1812.

Gylling, H, Radhakrishnan, R and Miettinen, TA (1997) Reduction of serum cholesterol in postmenopausal women with previous myocardial infarction and cholesterol malabsorption induced by dietary sitostanol ester margarine: women and dietary sitostanol. *Circulation*, 96, 4226–4231.

Gylling, H, Puska, P, Vartiainen, E and Miettinen, TA (1999) Retinol, vitamin D, carotenes and alpha-tocopherol in serum of a moderately hypercholesterolemic population consuming sitostanol ester margarine. *Atherosclerosis*, 145, 279–285.

Hallikainen, MA and Uusitupa, MI (1999) Effects of 2 low-fat stanol ester-containing margarines on serum cholesterol concentrations as part of a low-fat diet in hypercholesterolemic subjects. *Am. J. Clin. Nutr.*, 69, 403–410.

Hallikainen, MA, Sarkkinen, ES, Gylling, H, Erkkila, AT and Uusitupa, MI (2000a) Comparison of the effects of plant sterol ester and plant stanol ester-enriched margarines in lowering serum cholesterol concentrations in hypercholesterolaemic subjects on a low-fat diet. *Eur. J. Clin. Nutr.*, 54, 715–725.

Hallikainen, MA, Sarkkinen, ES and Uusitupa, MI (2000b) Plant stanol esters affect serum cholesterol concentrations of hypercholesterolemic men and women in a dose-dependent manner. *J. Nutr.*, 130, 767–776.

Heinemann, T, Leiss, O and von Bergmann, K (1986) Effect of low-dose sitostanol on serum cholesterol in patients with hypercholesterolemia. *Atherosclerosis*, 61, 219–223.

Heinemann, T, Kullak-Ublick, GA, Pietruck, B and von Bergmann, K (1991) Mechanisms of action of plant sterols on inhibition of cholesterol absorption. Comparison of sitosterol and sitostanol. *Eur. J. Clin. Pharmacol.*, 40(Suppl. 1), S59–S63.

Heinemann, T, Axtmann, G and von Bergmann, K (1993) Comparison of intestinal absorption of cholesterol with different plant sterols in man. *J. Clin. Invest.*, 23, 827–831.

Hendriks, HF, Weststrate, JA, van Vliet, T and Meijer, GW (1999) Spreads enriched with three different levels of vegetable oil sterols and the degree of cholesterol lowering in normocholesterolaemic and mildly hypercholesterolaemic subjects. *Eur. J. Clin. Nutr.*, 53, 319–327.

Hendriks, HFJ, Ntanios, FY, Brink, EJ, Princen, HMG, Buytenhek, R and Meijer, GW (2001) One year follow-up study on the use of a low-fat spread enriched with plant stanol esters. *Ann. Nutr. Metab.*, 45(Suppl. 1), 100.

Holmbom, B and Erä, V (1978) Composition of tall oil pitch. *J. Am. Oil Chem. Soc.*, 55, 342–344.

Ikeda, I, Kawasaki, A, Samezima, K and Sugano, M (1981) Antihypercholesterolemic activity of beta-sitostanol in rabbits. *J. Nutr. Sci. Vitaminol. (Tokyo)*, 27, 243–251.

Ikeda, I, Tanabe, Y and Sugano, M (1989) Effects of sitosterol and sitostanol on micellar solubility of cholesterol. *J. Nutr. Sci. Vitaminol. (Tokyo)*, 35, 361–369.

Jones, PJ, Ntanios, FY, Raeini-Sarjaz, M and Vanstone, CA (1999) Cholesterol-lowering efficacy of a sitostanol-containing phytosterol mixture with a prudent diet in hyperlipidemic men. *Am. J. Clin. Nutr.*, 69, 1144–1150.

Jones, PJ, Raeini-Sarjaz, M, Ntanios, FY, Vanstone, CA, Feng, JY and Parsons, WE (2000) Modulation of plasma lipid levels and cholesterol kinetics by phytosterol versus phytostanol esters. *J. Lipid Res.*, 41, 697–705.

Kalliomäki, JL, Waris, EK and Vasama, R (1957) Sitosterolernas verkan på blodets kolesterol ach lipoproteiner. *Nordisk Medicin*, 57, 1–10.

Knopp, RH (1999) Drug treatment of lipid disorders. New Engl. J. Med., 341, 498–511.

Kritchevsky, D, Tepper, SA, Czanecki, SK and Kyle, DJ (1999) Effects of 4-methylsterols and of β-sitosterol on cholesterol metabolism in rat. *Nutr. Res.*, 19, 1649–1654.

Law, MR, Wald, NJ and Thompson, SG (1994) By how much and how quickly does reduction in serum cholesterol concentration lower risk of ischaemic heart disease? *Br. Med. J.*, 308, 367–372.

Law, M (2000) Plant sterol and stanol margarines and health. *Br. Med. J.*, 320, 861–864.

Lees, AM, Mok, HY, Lees, RS, McCluskey, MA and Grundy, SM (1977) Plant sterols as cholesterol-lowering agents: clinical trials in patients with hypercholesterolemia and studies of sterol balance. *Atherosclerosis*, 28, 325–338.

Ling, WH and Jones, PJ (1995) Dietary phytosterols: a review of metabolism, benefits and side effects. *Life Sci.*, 57, 195–206.
Maki, KC, Davidson, MH, Umporowicz, DM, Schaefer, EJ, Dicklin, MR, Ingram, KA, Chen, S, McNamara, JR, Gebhart, BW, Ribaya-Mercado, JD, Perrone, G, Robins, SJ and Franke, WC (2001) Lipid responses to plant-sterol-enriched reduced-fat spreads incorporated into a National Cholesterol Education Program Step I diet. *Am. J. Clin. Nutr.*, 74, 33–43.
Mattson, FH, Volpenhein, RA and Erickson, BA (1977) Effect of plant stanol esters on the absorption of dietary cholesterol. *J. Nutr.*, 107, 1139–1146.
Mattson, FH, Grundy, SM and Crouse, JR (1982) Optimizing the effect of plant sterols on cholesterol absorption in man. *Am. J. Clin. Nutr.*, 35, 697–700.
Meguro, S, Higashi, K, Hase, T, Honda, Y, Otsuka, A, Tokimitsu, I and Itakura, H (2001) Solubilization of phytosterols in diacylglycerol versus triacylglycerol improves the serum cholesterol-lowering effect. *Eur. J. Clin. Nutr.*, 55, 513–517.
Mensink, RP, Ebbing, S, Lindhout, M, Plat, J and van Heugten, MM (2002) Effects of plant stanol esters supplied in low-fat yoghurt on serum lipids and lipoproteins, non-cholesterol sterols and fat soluble antioxidant concentrations. *Atherosclerosis*, 160, 205–213.
Miettinen, TA (2001) Phytosterols – what plant breeders should focus on. *J. Sci. Food Agric.*, 81, 895–903.
Miettinen, TA and Gylling, H (1999) Regulation of cholesterol metabolism by dietary plant sterols. *Curr. Opin. Lipidol.*, 10, 9–14.
Miettinen, TA and Vanhanen, H (1994) Dietary sitostanol related to absorption, synthesis and serum level of cholesterol in different apolipoprotein E phenotypes. *Atherosclerosis*, 105, 217–226.
Miettinen, TA, Tilvis, RS and Kesaniemi, YA (1990) Serum plant sterols and cholesterol precursors reflect cholesterol absorption and synthesis in volunteers of a randomly selected male population. *Am. J. Epidemiol.*, 131, 20–31.
Miettinen, TA, Gylling, H and Vanhanen, H (1993) Lipoprotein and cholesterol metabolism altered by rapeseed oil without and with added plant sterols. In: *17th Nordic Lipid Symposium, Imatra, Finland* (Y Mälkki and G Lambertsen, eds), Lipidforum, Berger, pp.14–19.
Miettinen, TA, Puska, P, Gylling, H, Vanhanen, H and Vartiainen, E (1995) Reduction of serum cholesterol with sitostanol-ester margarine in a mildly hypercholesterolemic population. *New Engl. J. Med.*, 333, 1308–1312.
Miettinen, TA, Vuoristo, M, Nissinen, M, Jarvinen, HJ and Gylling, H (2000) Serum, biliary and faecal cholesterol and plant sterols in colectomized patients before and during consumption of stanol ester margarine. *Am. J. Clin. Nutr.*, 71, 1095–1102.
NCEP Expert Panel (2001) Executive summary of the third report of the National Cholesterol Education Program (NCEP) expert panel on detection, evaluation and treatment of high blood cholesterol in adults (Adult treatment panel III). *J. Am. Med. Assoc.*, 16, 2486–2497.
Neil, HA, Meijer, GW and Roe, LS (2001) Randomised controlled trial of use by hypercholesterolaemic patients of a vegetable oil sterol-enriched fat spread. *Atherosclerosis*, 156, 329–337.
Nes, WR, Krevitz, K, Joseph, J, Nes, WD, Harris, B, Gibbons, GF and Patterson, GW (1977) The phylogenetic distribution of sterols in tracheophytes. *Lipids*, 12, 511–527.
Nes, WR and Parish, EJ (eds) (1989) *Analysis of Sterols and Other Biologically Significant Steroids*, Academic Press, New York.
Nestel, P, Cehun, M, Pomeroy, S, Abbey, M and Weldon, G (2001) Cholesterol-lowering effects of plant sterol esters and non-esterified stanols in margarine, butter and low-fat foods. *Eur. J. Clin. Nutr.*, 55, 1084–1090.
Nguyen, TT (1999) The cholesterol-lowering action of plant stanol ester. *J. Nutr. Sci.*, 12, 2109–2112.

Nguyen, TT, Dale, L, von Bergmann, K and Croghan, IT (1999) Cholesterol-lowering effect of stanol ester in a US population of mildly hypercholesterolemic men and women: a randomized, controlled trial. *Mayo Clin. Proc.*, 74, 1198–1206.

Niinikoski, H, Viikari, J and Palmu,T (1997) Cholesterol-lowering effect and sensory properties of sitostanol ester margarine in normocholesterolemic adults. *Scand. J. Nutr.*, 41, 9–12.

Normén, L, Johnsson, M, Andersson, H, van Gameren, Y and Dutta, P (1999) Plant sterols in vegetables and fruits commonly consumed in Sweden. *Eur. J. Nutr.*, 38, 84–89.

Ntanios, FY and Jones, PJ (1998) Effects of variable dietary sitostanol concentrations on plasma lipid profile and phytosterol metabolism in hamsters. *Biochim. Biophys. Acta*, 1390, 237–244.

Ntanios, FY, Jones, PJ and Frohlich, JJ (1998a) Dietary sitostanol reduces plaque formation but not lecithin cholesterol acyl transferase activity in rabbits. *Atherosclerosis*, 138, 101–110.

Ntanios, FY, MacDougall, DE and Jones, PJ (1998b) Gender effects of tall oil versus soybean phytosterols as cholesterol-lowering agents in hamsters. *Can. J. Physiol. Pharmacol.*, 76, 780–787.

Ong, JTH (1978) Taste-stable aqueous pharmaceutical suspension of tall oil sitosterols and a method for the preparation thereof. US Patent 4,195,084.

Ostlund, RE (1999) Sitostanol formulation to reduce cholesterol absorption and method for preparing and use of same. US Patent 5,932,562.

Ostlund, RE, Spilburg, CA and Stenson, WF (1999) Sitostanol administered in lecithin micelles potently reduces cholesterol absorption in humans. *Am. J. Clin. Nutr.*, 70, 826–831.

Ostlund, RE, Jr, Racette, SB, Okeke, A and Stenson, WF (2002) Phytosterols that are naturally present in commercial corn oil significantly reduce cholesterol absorption in humans. *Am. J.Clin. Nutr.*, 75(6), 1000–1004.

Peers, KE (1977) The non-glyceride saponifiables of shea butter. *J. Sci. Food Agric.*, 28, 1000–1009.

Pelletier, X, Belbraouet, S, Mirabel, D, Mordret, F, Perrin, JL, Pages, X and Debry, G (1995) A diet moderately enriched in phytosterols lowers plasma cholesterol concentrations in normocholesterolemic humans. *Ann. Nutr. Metab.*, 39, 291–295.

Peterson, DW (1951) Effect of soybean sterols in the diet on plasma and liver cholesterol in chicks. *Proc. Soc. Exp. Biol. Med.*, 78, 143–147.

Phillips, KM, Tarragó-Trani, MT and Stewart, KK (1999) Phytosterol content of experimental diets differing in fatty acid composition. *Food Chem.*, 64, 415–422.

Piironen, V, Lindsay, DG, Miettinen, TA, Toivo, J and Lampi, A-M (2000a) Plant sterols: biosynthesis, biological function and their importance to human nutrition. *J. Sci. Food Agric.*, 80, 939–966.

Piironen, V, Toivo, J and Lampi, A-M (2000b) Natural sources of dietary plant sterols. *J. Food Comp. Anal.*, 13, 619–624.

Plat, J (2001) *Plant Stanol Esters: Effects on Cardiovascular Risk Markers and Cholesterol Metabolism*, Datawyse, Maastricht.

Plat, J, van Onselen, ENM, van Heugten, MMA and Mensink, RP (2000) Effects on serum lipids, lipoproteins and fat soluble antioxidant concentrations of consumption frequency of margarines and shortenings enriched with plant stanol esters. *Eur. J. Clin. Nutr.*, 54(9), 671–677.

Plat, J and Mensink, RP (2000) Vegetable oil based versus wood based stanol ester mixtures: effects on serum lipids and hemostatic factors in non-hypercholesterolemic subjects. *Atherosclerosis*, 148, 101–112.

Plat, J and Mensink, RP (2001) Effects of plant sterols and stanols on lipid metabolism and cardiovascular risk. *Nutr. Metab. Cardiovasc. Dis.*, 11, 31–40.

Pollak, OJ (1953) Reduction of blood cholesterol in man. *Circulation*, 7, 702–706.

Pollak, OJ (1985) Effect of plant sterols on serum lipids and atherosclerosis. *Pharmaceut. Ther.*, 31, 177–208.

Pollak, OJ and Kritchevsky, D (1981) *Sitosterol (Monographs in Atherosclerosis 10)*, S Karger AG, Basel.

Rudd, EA and Brockman, HL (1984) Pancreatic carboxyl ester lipase (cholesterol esterase). In: *Lipases* (B Borgström and HL Brockman, eds), Elsevier, Amsterdam, pp.185–204.

Salen, G, Ahrens, EH and Grundy, SM (1970) Metabolism of beta-sitosterol in man. *J. Clin. Invest.*, 49, 952–967.

Sanders, DJ, Minter, HJ, Howes, D and Hepburn, PA (2000) The safety evaluation of phytosterol esters. Part 6. The comparative absorption and tissue distribution of phytosterols in the rat. *Food Chem. Toxicol.*, 38, 485–491.

Schlierf, G, Oster, P, Heuck, CC, Raetzer, H and Schellenberg, B (1978) Sitosterol in juvenile type II hyperlipoproteinemia. *Atherosclerosis*, 30, 245–248.

Schothorts, RC and Jekel, AA (1999) Oral sterol intake in the Netherlands: evaluation of the results obtained by GC analysis of duplicate 24-h diet samples collected in 1994. *Food Chem.*, 64, 561–566.

Schuhmann, P and Scheller, R (1996) Methode zur qualitativen und quantitativen bestimmung von phytosterinen in pflanzenölen mittels LC-GC off-line. *Mitt. Gebiete Lebensm. Hyg.*, 87, 708–715.

Sierksma, A, Weststrate, JA and Meijer, GW (1999) Spreads enriched with plant sterols, either esterfied 4,4,-dimethylsterols or free 4-desmetylsterols, and plasma total- and LDL-cholesterol concentrations. *Br. J. Nutr.*, 82, 273–282.

Sugano, M, Kamo, F, Ikeda, I and Morioka, H (1976) Lipid-lowering activity of phytostanols in rats. *Atherosclerosis*, 24, 301–309.

Sugano, M, Morioka, H and Ikeda, I (1977) A comparison of hypocholesterolemic activity of beta-sitosterol and beta-sitostanol in rats. *J. Nutr.*, 107, 2011–2019.

Swahn, R, Feaster, R and Quinn, DM (1997) Mechanism-based inhibitors of mammal cholesterol esterase. *Meth. Enzymol.*, 286, 231–252.

Tammi, A, Ronnemaa, T, Gylling, H, Rask-Nissila, L, Viikari, J, Tuominen, J, Pulkki, K and Simell, O (2000) Plant stanol ester margarine lowers serum total and low-density lipoprotein cholesterol concentrations of healthy children: the STRIP project. (Special Turku Coronary Risk Factors Intervention Project). *J. Pediatr.*, 136, 503–510.

Tang, JL, Armitage, JM, Lancaster, T, Silagy, CA, Fowler, GH, Neil, HAW, Smith, GD and Ebrahim, S (1998) Systematic review of dietary intervention trials to lower blood total cholesterol in free-living subjects. *Br. Med. J.*, 316, 1213–1220.

Tiainen, I, Nurmi, J and Klasi, L (1998) A product, a method for its production, and its use. World Patent Application WO 98/13023.

Tikkanen, I, Nurmi, J and Klasi, L (1998) A product, a method for its production, and its use. World Patent Application WO 98/13023.

Tikkanen, MJ, Hogstrom, P, Tuomilehto, J, Keinanen-Kiukaanniemi, S, Sundvall, J and Karppanen, H (2001) Effect of a diet based on low-fat foods enriched with nonesterified plant sterols and mineral nutrients on serum cholesterol. *Am. J. Cardiol.*, 88, 1157–1162.

Thompson, RH, Jr, Patterson, G, Thompson, MJ and Slover, HT (1981) Separation of pairs of C-24 epimeric sterols by glass capillary gas liquid chromatography. *Lipids*, 16, 694–699.

Vanhanen, HT and Miettinen, TA (1992) Effects of unsaturated and saturated dietary plant sterols on their serum contents. *Clin. Chim. Acta*, 205, 97–107.

Vanhanen, HT, Blomqvist, S, Ehnholm, C, Hyvonen, M, Jauhiainen, M, Torstila, I and Miettinen, TA (1993) Serum cholesterol, cholesterol precursors, and plant sterols in hypercholesterolemic subjects with different apoE phenotypes during dietary sitostanol ester treatment. *J. Lipid Res.*, 34, 1535–1544.

Vanhanen, HT, Kajander, J, Lehtovirta, H and Miettinen, TA (1994) Serum levels, absorption efficiency, faecal elimination and synthesis of cholesterol during increasing doses of dietary sitostanol esters in hypercholesterolaemic subjects. *Clin. Sci.*, 87, 61–67.

Van Horn, L and Ernst, N (2001) A summary of the science supporting the new National Cholesterol Education Program dietary recommendations: What dietitians should know. *J. Am. Diet. Assoc.*, 101, 1148–1154.

Vissers, MN, Zock, PL, Meijer, GW and Katan, MB (2000) Effect of plant sterols from rice bran oil and triterpene alcohols from sheanut oil on serum lipoprotein concentrations in humans. *Am. J. Clin. Nutr.*, 72, 1510–1515.

Volpe, R, Niittynen, L, Korpela, R, Sirtori, C, Bucci, A, Fraone, N and Pazzucconi, F (2001) Effects of yoghurt enriched with plant sterols on serum lipids in patients with moderate hypercholesterolaemia. *Br. J. Nutr.*, 86, 233–239.

Wester, I (1998) Texturizing compositions for use in fat blends in food. World Patent Application WO 98/19556.

Wester, I (1999) Dose responsiveness to plant stanol esters. *Eur. Heart J.*, Suppl. 1, S104–S108.

Weststrate, JA and Meijer, GW (1998) Plant sterol-enriched margarines and reduction of plasma total- and LDL-cholesterol concentrations in normocholesterolaemic and mildly hypercholesterolaemic subjects. *Eur. J. Clin. Nutr.*, 52, 334–343.

Williams, CL, Bollella, MC, Strobino, BA, Boccia, L and Campanaro, L (1999) Plant stanol ester and bran fiber in childhood: effects on lipids, stool weight and stool frequency in preschool children. *J. Am. Coll. Nutr.*, 18, 572–581.

Wright, JLC, McInnes, AG, Shimizu, S, Smith, DG and Walter, JA (1978) Identification of C-24 alkyl epimers of marine sterols by ^{13}C nuclear magnetic resonance spectroscopy. *Can. J. Chem.*, 56, 1898–1903.

Vuorio, AF, Gylling, H, Turtola, H, Kontula, K, Ketonen, P and Miettinen, TA (2000) Stanol ester margarine alone and with simvastatin lowers serum cholesterol in families with familial hypercholesterolemia caused by the FH-North Karelia mutation. *Arterioscler. Thromb. Vasc. Biol.*, 20, 500–506.

Yliruusi, J, Hiltunen, R and Petäjä, L (1999) Method for producing a fat mixture. World Patent Application WO 99/4321.

CHAPTER 8

Omega-3 (*n*-3) fatty acids

DUO LI[1], ORTWIN BODE[2], HAMISH DRUMMOND[2] and ANDREW J. SINCLAIR[3]

[1]*Hangzhou University of Commerce, Hangzhou, China;* [2]*Clover Corp Ltd, Sydney, New South Wales, Australia;* [3]*RMIT University, Melbourne, Victoria, Australia*

A. Introduction

Omega-3 (also referred to as ω-3 or *n*-3) fatty acids are a family of compounds, widely distributed in animal tissues and plants, which should be used in functional foods since they play a critical role in human health. The main *n*-3 fatty acids in food sources are α-linolenic acid (ALA; 18:3*n*-3), docosahexaenoic acid (DHA; 22:6*n*-3), eicosapentaenoic acid (EPA; 20:5*n*-3) and docosapentaenoic acid (DPA; 22:5*n*-3).

ALA, an essential fatty acid present as a relatively high proportion of the total fatty acids in some vegetable oils such as perilla and chia (60–70%), flaxseed (55–60%), canola (~10%), soybean (~7%), and walnut (~13%) oils, is the precursor of C_{20} and C_{22} long-chain *n*-3 polyunsaturated fatty acids (PUFA). DHA and EPA are the predominant long-chain *n*-3 PUFA found in fish, fish oils, and other marine organisms. DPA is a major long-chain *n*-3 PUFA found in fish, meat, and meat products.

DHA is specifically localized in the retina and the brain in humans and other mammals, where it has important effects on membrane order (fluidity), the activity of membrane-bound enzymes, ion channels, and signal transduction. *n*-3 PUFA affect the regulation of gene expression, such as the down-regulation of aggrecanases, interleukin-1α, tumor necrosis factor α (TNF-α), cyclooxygenase (COX), fatty acid synthase, acetyl-coenzyme A (CoA) carboxylase, S14 protein, and stearoyl-CoA desaturase; and the up-regulation of cytochrome P-450 4A2 and peroxisome proliferator-activated receptor α. The marine long-chain *n*-3 PUFA show beneficial effects by increasing heart rate variability, decreasing the risk of stroke, reducing serum triacylglycerol (TAG) levels, reducing systolic and diastolic blood pressure, reducing insulin resistance, and modulating glucose metabolism. *n*-3 PUFA also have anti-cancer and anti-inflammatory activities. They have been reported to have a beneficial effect on patients with attention-deficit/hyperactivity disorder (ADHD) and schizophrenia, and may be effective in managing depression in adults. Like linoleic acid (LA; 18:2*n*-6), ALA from plant sources is able to decrease plasma/serum total and LDL cholesterol levels. Consequently, *n*-3 fatty acids have been added to a range of food products, such as margarines, mayonnaise, milk, milk shakes, soy milk, yoghurt, ice cream, bread, and cereal bars in various countries.

B. Physiological functions of *n*-3 fatty acids

Fatty acids are major components of the phospholipids in most biological membranes, and long-chain *n*-3 and *n*-6 PUFA are important in membrane structure and function (Holman, 1968; Mitchell *et al.*, 1998). DHA is specifically highly concentrated in the retina and brain in humans and other mammals. DHA is essential for visual (Salem and Ward, 1993; Weisinger *et al.*, 1999) and brain function (Salem and Ward, 1993; Greiner *et al.*, 1999; Salem *et al.*, 2001), where it exerts a primary role through its effect on membrane fluidity (membrane order). This, in turn, influences the function of membrane receptors such as rhodopsin (Mitchell *et al.*, 1998; Salem *et al.*, 2001), the regulation of membrane-bound enzymes such as Na/K-dependent ATPase (Kurlack and Stephenson, 1999), and signal transduction through effects on inositol phosphates, diacylglycerols (DAG), and protein kinase C (Kurlack and Stephenson, 1999). DHA directly influences neurotransmitter biosynthesis, signal transduction, uptake of serotonin, binding of β-adrenergic and serotonergic receptors, and monoamine oxidase activity (Hibbeln and Salem, 1995; Salem *et al.*, 2001).

n-3 PUFA can alter gene expression. They down-regulate proteoglycan-degrading enzymes (aggrecanases), inflammation-inducible cytokines (interleukin-1α and TNF-α), COX-2 (Curtis *et al.*, 2000), fatty acid synthase, acetyl-CoA carboxylase, S14 protein, and stearoyl-CoA desaturase (Clarke and Jump, 1994; Pegorier, 1998). They up-regulate lipoprotein lipase fatty acid-binding protein, acyl-CoA synthetase, carnitine palmitoyltransferase 1, acyl-CoA dehydrogenase, acyl-CoA oxidase, cytochrome P-450 4A2, and peroxisome proliferator-activated receptor α (Clarke and Jump, 1994; Ren *et al.*, 1997; Pegorier, 1998; Price *et al.*, 2000).

The roles of EPA and DHA in cells include the regulation of eicosanoid production from arachidonic acid (AA; 20:4*n*-6), in that EPA competes with AA to produce various eicosanoids such as the 3-series prostaglandins, prostacyclin, and thromboxane, and the 5-series leukotrienes (LT) and lipoxins. EPA and DHA lower plasma TAG levels and blood pressure (De Deckere *et al.*, 1998; Mori *et al.*, 1999) and regulate ion flux in cardiac cells (Kang and Leaf, 2000). There have been successful trials of secondary prevention of myocardial infarction with both ALA-rich vegetable oils (Singh *et al.*, 1997; De Lorgeril *et al.*, 1999) and oils rich in EPA and DHA (Singh *et al.*, 1997; GISSI-Prevenzione Investigators, 1999). In addition, ALA, EPA, and DHA all prevented fatal arrhythmias in dogs following infusion as free fatty acids (Billman *et al.*, 1999). These observations show that ALA may also play an important role in the cardiovascular system.

C. Metabolism of *n*-3 fatty acids

There has been considerable discussion about whether the C_{18} fatty acids (ALA and LA) or their longer-chain metabolites (AA in the case of the n-6 family, and EPA and DHA in the case of the n-3 family) are essential dietary components. Specific functions have been described for LA: there is evidence that it is required for optimal skin function in the formation of ceramides (*N*-acyl-sphingosines), which are involved in regulating permeability to water (Hansen and Jensen, 1985); and 15-hydroxyoctadecadienoic acid (a 15-lipoxygenase derivative of LA) is an anti-proliferative compound (Ziboh *et al.*, 2000). In contrast, no specific functions have yet been described for ALA apart from a putative role in animal fur (Rokkones, 1953; Fiennes *et al.*, 1973; Fu and Sinclair, 2000a). The critical issue of whether dietary ALA can supply tissues with all the EPA and DHA needed for optimal function will be addressed later.

1. ALA metabolic pathways

ALA was long considered to be important only as a precursor of longer-chain compounds; however, it is now recognized that ALA is also metabolized by other pathways which may be quantitatively more important than the formation of DHA. Studies in humans and animals have revealed that a major catabolic route of metabolism of ALA is β-oxidation. In rats, ~60% of an oral dose of labelled ALA was expired as CO_2 within 24 h, compared with less than 20% for palmitic acid, stearic acid, and AA (Leyton *et al.*, 1987). Pan and Storlien (1993) reported that the higher the proportion of PUFA in the diet, the higher the rate of oxidation of labelled ALA in rats. In humans, 16–20% of ALA was expired as CO_2 within 12 h (Vermunt *et al.*, 2000).

Another major route of metabolism of ALA in rats and primates involves carbon recycling into *de novo* lipogenesis in the brains of pregnant, lactating, fetal, and infant animals. In the brain, ALA is β-oxidized to acetyl-CoA and these carbon atoms are used for the synthesis of saturated and monounsaturated fatty acids and cholesterol (Sinclair, 1975; Menard *et al.*, 1998; Sheaff Greiner *et al.*, 1996). When rats were given [^{13}C]-labelled ALA orally, incorporation of the label into cholesterol was 4–16 fold higher than into DHA in the developing rat brain (Menard *et al.*, 1998).

A third metabolic pathway for ALA is as a precursor for longer-chain more

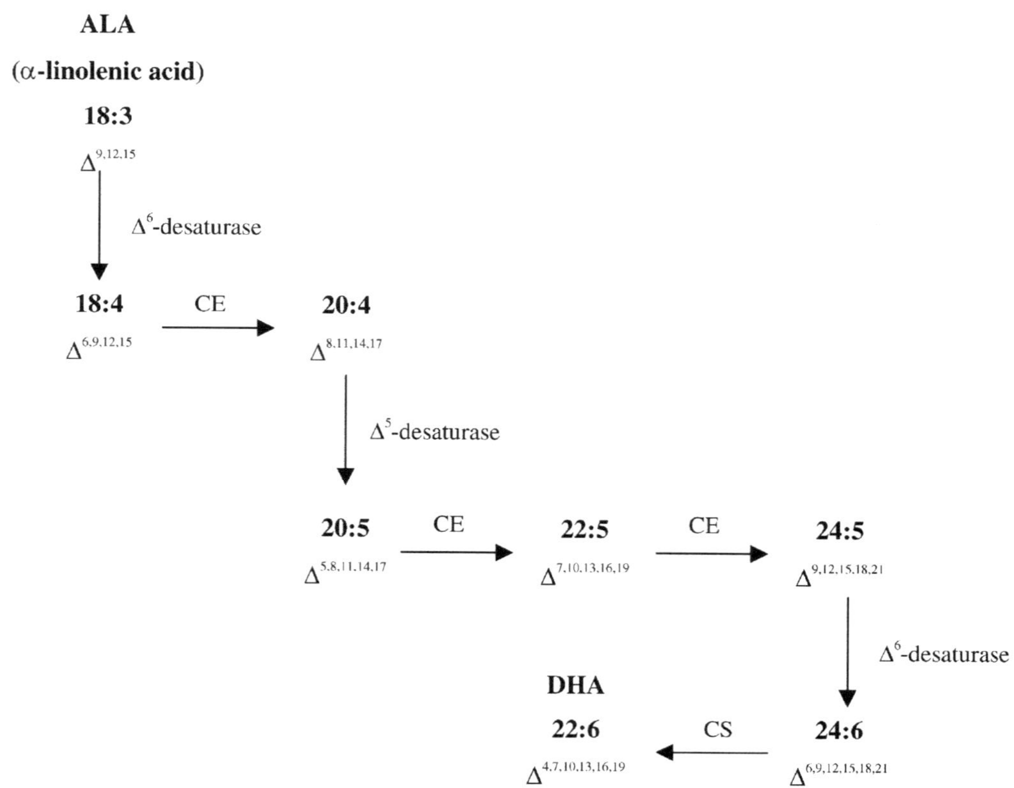

Figure 1. Metabolic steps involved in converting ALA (18:3*n*-3) to DHA (22:6*n*-3). CE, chain-elongation reaction. CS, chain-shortening reaction (peroxisomal oxidation).

unsaturated fatty acids such as EPA, DPA and DHA (Voss *et al.*, 1991). The pathway to DHA involves 7 different steps: three desaturations, three chain-elongations, and one chain-shortening (Figure 1). This pathway takes place mainly in the endoplasmic reticulum but the final chain-shortening step involves peroxisomal oxidation of 24:6*n*-3 to DHA (Sprecher *et al.*, 1998). Patients with a peroxisomal disorder known as Zellwegger's syndrome cannot synthesize DHA (Martinez *et al.*, 2000; Moore *et al.*, 1995).

Another pathway of ALA metabolism is deposition in skin (Fu and Sinclair, 2000a; Bowen and Clandinin, 2000) and secretion onto the fur of small mammals (Fu and Sinclair, 2000a). A recent study in guinea pigs reported that more than 46% of an oral dose of labelled ALA was incorporated into skin and fur lipids, with the highest proportion of the label being found in the fur (Fu and Sinclair, 2000b). Approximately 39% of the oral dose was oxidized to CO_2, indicating that nearly 90% of the ALA was unavailable for other metabolic pathways. The authors of the study proposed that the function of ALA on the fur was to protect the fur from damage by water, light, or other agents. Further preliminary data from our laboratory reveal that ALA is similarly metabolized onto the fur in rats and mice (Sinclair and Fu, unpublished observations). This observation may be related to the commonly reported use of flaxseed (linseed) oil by dog and horse owners to provide shine to the coats of their animals. While it has been widely reported that LA plays an important role in skin in preventing water loss and in maintaining epidermal integrity (Holman, 1968; Hansen and Jensen, 1985), some studies indicate that ALA might also play a role in skin and fur function. For example, an early study in rats showed that linseed oil (which is rich in ALA) contained a factor which promoted fur growth compared with a fat-deficient control group, and that ALA was more effective than LA in restoring fur growth (Rokkones, 1953). A later study in capuchin monkeys reported skin lesions, fur loss, and abnormal behavior following a diet very low in ALA, but rich in LA; normal skin and fur appearance were restored following the inclusion of linseed oil into the diet (Fiennes *et al.*, 1973). Ando *et al.* (1998) found that both LA and ALA lightened skin following ultraviolet light-induced hyperpigmentation of skin in guinea pigs, and suggested that the mechanism of action involved the suppression of melanin production and enhanced desquamation of the pigment from the epidermis. If a substantial disposal of ALA occurs via skin, fur and β-oxidation in other species, it might explain why ALA rarely accumulates in most tissues (Horrobin *et al.*, 1984).

It is well known that LA prevents abnormal water loss through the skin, which occurs in cases of essential fatty acid deficiency. However, ALA was not effective at reducing excessive water loss in rats (Hansen and Jensen, 1985). A case report on three elderly people revealed that a low ALA intake was associated with dry, scaly and atrophic skin (Bjerve, 1987), although there was no measurement of water loss in these subjects. In rhesus monkeys, ALA deficiency is associated with increased water intake, balanced by increased urination rather than by water loss through the skin (Reisbick *et al.*, 1992). A recent report indicated that perinatal *n*-3 fatty acid deficiency in rats altered water homeostasis and led to increased blood pressure by the time the rats were adults (Weisinger *et al.*, 2001).

2. *How effectively is ALA metabolized to DHA?*

There has been considerable discussion about the effectiveness of dietary ALA as a precursor for tissue EPA and DHA formation, especially in humans (Gerster, 1998). Several different approaches have been used to examine this question, including:

(1) feeding animals and humans with ALA and DHA, then comparing tissue levels of DHA;
(2) administering isotopically labelled ALA, then determining the fractional metabolism to DHA in available tissue compartments (such as plasma), or in various tissues in the case of animal studies;
(3) more recently, measuring the amount of mRNA coding for the desaturase enzymes in various tissues.

Feeding and labelling studies indicate that dietary DHA is a more effective source of tissue DHA than is ALA, on a gram-for-gram basis. Early studies in weanling rats revealed that 0.16% of a radiolabelled oral dose of DHA was recovered in brain DHA, while only 0.008% of a radiolabelled oral dose of ALA was recovered in brain DHA: a 20 fold difference (Sinclair, 1975). It has been estimated that 10 times more dietary ALA than DHA is required to maintain a given DHA level in guinea pig brains (Abedin *et al.*, 1999). Dietary DHA was found to be 7 times more effective than dietary ALA as a substrate for neural DHA in baboons (Greiner *et al.*, 1997).

Diets rich in ALA do not necessarily lead to high levels of tissue DHA. If animals are fed diets increasingly rich in ALA there is an increase in ALA levels in most tissues. This is accompanied by a proportionately smaller increase in EPA and DPA levels, and a still smaller increase in DHA levels – except in the case of nervous tissues, where the levels of DHA rise more than those of the other *n*-3 PUFA (Fu and Sinclair, 2000b). However, even very high levels of ALA fed to neonatal rats cannot maintain the same level of DHA in the nervous system as a well-nourished dam providing preformed DHA in her milk (Woods *et al.*, 1996). A similar situation has also been noted in ruminants, where tissue DHA levels are substantially lower than ALA, EPA and DPA levels in all tissues except nervous and reproductive tissues (Christie, 1981).

In human studies, ALA intakes of up to 15 g/day for 4 weeks lead to significant increases in ALA, EPA, and DPA contents in plasma TAG and phospholipids. But there is very little if any detectable increase in DHA levels in plasma, platelets, white cells, or red blood cells (Gerster, 1998; Li *et al.*,

1999a; Mantzioris *et al.*, 1994). At still greater intakes of ALA there has been one report of an increase in DHA as well: Mest *et al.* (1983) reported that 30 ml of flaxseed oil consumed daily with the usual diet by 10 healthy volunteers for 4 weeks led to a significant rise in EPA and DHA contents, and a significant decrease in AA content, in plasma phospholipids. In addition, longer-term studies in adults who increased their ALA intake by 3 g/day for 10 months found an increase in plasma DHA levels of >20% (Ezaki *et al.*, 1999). This dietary change was effected by substituting perilla oil (which is rich in ALA) for soy oil (which has only 10% the ALA level of perilla oil). In cross-sectional studies, DHA was found in both platelet and plasma phospholipid fractions of vegans, despite the lack of DHA in their diet (Phinney *et al.*, 1990; Li *et al.*, 1999b). These results suggest that DHA can be endogenously synthesized from the parent fatty acid ALA in humans to meet physiological needs, independent of dietary sources of EPA and DHA, especially for vegans. However, the *in vivo* conversion of dietary ALA to EPA and DHA appears to be a slow process. It is not as effective as direct consumption of EPA and DHA from fish or fish oil in raising the EPA and DHA contents of platelet and plasma lipids (Li *et al.*, 1999a). For example, Mann *et al.* (1997) reported that the EPA content of platelet phospholipids in 29 healthy adult omnivores increased from 0.4% to 1.9% of total fatty acids after 2 weeks on an Atlantic salmon diet (mean EPA intake: 847 mg/day). Li *et al.* (1999a) reported that a high-ALA diet (15.4 g/day) only increased the EPA content of platelet phospholipids from 0.2% to 0.5%. The EPA content of platelet phospholipids is increased markedly by dietary fish or fish oil supplements, moderately by linseed oil or lean red meat, and only slightly by canola oil.

Reasons proposed for the relatively poor conversion efficiency of ALA to EPA and DHA are that a substantial proportion of the ALA is diverted to β-oxidation (discussed above), that ALA is found distributed throughout all major tissue lipid pools (adipose, carcass, skin), and that in animals ALA might be excreted onto the fur (as discussed above). In addition, LA is the major dietary PUFA and is a competitive inhibitor of the metabolism of ALA to 18:4*n*-3 (commonly known as stearidonic acid; see Chapter 9), a precursor of long-chain *n*-3 PUFA (see Figure 1). Furthermore, diets rich in LA decrease expression of the hepatic Δ^6-desaturase compared with fat-free diets; this presumably also reduces the possibility of conversion of ALA to 18:4*n*-3 and of 24:5*n*-3 to 24:6*n*-3 (Cho *et al.*, 1999a).

A major rate-limiting step in the conversion of ALA to EPA and other long-chain *n*-3 PUFA is considered to be the first Δ^6-desaturation. This is demonstrated effectively by consideration of data from studies involving feeding 18:4*n*-3, the product of the Δ^6-desaturation of ALA, to humans or animals. In one study, a group of 32 stroke patients supplemented their diets with one of three diets for 3 weeks: (1) a mixture of 7.5% blackcurrant seed oil [rich in 18:4*n*-3 and γ-linolenic acid (GLA; 18:3*n*-6)], 50% soybean oil, and 42.5% medium-chain

TAG; (2) 100% soybean oil; or (3) 50% soybean oil plus 50% medium-chain TAG. The EPA content of red blood cell membrane phospholipids increased significantly (from 0.8% to 1.1%) in those consuming the blackcurrant seed oil (Diboune *et al.*, 1992). In another study, the EPA contents of TAG, cholesterol esters, and phospholipids of guinea pig liver were significantly higher after the guinea pigs were fed a diet containing 10% blackcurrant seed oil for 40 days compared with a diet containing 10% walnut oil (which contains ALA) (Crozier *et al.*, 1989). A similar result was found in rats by Yamazaki *et al.* (1992). Dietary sources of 18:4*n*-3 (blackcurrant seed oil, alpine currant seed oil, and oil from *Echium* sp.) might therefore be a viable source of long-chain *n*-3 PUFA, particularly for vegetarians. Because the formation of DHA from DPA also requires a Δ^6-desaturase, it is unlikely that any of the intermediate *n*-3 PUFA will be as effective a source of tissue DHA as dietary DHA itself (on a gram-for-gram basis).

Plasma DHA levels may not reflect DHA status in other tissues. In studies in humans (infants and adults) with either [^{13}C]- or [^{2}H]-labelled ALA, the main compartments that can be accessed are plasma, red blood cells, and white blood cells. All these studies report that labelled DHA is incorporated into plasma lipids over time, with substantial variability in the formation of isotope-labelled DHA being reported in infants (Salem *et al.*, 1996). It is presumed that most of the DHA found in these studies is derived from synthesis in the liver since that tissue is considered a major site of PUFA synthesis. If other tissues, such as the brain and testes, have specific capacities for PUFA formation, then examining the plasma might not reveal this capacity. For example, in guinea pigs fed 0.7% (w/w of total diet) ALA, DHA levels of <1% of liver and heart phospholipid fatty acids suggested little DHA synthesis, whereas DHA levels of 16% and 11% in retina and brain, respectively, implied DHA synthesis in these tissues (Abedin *et al.*, 1999). In other words, there are tissue-specific levels of PUFA, which presumably depend on a number of factors including the activity of the desaturase and elongase enzymes in the specific tissues, preferential transport, decreased degradation, etc. In cats, the liver apparently has the capacity to synthesize DPA from ALA. This is then transported to the brain, where DHA can be synthesized, presumably because of the presence of brain-specific enzymes for DHA synthesis (Pawlosky *et al.*, 1994). Recent studies on the cloning and expression of mammalian Δ^6- and Δ^5-desaturases support the idea that tissues other than liver have high levels of desaturase activity. For example, Cho *et al.* (1999a, 1999b) showed that the brain has a higher Δ^6-desaturase mRNA level than liver, lung, heart, or skeletal muscle in mice; in humans, the liver, brain and heart have a higher Δ^6-desaturase mRNA levels than other tissues.

In spite of the apparent differences between ALA and DHA as a source of tissue DHA, it is relevant to ask whether it really matters if all that is needed is a larger dose of ALA than would be required if only DHA were supplied. Based on the finding that LA reduces Δ^6-desaturase levels (Cho *et al.*, 1999a), an obvious mechanism to promote EPA and DHA synthesis would be to reduce the dietary LA level at the same time as increasing the ALA level (Woods *et al.*, 1996). This strategy has been used successfully in human dietary intervention studies (Mantzioris *et al.*, 1994). This question can also be answered by examining the function of the retina, since the physiological response of this tissue to light is positively correlated with the supply of *n*-3 fatty acids (Sinclair, 2000). One study has compared the effects of dietary ALA and dietary DHA on electroretinographic amplitude responses in guinea pigs (Weisinger *et al.*, 1996). This study used three dietary ALA levels (low, medium, and high) and a fourth diet containing EPA and DHA in addition to ALA. The maximum electroretinographic amplitude was found in the group given the highest ALA level (where ALA provided 2% of the total energy in the diet). The diet containing ALA together with long-chain *n*-3 PUFA, which gave the highest total *n*-3 PUFA intake (3.1% of energy), gave a lower electroretinographic signal than the high-ALA-only diet. Thus, dietary ALA was able to provide sufficient retinal DHA to support optimal electroretinographic function in the guinea pig (a species with rod-dominated vision).

3. *EPA metabolic pathways*

One major catabolic pathway for EPA metabolism is via β-oxidation (Ikeda *et al.*, 1998a). Another is incorporation into tissue membrane lipids. In tissues, EPA can be metabolized to DHA via chain-elongation and chain-shortening reactions (Figure 1). After EPA is released from membrane phospholipids by the action of phospholipase A_2, free EPA can be metabolized to biologically active compounds by three enzyme systems: COX, lipoxygenase, and cytochrome P-450. Lipoxygenases are cytosolic enzymes, while both COX and cytochrome P-450 are membrane-bound (Harder *et al.*, 1995). Both EPA and its metabolites may act as competitive antagonists of AA and its products at different levels of the metabolic pathways and receptor sites. The main pathways are summarized below.

In the cyclooxygenase-1 (COX-1) pathway (see the left-most pathway on Figure 2), molecular oxygen is added to free EPA by COX-1 to form prostaglandin G_3 (PGG_3), which is quickly converted to PGH_3 by a peroxidase. Both PGG_3 and PGH_3 are unstable, biologically active molecules (endoperoxides), which are intermediates in the transformation of EPA to 3-series prostanoids such as prostaglandins (PGD_3, PGE_3 and $PGF_{3\alpha}$), prostacyclin (PGI_3), and thromboxane A_3, all of which are anti-thrombotic agents (De Caterina *et al.*, 1993).

The cyclooxygenase-2 (COX-2) pathway (the middle pathway shown on Figure 2) is activated when NSAIDs (non-steroidal anti-inflammatory drugs) such as aspirin (acetylsalicylic acid; ASA) selectively acetylate the hydroxyl

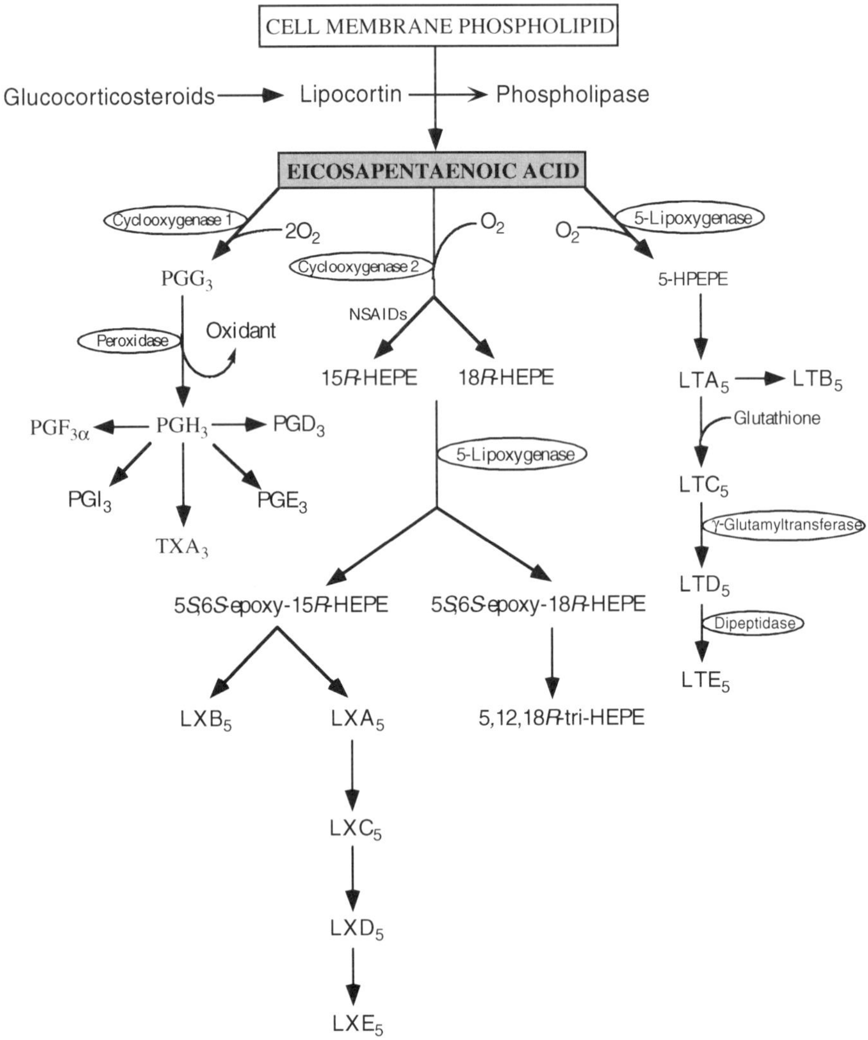

Figure 2. Biosynthetic pathways of EPA-derived eicosanoids. The binding of a stimulant (e.g. a glucocorticosteroid) to a membrane receptor results in the activation of phospholipase A_2, which cleaves EPA from membrane phosphatidylcholine. After its release from membrane phospholipids, free EPA can be metabolized by various enzyme systems to form a range of biologically active compounds. These are generally less potent than the AA-derived metabolites that compete with EPA for the enzymes. HEPE, hydroxy-EPA. HPEPE, hydroperoxy-EPA. NSAIDS, non-steroidal anti-inflammatory drugs. PGD_3, PGE_3, PGF_3, PGG_3, PGH_3, prostaglandins. PGI_3, prostacyclin. LXA_5, LXB_5, LXC_5, LXD_5, LXE_5, lipoxins. LTA_5, LTB_5, LTC_5, LTD_5, LTE_5, leukotrienes.

group of one serine residue within the polypeptide chain of COX-1, resulting in a reduction in all cyclooxygenase metabolites such as prostaglandins, thromboxane, and prostacyclin. However, the ASA-acetylated form of the enzyme, COX-2, is still active. Following treatment with *n*-3 PUFA and ASA, human cells generate several novel 18*R*- and 15*R*-hydroxy-EPA compounds (HEPE) from EPA (Serhan and Oliw, 2001). Polymorphonuclear leukocytes take up 18*R*-HEPE and convert it via 5-lipoxygenase to an unstable 5,6-epoxide that gives rise to 5,12,18*R*-tri-HEPE. In an analogous biosynthetic pathway, 15*R*-HEPE released by endothelial cells is converted by activated polymorphonuclear leukocytes via 5-lipoxygenase to a 5-series lipoxin (LX) analog that retains its C-15 *R* configuration, namely 15*R*-LXA_5. Trout macrophages and human leukocytes can convert endogenous EPA to 15*S*-containing LX or 5-series LX_5 compounds (Hill *et al.*, 1999). It has been suggested that stable synthetic analogues of lipoxins and the ASA-triggered 15-epi-lipoxins can mimic many of the desirable anti-inflammatory, "pro-resolution" actions of native lipoxins (McMahon *et al.*, 2001).

The biosynthesis of LT involves EPA being converted to 5-HPEPE (5*S*-hydroperoxy-EPA) by 5-lipoxygenase (see the right-most pathway on Figure 2). 5-HPEPE is converted by LTA synthetase to the labile intermediate 5,6-epoxide LTA_5, which may then be transformed to 5-series LT. LT are mainly produced by macrophages, monocytes, neutrophils, eosinophils, mast cells, and basophils. Additionally, transcellular synthesis of LTB_5 and LTC_5 from the 5,6-epoxide LTA_5 occurs in endothelial cells, platelets, mast cells, lymphocytes, and erythrocytes. The 5-series LT are less active than the potent AA-derived 4-series LT (De Caterina *et al.*, 1993).

The cytochrome P-450 pathway enzymes (not shown on Figure 2) are monooxygenases, whereas COX is a dioxygenase (Estabrook *et al.*, 1982). Unlike COX and lipoxygenases, cytochrome P-450 enzymes require several cofactors to metabolize fatty acids, including P-450 reductase and NADPH (Fitzpatrick and Murphy, 1989). In the presence of these cofactors and molecular oxygen, cytochrome P-450 can serve as the catalyst for the biotransformation of EPA to a variety of oxygenated metabolites (Fitzpatrick and Murphy, 1989). These include epoxides and hydroxy fatty acids such as epoxy-EPA (EEP) and HEPE (Kiss *et al.*, 1998).

D. Biological importance of *n*-3 PUFA in human health

1. Blood pressure

The effect of long-chain *n*-3 PUFA from fish on blood pressure has been evaluated in a meta-analysis of 31 placebo-controlled trials in 1356 subjects. The results indicated that in a group of studies on people with hypertension, average systolic blood pressures fell by 3.4 mm Hg and diastolic blood pressures fell 2.0 mm Hg following ingestion of fish oil (5.6 g/day) (Morris *et al.*, 1993). Two human studies recently provided further evidence of the importance of long-chain *n*-3 PUFA on blood pressure. In a dietary invention study (Bao *et al.*, 1998), 69 overweight (body mass index (BMI) of >25 kg/m^2) medication-treated hypertensive subjects were randomized among 4 regimens for 16 weeks. These were: (1) a daily meal of fish (providing ~3.65 g/day of long-chain *n*-3 PUFA), (2) a weight reduction programme, (3) regimens (1) and (2) combined, or (4) a control regimen. Sixty-three subjects completed the study. Systolic and diastolic blood pressure, body weight, and heart rate were significantly lower in the fish diet group than in the control group, even after adjustment for changes in urinary sodium and potassium levels, the sodium:potassium ratio, and dietary macronutrients. The second study was an observational study in which the effects of *n*-3 PUFA from fish on blood pressure, platelet fatty acid levels, and heart rate variability were investigated. This was conducted with 24 male and 19 female subjects aged 18–62 with type I diabetes mellitus, and 24 male and 14 female subjects aged 37–77 with type II diabetes mellitus (Christensen *et al.*, 2001). Each patient completed a food frequency questionnaire and gave a blood sample. Blood pressure, heart rate variability, plasma and lipoprotein lipids, and platelet fatty acid composition were measured. The study found that fish intake was significantly positively associated with platelet membrane DHA levels. According to their platelet DHA level, patients were divided into three tertiles: the 1st tertile (n=14) with the lowest DHA levels, the 3rd tertile (n=15) with the highest DHA levels, and the remainder with moderate DHA levels in the 2nd tertile. Compared with the 1st tertile, the 3rd tertile had significantly lower diastolic blood pressure and higher 24-h heart rate variability (increased heart rate variability has a beneficial effect on dysrhythmia). Platelet DHA level was significantly positively associated with 24-h heart rate variability in patients with type I diabetes mellitus, but this association was not significant in patients with type II diabetes mellitus. Mori *et al.* (1999) found that EPA and DHA differed in their effects on blood pressure and heart rate in a double-blind, randomized placebo-controlled human study. In this study, 55 overweight (BMI 25–30) subjects aged 20–65 were randomized to groups taking 4 g/day of either purified EPA, DHA, or placebo (olive oil) capsules for 6 weeks. Compared with the placebo group, DHA significantly reduced both the sleeping systolic and diastolic blood pressures (measured over 24 h), by 5.8 mm and 3.3 mm Hg, respectively, and the waking systolic and diastolic blood pressures by 3.5 mm and 2.0 mm Hg, respectively ($p<0.05$). Relative to the placebo group, the DHA group had an average 3.5±0.8 bpm (beats per minute) reduction in 24-h heart rate: 3.7±1.2 bpm when awake and 2.8±1.2 bpm when asleep.

However, EPA had no significant effect on blood pressure or heart rate. A recent study in rats found that *n*-3 PUFA deficiency in the perinatal period resulted in increased blood pressure later in life (Weisinger *et al.*, 2001). Forty

Sprague–Dawley rats were randomly divided into two groups receiving either a diet deficient in *n*-3 PUFA, or a diet adequate in *n*-3 PUFA and containing ALA, for 64 days. Half the animals in each group then crossed-over to the other group for 150 days. The results showed that the mean blood pressure was significantly higher in the group that was fed the PUFA-deficient diet throughout the study than in the other 3 groups, and also significantly higher in the group that switched from an initially PUFA-deficient diet to a PUFA-containing diet than in the group that was fed the PUFA-containing diet from the beginning. Although the study was carried out on rats, the implication of these findings is that an adequate *n*-3 PUFA intake at an early age may prevent increased blood pressure in later life in humans.

2. *Plasma/serum and lipoprotein lipids*

Based on early studies (Keys *et al.*, 1965; Hegsted *et al.*, 1965) it has been claimed that LA is the only fatty acid to appreciably lower plasma/serum total and LDL cholesterol levels when substituted for carbohydrate in the diet. However, more recent dietary intervention studies have found that ALA from plant sources has a similar effect on plasma/serum total and LDL cholesterol levels as LA (McDonald *et al.*, 1989, Li *et al.*, 1999a). Furthermore, *n*-3 PUFA, especially those from marine oils, reduce serum TAG levels (Svaneborg *et al.*, 1994; Zampelas *et al.*, 1994); this is considered to be important since raised serum TAG levels are now considered to be an independent risk factor for coronary heart disease (CHD) (Frontini *et al.*, 2001).

3. *Thrombosis*

Arterial thrombosis is generally considered to play a major role in the transition from stable to acute ischemic heart and cerebral diseases, manifested by unstable angina, acute thrombotic infarction, and sudden death. Platelet aggregation is an early event in the development of thrombosis. It is initiated by thromboxane A_2 (TXA_2), a potent platelet aggregation agent and vascular contractor which is produced from AA in the platelet membrane (Hamberg *et al.*, 1975; Moncada and Vane, 1979). EPA released from phospholipids of the platelet membrane competes with AA for access to COX, producing an alternative form of thromboxane, TXA_3 (see Figure 2), which is relatively inactive in promoting platelet aggregation and vasoconstriction (Raz *et al.*, 1977). This situation can lead to reduced production of TXA_2 and thus a decreased tendency to thrombosis (Dyerberg, 1986; Lands, 1986).

A diet with a high *n*-6 to *n*-3 PUFA ratio can cause a high tissue *n*-6 to *n*-3 PUFA ratio (i.e. a high AA to EPA ratio), which may promote the production of TXA_2, leading to increased tendency to thrombosis (Kinsella *et al.*, 1990). Evidence from dietary intervention studies suggests that the production of

TXA_2 was decreased by long-chain *n*-3 PUFA in humans (von Schacky *et al.*, 1985; Ferretti *et al.*, 1998) and in animals (Ikeda *et al.*, 1998b; Abeywardena *et al.*, 1991), and also by plant *n*-3 PUFA (ALA) in humans (Bjerve *et al.*, 1987) and in animals (Budowski *et al.*, 1980; Lee *et al.*, 1988). Two prospective cohort studies from the US Nurses' Health Study examined the effect of the intake of *n*-3 PUFA from plants and fish on ischemic heart and cerebrovascular diseases in women. In the first study (Hu *et al.*, 1999), the dietary intake of ALA was calculated from a food frequency questionnaire completed in 1984 by 76 283 nurses aged 38–63 and free from previously diagnosed cardiovascular disease and cancer. There were 597 cases of non-fatal myocardial infarction and 232 cases of fatal ischemic heart disease documented during 10 years of follow-up. After adjustment for confounding factors, the results showed that higher ALA intakes were significantly associated with reduced risk of fatal ischemic heart disease in this study population. In the second study (Iso *et al.*, 2001), 79 839 female nurses, aged 34–59 in 1980 and without prior diagnosed cardiovascular disease, cancer, or history of diabetes or hypercholesterolemia, completed a food frequency questionnaire. After 14 years of follow-up, 574 stroke incidents were documented, including 303 ischemic strokes (264 thrombotic and 39 embolic infarctions). Compared with women who ate fish less than once per month, those with higher intakes of long-chain *n*-3 PUFA from fish had a lower risk of total strokes. Among stroke sub-types, a significantly reduced risk of thrombotic infarction was found among women who ate fish two or more times per week. These data indicate that higher fish consumption is associated with reduced risk of thrombotic infarction. These results confirm that an increased dietary *n*-3 PUFA intake protects against cardiovascular thrombotic infarction. A recent cross-sectional study of 4584 participants in the National Heart, Lung and Blood Institute "Family Heart" study revealed that higher intakes of either ALA or LA, assessed using a semi-quantitative food frequency questionnaire, were inversely related to the prevalence of coronary artery disease (Djousse *et al.*, 2001). In this study, the two fatty acids had synergistic effects on reducing the prevalence of coronary artery disease.

4. Secondary prevention of cardiovascular disease

Do marine *n*-3 oils make any real difference as to whether we will suffer from fewer adverse cardiac events? The first prospective study to explore the cardioprotective effect of *n*-3 fatty acids in a secondary prevention trial was the Diet and Reinfarction Trial. Burr *et al.* (1989) studied 2013 men who had survived a heart attack. Half were advised to eat oily fish twice a week or to take marine *n*-3 oil capsules giving an equivalent amount of *n*-3 fatty acid as the fish diet, while the other half were advised only to eat a "prudent" diet. Survival over the following 2 years was monitored. The group advised to consume oily fish had a 29% lower overall 2-year mortality than the control group. These results point to a possible protective effect for *n*-3 fatty acids during ischemia and reperfusion, as already indicated by animal studies. Another secondary prevention trial, the Lyon Diet Heart Study, aimed to reduce the risk of death from cardiovascular disease by dietary modification (de Lorgeril *et al.*, 1999). The experimental group (n=302) who consumed a "Mediterranean" diet rich in ALA had a significantly lower incidence of cardiac deaths and non-fatal myocardial infarctions than the control group.

Another prospective, randomized clinical trial using *n*-3 fatty acids was reported by Singh *et al.* (1997). Patients presenting with suspected myocardial infarctions (n=360) were randomized to three groups receiving either a placebo, a marine *n*-3 oil (providing 2 g EPA+DHA per day), or mustard seed oil (providing 2.9 g ALA per day). After one year, CHD events were significantly lower in both groups receiving *n*-3 fatty acids. Von Schacky *et al.* (1999) recently reported a small but statistically significant ($p<0.041$) reduction in angiographically determined CHD progression in a study in which 223 patients were provided with either a placebo or long-chain *n*-3 fatty acids (6 g/day for the first 3 months, followed by 3 g/day for a further 21 months). During the study there were 7 cardiovascular events in the placebo group and 2 in the *n*-3 group, although the difference was not statistically significant (p=0.10).

The most recent test of the effects of marine *n*-3 fatty acids on CHD morbidity and mortality was the GISSI-Prevention Trial (GISSI-Prevenzione Investigators, 1999). This study was conducted in Italy and included 11 324 patients with known CHD. In a factorial design, 2830 of the patients were assigned to take vitamin E (300 mg/day), another 2836 were given 850 mg of long-chain *n*-3 fatty acids daily, another 2830 were given both, and the final group of 2828 received neither. After 3.5 years of follow up, analysis revealed that total mortality in the patients given long-chain *n*-3 fatty acids was 20% lower than in those patients not so treated, and the incidence of sudden cardiac death was 45% lower. Vitamin E showed a slight benefit but this was not statistically significant. These results were achieved despite the fact that over 25% of patients reported that they stopped taking the *n*-3 fatty acid supplements partway through the trial. The findings of the GISSI-Prevention study provides strong support for the use of marine *n*-3 fatty acids in the secondary prevention of acute coronary syndromes. The mechanism by which marine *n*-3 fatty acids protect against cardiac death is not known with certainty, but it may relate to their ability to prevent cellular damage during periods of ischemic stress.

5. *Cancers*

Numerous studies have investigated the effects of individual dietary fatty acids on cancers in experimental models. In general, high intakes of *n*-6 PUFA favored the development of tumors, whereas equivalent levels of *n*-3 PUFA reduced or protected against tumor development, as reviewed by Cave (1997). Compared with *n*-6 PUFA, long-chain *n*-3 PUFA from fish oil had protected against the development of colon cancer in rats (Collett *et al.*, 2001). A recent study by Igarashi and Miyazawa (2000), which compared the effects of conjugated linoleic acid (CLA) and conjugated linolenic acid (CALA) on cultured human tumor cells, found that CALA was more cytotoxic to human tumor cells than CLA. DHA was more effective than EPA at inhibiting transcription factor activator protein 1 (AP-1), which has been implicated in the development of cancer in an experimental animal model (Liu *et al.*, 2001a).

However, there are few publications on human studies. The consumption of fatty fish, high in long-chain *n*-3 PUFA, decreased prostate cancer risk in a population-based prospective study from Sweden in 6272 men (Terry *et al.*, 2001). Men who ate moderate or high amounts of fish had a 2–3 fold lower frequency of prostate cancer than men who ate no fish. A second recent prospective cohort study from Finland reported the effects of dietary fat and cholesterol levels on colorectal cancer risk in 9959 men and women initially free of diagnosed cancer (Jarvinen *et al.*, 2001). Baseline information was collected from the subjects during 1967–1972, and 109 cases of colorectal cancer were documented in late 1999. After adjustment for other confounding factors, a high cholesterol intake was associated with increased colorectal cancer risk, while intakes of total fat, saturated fat, monounsaturated fat, and PUFA were not significantly associated with colorectal cancer risk.

6. *Inflammation*

AA is a substrate for pro-inflammatory eicosanoids such as PGE_2 and LTB_4, while long-chain *n*-3 PUFA such as DHA and EPA, and *n*-6 PUFA such GLA are substrates for eicosanoids that are antagonistic towards those produced from AA (Ziboh *et al.*, 2000). Long-chain *n*-3 PUFA and GLA are claimed to have anti-inflammatory effects, and the long-chain *n*-3 PUFA especially are thought to ameliorate rheumatoid arthritis and related disorders (James and Cleland, 1997). AA-derived LT, initially recognized as "slow-reacting substances of anaphylaxis", are mediators of allergic responses and inflammation (Mayatepek and Hoffmann, 1995). AA can also increase the formation of other pro-inflammatory substances, such as cytokines, TNF-α, interleukin-6, and reactive oxygen species, as reviewed by Darlington and Stone (2001). EPA and DHA compete at the COX and lipoxygenase level with AA, resulting in decreased production of AA-derived pro-inflammatory eicosanoids, cytokines, interleukin-6, and reactive oxygen species (Calder and Zurier, 2001). The 5-series LT, produced from EPA, are inactive relative to the AA-derived 4-series LT (Hammarstrom, 1982; James and Cleland, 1997). GLA can be elongated *in vivo* to 20:3*n*-6, an immediate precursor of prostaglandin E_1, an eicosanoid with known anti-inflammatory and immunoregulatory properties (Fantone *et al.*,

1983; Rothman *et al.*, 1995). A recent study found that GLA reduces the production of interleukin 1β (IL-1β), an important mediator of joint tissue injury and inflammation, by human monocytes *in vitro* (Furse *et al.*, 2001).

Clinical trials of the ability of marine *n*-3 oils to prevent inflammatory conditions such as arthritis have led to improvements in the condition of the subjects in all trials (Adam, 1995; James and Cleland, 1997). The clinical signs most affected were improvements in joint tenderness, morning stiffness, grip strength, and the interval to fatigue onset. In these studies, *n*-3 oils significantly decreased the production of LTB_4 from stimulated neutrophils. Studies of rheumatoid arthritis patients have shown that marine *n*-3 oils also decrease the production of other pro-inflammatory mediators (probably as a result of the reduced levels of LT), such as interleukin-1, platelet-activating factor, and TNF-α (Adam, 1995; James and Cleland, 1997). Although the dose of marine *n*-3 oil has to be quite high to achieve results (~5 g EPA+DHA per day), there is considerable interest in the combined use of marine *n*-3 oil and conventional anti-inflammatory drugs to reduce the pain and inconvenience caused by rheumatoid arthritis (James and Cleland, 1997). Vegetarian diets have also been reported to ameliorate the symptoms of arthritis (Adam, 1995), and diets containing up to 4 g/day of ALA have been recommended (Adam, 1995).

Donadio *et al.* (1994) showed that long-chain *n*-3 PUFA can lead to clinical benefits in immunoglobulin A (IgA) nephropathy, a glomerulonephritis which results in chronic renal insufficiency. A multi-center, placebo-controlled, randomized trial involving more than 100 patients given 12 g of long-chain *n*-3 PUFA daily for 2 years showed that only 6% of patients in the treatment group developed progressive renal failure, compared with 33% in the placebo group (p=0.002). Belluzzi *et al.* (1996) showed that long-chain *n*-3 PUFA were effective in maintaining remission in 78 patients with Crohn's disease, an inflammatory bowel disease. After 1 year of treatment, 59% of patients in the group receiving 2.7 g/day of long-chain *n*-3 PUFA were still in remission, compared with only 26% in the placebo group (p=0.006).

7. *Neuropsychiatric disorders*

Manic–depressive illness (bipolar disorder), depression, and schizophrenia are common neuropsychiatric disorders. Results from case–control studies and clinical trails have shown that long-chain *n*-3 PUFA play a regulatory role in neuropsychiatry performance. It has been suggested that a lower serum/plasma cholesterol concentration is associated with an increased risk of suicide and depression (Hibbeln and Salem, 1995). Results from a recent 4-month, double-blind, placebo-controlled trial with 30 patients aged 18–65 with type I or type II bipolar disorder showed that episodes of severe mania and depression were significantly lower in the long-chain *n*-3 PUFA supplementation group (n=14) (9.6 g/day) than in the placebo group (n=16) (Stoll *et al.*, 1999).

Decreased long-chain *n*-3 PUFA levels have been reported in the serum phospholipids and cholesterol esters of depressive patients (Adams *et al.*, 1996; Maes *et al.*, 1999), and in the erythrocyte membranes of schizophrenic patients (Richardson *et al.*, 2000; Assies *et al.*, 2001). As reviewed by Richardson and Puri (2000) and Kidd (2000), abnormalities of C_{20} and C_{22} PUFA such as AA, EPA, and DHA are associated with ADHD. A recent prospective 5–8 year follow-up study in 29 133 men aged 50–69 showed that low serum cholesterol concentrations were associated with low mood, resulting in an increased risk of hospital treatment due to major depressive disorder and attempted suicide (Partonen *et al.*, 1999). A recent case–control study found that serum cholesterol concentrations were significantly lower in parasuicide subjects than in 331 sex- and age-matched controls ($p<0.001$) (Gallerani *et al.*, 1995). Mufti *et al.* (1998) reported that low serum cholesterol levels were strongly associated with violent behavior in 20 psychiatric patients. These results suggest that the widespread use of cholesterol-lowering drugs could lead to changes in mood.

The proposed mechanisms of action of *n*-3 PUFA in neuropsychiatric disorders include effects on neurotransmitter receptors and G-proteins via effects on the biophysical properties of cell membranes, secondary messengers, and protein kinases (Hibbeln and Salem, 1995; Edwards and Peet, 1999). The biophysical properties of synaptic membranes directly affect neurotransmitter biosynthesis, signal transduction, serotonin uptake, the binding of α-adrenergic and serotonergic receptors, and monoamine oxidase activity. These factors are all implicated in the role of long-chain *n*-3 PUFA in neurobiology. It has been reported that EPA is able to reverse the phospholipid abnormalities in patients with schizophrenia via inhibition of PUFA-specific phospholipase A_2, an enzyme that removes PUFA from the *sn*-2 position of membrane phospholipids, or by activation of a fatty acid CoA ligase (Richardson *et al.*, 2000). It may also be possible that *n*-3 PUFA operate through *N*-acyl-ethanolamines (NAEs) and 2-acylglycerols, since these lipids are endogenous ligands for the cannabinoid receptors found predominantly in the brain (Hillard and Campbell, 1997). Brain levels of bioactive NAEs increased when piglets were fed a diet containing added AA and DHA for 18 days: levels of AA-NAE increased 4 fold, EPA-NAE increased 5 fold, DPA-NAE increased 9 fold, and DHA-NAE increased 10 fold, compared with a diet without added AA and DHA (Berger *et al.*, 2001).

8. *Obesity*

The consumption of 1,3-DAG results in a smaller postprandial increase in plasma TAG concentrations than the consumption of TAG (Taguchi *et al.*, 2000) (see also Chapter 5). Furthermore, short-term and long-term consumption of DAG in adults is associated with weight loss (Nagao *et al.*, 2000). A recent study in mice fed a high-sucrose diet examined the effects of replacing dietary TAG with an ALA-rich 1,3-DAG (in which ALA provided 61% of the fatty acyl groups). Serum leptin and insulin levels increased in the high-TAG group over a 20-week period. However, replacement of 3% (w/w) of the dietary TAG by the ALA-rich 1,3-DAG resulted in reduced body weight and visceral fat, and lower serum insulin and leptin levels (Hase *et al.*, 2001). A study in humans on calorie-restricted diets found that those who consumed 2.5–3.7 g/day of an ALA-rich DAG (containing 49% ALA) for 12–16 weeks had a significant reduction in body fat area (Takei *et al.*, 2001). A second study in humans, using 2 g/day of an ALA-rich DAG (containing 59% ALA) for 6–12 weeks, showed a significant reduction in visceral fat area, a significant reduction in very-low-density lipoprotein (VLDL) TAG levels, and a significant increase in resting oxygen consumption rates. The authors concluded that the reduction in visceral fat area was the result of reduced TAG synthesis and increased oxidation of fatty acids (Katsuragi *et al.*, 2001).

E. Why is there a need for *n*-3 fatty acids in functional foods?

1. Current intakes of n-3 PUFA

The rationale for including *n*-3 fatty acids in foods is based on various studies which have shown that the intakes of *n*-3 PUFA in various populations (the UK, USA, Canada, Australia, and presumably also other countries) are below the lowest of the recommended intakes. This is 1.5 g/week (215 mg/day) of long-chain *n*-3 PUFA (UK Department of Health, 1991; Newton, 1998). To put this figure into perspective, it equates to consuming one 100 g can of tuna in water daily throughout the year, and for every year of life. This intake is certainly not achieved in Australia and the USA, where estimated intakes are in the range of <100–200 mg/day (Kris-Etherton *et al.*, 1998; Sinclair and Vingrys, 1998). Other groups have recommended even higher intakes of long-chain *n*-3 PUFA: the British Nutrition Foundation (1992) recommended an intake of 1.25 g/day, and Simopoulos *et al.* (2000) recommended 650 mg/day. Whatever the recommended intake chosen, it is clear that there is a substantial gap between currently achieved and recommended intakes. This gap leaves an opportunity for functional foods containing *n*-3 PUFA.

2. Dietary sources of n-3 PUFA

LA, which is commonly found in vegetable seed oils, is the most predominant PUFA in our diet. ALA is less abundant than LA; however, ALA is present in some vegetable oils such as perilla, flaxseed, canola, soybean and walnut oils (Table 1) and in green vegetables (Table 2). Fish and fish oils are the main sources of long-chain *n*-3 PUFA (e.g. DHA, EPA, and DPA) in the human diet (Tables 3 and 4). Other dietary sources of long-chain *n*-3 PUFA include lean

Table 1. ALA contents of vegetable oils/fats[a]

Oil	ALA content (g/100g oil)
Perilla oil	>60
Flaxseed oil	60
Canola oil	8
Walnut oil	7
Soybean oil	6
Corn (maize) oil	0.7
Olive oil	0.6
Safflower oil	0.4
Sesame oil	0.3
Peanut oil	0.2
Palm oil	0.2
Evening primrose oil	0.2
Cocoa butter	0.1

[a]Data taken from Gunstone *et al.* (1994).

Table 2. *n*-3 Fatty acid contents of common fresh green vegetables[a]

	Fatty acid content (mg/100 g fresh product)		
Vegetable	16:3*n*-3	18:3*n*-3 (ALA)	Total
Spinach	21 ± 2	129 ± 16	150 ± 19
Watercress	45 ± 14	180 ± 71	225 ± 83
Parsley	44 ± 12	125 ± 34	169 ± 46
Chinese cabbage	0.6 ± 0.3	22 ± 3	23 ± 4
Brussels sprouts	3 ± 2	98 ± 25	101 ± 24
Bok choy	22 ± 4	103 ± 16	125 ± 20
Cos lettuce	ND	94 ± 19	94 ± 19
Broccoli	4 ± 0.6	110 ± 16	114 ± 17
Chinese broccoli	15 ± 2	68 ± 10	83 ±12
Baby bok choy	3 ± 0.7	28 ± 6	31 ± 6
Mint	ND	195 ± 39	195 ± 39

[a]Data adapted from Pereira *et al.* (2001). Values are means ± SD. ND, not detected.

meat and meat products, offal, egg yolk, and milk and dairy products (Tables 5 and 6). Omnivores can obtain their long-chain *n*-3 PUFA either from dietary ALA or directly from the consumption of fish, eggs, or animal products. Lacto-ovo vegetarians can gain a limited amount of long-chain *n*-3 PUFA from milk, dairy products, and eggs. In contrast, vegans must rely totally on endogenous synthesis from ALA by desaturation and elongation reactions. Because plants cannot normally convert ALA to long-chain *n*-3 PUFA, there are no long-chain *n*-3 PUFA in plant-based vegan diets.

Table 3. n-3 Fatty acid contents of fresh fish[a]

Fish	Fatty acid content (mg/100 g edible flesh)				
	18:3n-3 (ALA)	20:5n-3 (EPA)	22:5n-3 (DPA)	22:6n-3 (DHA)	Total[b]
Barracouta	2	22	10	171	205
Bream, black	2	34	24	155	215
Bream, golden	3	43	34	168	247
Flathead, rock	1	41	16	105	163
Flathead, sand	22	51	28	233	334
Flathead, tiger	1	19	6	155	187
Flounder, greenback	9	111	72	55	247
Garfish, southern sea	27	46	47	276	396
Gemfish	13	44	38	390	485
Grenadier, blue	3	142	118	391	654
Groper, blue	2	17	20	143	182
Gurnard, red	10	74	48	336	468
Gurnard, perch	11	15	8	130	164
Jackass Fish	1	39	15	161	216
John Dory	1	20	10	171	202
Leatherjacket, Chinaman	2	46	12	144	204
Leatherjacket, Scaber	3	39	17	151	210
Ling, pink	1	21	10	168	200
Ling, rock	2	17	8	161	188
Luderick	35	72	55	51	213
Mackerel, blue	53	251	129	787	1220
Mullet, red	3	87	28	220	338
Mullet, sea	120	1318	635	286	2359
Nannygai	3	47	23	177	250
Orange roughy	16	22	8	69	115
Oyster, Sydney rock	313	494	108	502	1417
Perch, golden	111	79	124	302	616
Pike, long-finned	7	27	34	123	191
Salmon, Australian	14	96	28	512	650
Salmon, Atlantic	80	472	439	1147	2138
Scad, yellowtail	83	240	250	1101	1674
Scallop, bay	21	140	26	138	325
Snapper	5	104	52	295	456
Tailor	262	22	251	827	1360
Tarwhine	66	381	192	643	1282
Trevally	72	444	248	1168	1932
Trout, rainbow (cultivated)	35	53	26	442	556
Trout, rainbow (wild)	162	97	55	336	650
Tuna, southern bluefin	3	230	115	804	1152
Whiting, King George	22	225	136	194	577
Whiting, sand	3	57	22	176	258
Whiting, school	3	60	21	144	228

[a]Data adapted from Sinclair *et al.* (1992). [b]Note that the totals include other n-3 fatty acids, such as 20:4n-3, in addition to the main ones listed, and so not all rows will add up exactly.

Table 4. *n*-3 Fatty acid contents of commercial canned fish products[a]

Brand	Fatty acid content (mg/100 g edible flesh)					
	18:3*n*-3 (ALA)	18:4*n*-3	20:5*n*-3 (EPA)	22:5*n*-3 (DPA)	22:6*n*-3 (DHA)	Total
Mackerel:						
Assi	54	74	558	86	1148	1920
John West	217	639	937	147	1630	3570
Admiral	178	431	700	161	1606	3075
Seawall	34	52	407	163	753	1409
Salmon:						
Greenseas (Australian)	9	13	69	25	381	497
Savings (Australian)	29	35	168	69	974	1275
Safcol (Australian)	36	72	319	74	588	1089
Rex (pink)	22	24	268	153	443	909
John West (pink)	44	138	263	60	610	1114
Parramount (pink)	76	154	552	157	956	1895
Farmland (pink)	44	90	273	80	443	930
Safcol (silver)	16	15	158	92	672	953
John West (red)	108	236	888	206	1301	2738
Parramount (red)	89	304	618	137	985	2133
Farmland (red)	34	67	296	72	490	960
Sardine:						
John West	175	326	816	129	1391	2837
King Oscar	144	298	1255	92	2256	4044
Admiral	55	115	309	41	488	1009
Tuna:						
Ayam	1	1	23	6	180	212
Rex	2	2	52	15	209	281
Greenseas	1	2	28	4	207	242
Farmland	4	5	51	16	293	370
Safcol	1	1	18	4	153	177

[a]Data adapted from Sinclair *et al.* (1998).

Some novel seed oils contain a relatively high proportion of 18:4*n*-3, such as oils from blackcurrant seeds, redcurrant seeds, alpine currant seeds, and *Echium* species (Table 7). As a precursor of 20:4*n*-3 (see Figure 1), 18:4*n*-3 is regarded as a good source of long-chain *n*-3 PUFA and appears to possess a higher biological activity than ALA (Petrick *et al.*, 2000).

F. Novel sources of long-chain *n*-3 PUFA

Several groups around the world have investigated the fermentation of microalgae and algae-like microorganisms as sources of PUFA, since in the marine environment these are rich sources of long-chain *n*-3 PUFA (Barclay *et*

Table 5. n-3 Fatty acid contents of lean meat and visible fat of common meats, milk, and egg[a]

	Fatty acid content (mg/100 g product)				
Product	**18:3n-3 (ALA)**	**20:5 n-3 (EPA)**	**22:5 n-3 (DPA)**	**22:6 n-3 (DHA)**	**Total[b]**
Lean meat:					
Beef	16 ± 2	16 ± 2	23 ± 4	3 ± 1	58 ± 9
Lamb	65 ± 18	23 ± 4	22 ± 1	8 ± 2	118 ± 24
Chicken	16 ± 8	4 ± 0	8 ± 2	6 ± 2	34 ± 11
Pork	8 ± 2	3 ± 1	7 ± 1	3 ± 1	21 ± 4
Duck	6 ± 3	2 ± 0	12 ± 3	10 ± 3	30 ± 7
Turkey	4 ± 1	2 ± 1	9 ± 0	8 ± 1	22 ± 2
Visible fat:					
Beef	441 ± 50	24 ± 7	ND	ND	467 ± 40
Lamb	850 ± 23	23 ± 9	ND	ND	889 ± 103
Chicken	646 ± 96	20 ± 3	ND	ND	666 ± 93
Pork	726 ± 143	ND	ND	ND	822 ± 158
Duck	564 ± 138	29 ± 0	ND	ND	574 ± 148
Turkey	508 ± 214	6 ± 0	ND	ND	510 ± 211
Milk	22	Tr	5	5	32
Egg	40 ± 8	Tr	32 ± 7	62 ± 9	134 ± 24

[a]Data adapted from Li *et al.* (1998). Most values are means ± SD. [b]Note that the totals include other n-3 fatty acids, such as 20:3n-3, in addition to the main ones listed, and so not all rows will add up exactly. ND, not detected. Tr, trace.

al., 1998; Kyle, 2001; Zeller *et al.*, 2001). Driving forces are the possibly inadequate supply of fish oils obtained from the natural harvesting of ocean fish, and the pressure to find pure sources of PUFA such as AA and DHA, particularly for inclusion in infant formulas (Kyle, 2001). Additional benefits of these newer sources include reduced taste and odor problems, higher levels of DHA in the oils produced, the use of environmentally sustainable resources, and the minimal content of environmental contaminants. The two organisms which have been used to produce DHA-rich oils in commercial quantities are *Crypthecodinium cohnii* (a member of the Dinophyta) and *Schizochytrium* sp. (a member of the kingdom Chromista) (Kyle, 2001; Zeller *et al.*, 2001). Fermentation of these microorganisms yields oils rich in specific PUFA that can be included in nutritional supplements or in foods; alternatively the microalgae can be used directly as animal feed ingredients to produce eggs, chicken, and pork meat enriched in DHA (Abril and Barclay, 1998; Zeller *et al.*, 2001).

Major advances in the cloning and manipulation of fatty acid desaturase and elongase genes from various organisms have been made during the past several years, and these recombinant enzymes offer the prospect of producing a desired

Table 6. *n*-3 Fatty acid contents of canned meats commonly available in Australia[a]

	Fatty acid content (mg/100 g product)				
Product (country of origin)	**18:3*n*-3 (ALA)**	**20:5*n*-3 (EPA)**	**22:5*n*-3 (DPA)**	**22:6*n*-3 (DHA)**	**Total[b]**
Corned beef (France)	53 ± 2	8 ± 0	19 ± 1	ND	82 ± 3
Corned beef (Australia) sample 1	67 ± 2	14 ± 0	24 ± 0	ND	105 ± 2
Corned beef (Australia) sample 2	97 ± 2	23 ± 0	34 ± 1	ND	154 ± 3
'Camp Pie' beef (Australia)	71 ± 0	7 ± 1	14 ± 0	ND	92 ± 1
Beef goulash (Croatia)	27 ± 4	6 ± 1	12 ± 1	ND	44 ± 6
Corned mutton (Australia) sample 1	191 ± 1	20 ± 1	38 ± 0	10 ± 0	262 ± 2
Corned mutton (Australia) sample 2	141 ± 3	18 ± 0	28 ± 0	8 ± 1	195 ± 3
Shoulder ham (Australia)	59 ± 3	4 ± 1	12 ± 0	6 ± 0	91 ± 3
Leg ham (Australia)	71 ± 2	5 ± 0	18 ± 0	13 ± 1	120 ± 3
Stewed pork leg (China)	283 ± 11	18 ± 1	40 ± 1	8 ± 0	436 ± 17
Stewed pork (China)	350 ± 56	18 ± 2	45 ± 7	12 ± 2	541 ± 84
Leg ham (the Netherlands)	39 ± 5	4 ± 1	10 ± 1	5 ± 1	64 ± 8
Leg ham (USA)	52 ± 4	ND	9 ± 1	ND	70 ± 8
Pantry ham (USA)	78 ± 6	ND	9 ± 1	ND	98 ± 7
Spiced ham (USA)	98 ± 7	ND	14 ± 2	ND	130 ± 8
Chicken hot dog (Denmark)	590 ± 25	24 ± 0	21 ± 1	18 ± 1	659 ± 25
Chicken luncheon meat (Denmark)	580 ± 30	20 ± 3	20 ± 2	17 ± 1	643 ± 35
Chicken luncheon meat (Jordan)	63 ± 1	ND	5 ± 0	4 ± 0	73 ± 1
Chicken (USA)	14 ± 0	1 ± 0	5 ± 0	10 ± 1	30 ± 1
Roasted goose (China)	486 ± 10	5 ± 0	19 ± 0	11 ± 0	525 ± 10

[a]Data adapted from Li *et al.* (2002). Values are means ± SD (*n*=3 samples). [b]Note that the totals include other *n*-3 fatty acids, such as 20:3*n*-3, in addition to the main ones listed, and so not all rows will add up exactly. ND, not detected.

Table 7. *n*-3 Fatty acid contents of novel seeds oils (g/100 g oil)[a]

	Fatty acid content (g/100 g oil)	
Oil	**18:3*n*-3 (ALA)**	**18:4*n*-3**
Echium sp.	35	14
Blackcurrant	18	3
Redcurrant	19	6
Alpine currant	18	5

[a]Data taken from Johansson (1999).

fatty acid in oilseed crops (Napier *et al.*, 1999; Parker-Barnes *et al.*, 2000). In particular, the use of genetically engineered plants has been considered a possibility for enriching plant oils (such as canola oil) with long-chain *n*-3 PUFA such as 18:4*n*-3, EPA, and even DHA (Napier *et al.*, 1999).

G. Practical aspects of introducing *n*-3 fatty acids into foods

The fortification of food groups or food products with *n*-3 PUFA can be accomplished in two ways: (1) by feeding animals used in food production with *n*-3 PUFA from vegetable or aquatic sources to generate eggs, meat, and milk enriched in *n*-3 PUFA; or (2) by adding highly refined odorless fish oils or *n*-3 plant oils directly, or in a protected microencapsulated form, to the food product.

1. Stability of n-3 fatty acids

Fish oils are an ideal way of enriching food products with long-chain *n*-3 fatty acids since they contain a high percentage of DHA and EPA. However, the highly unsaturated fish oils are notoriously unstable and oxidize spontaneously if not protected. This oxidation is the major cause of rancidity in foods to which fish oils have been added, either directly or in a protected form such as a microencapsulated product.

Organoleptically detectable lipid oxidation can occur in foods containing even minute traces of long-chain *n*-3 PUFA. One difficulty is in how to quantify the degree of rancidity. As the total amount of long-chain *n*-3 PUFA decreases in a food product, so does the analytical ability to measure and quantify degrees of oxidation. The most effective way of measuring the state of rancidity is by taking advantage of the objectionable odor and off-flavors, and use of subjective sensory evaluation by a statistically relevant sensory panel (Warner and Eskin, 1995).

This is considered by many as the ultimate test of quality control or assurance. The only drawback with this approach is the cost of using such a panel to get reliable shelf life stability data points at every testing time interval in order to obtain a representative sensory shelf life stability profile which reflects changes in the rancidity of the product over time. Lipid oxidation and rancidity represent two sides of the same coin: oxidation is the process, and rancidity is the perceived result (Warner and Eskin, 1995).

There are many accepted analytical methods which give good insight into the oxidative status of foods (Warner and Eskin, 1995; Frankel, 1998). From a practical point of view, however, there are some simple analytical methods that give a good indication of the extent of oxidative degradation or rancidity. It is important to understand that each analytical method gives a measure of only a small part of the complex cascade of reactions that occur during the oxidation of long-chain PUFA, and that several methods may need to be used to obtain a good indication of the oxidative status of a sample (Warner and Eskin, 1995; Frankel, 1998).

Widely used analytical methods that are useful for measuring the state of oxidative decomposition in lipid-containing foods include the determination of the peroxide value, anisidine value, percent free fatty acids, lipid class (total TAG remaining in the lipid phase), and headspace analysis (Warner and Eskin, 1995; Frankel, 1998).

A disadvantage of using such methods is that the fat phase has to be extracted from the food product before the analysis can be performed, and so it can be argued that the extracted fat phase does not fully represent the lipid phase as present within the food product. This applies especially to measurements of the decomposition products of the lipid phase by the peroxide and anisidine values, which quantify the more volatile and/or polar decomposition products. In our experience, the percent free fatty acids value and the quantification of total TAG in conjunction with headspace analysis mimics extremely well the results of evaluation by a sensory panel (O. Bode, personal communication).

2. *Antioxidants*

A number of natural and synthetic antioxidants are available to the food industry. These can be used on their own or in blends acting in synergy with each other to inhibit oxidation and to slow down the eventual onset of the oxidizing process. However, the only certain way to stop oxidation is to remove the oxygen supply completely (Löliger, 1983). Enzymatic hydrolysis of the lipid phase in the product and breakdown of the TAG structure followed by oxidation of long-chain *n*-3 PUFA is another issue to be aware of.

3. *Direct addition of n-3 PUFA oils to food products*

Edible fish oils containing long-chain *n*-3 PUFA can be added directly to food products, blended with other more stable vegetable oils, or be protected from oxidation by entrapment in a preformed microcapsule which is then used as an ingredient in the form of a powder.

If the oil is added directly to the final product, care has to be taken to package the product under vacuum or introduce an inert gas into the package before final sealing. The inert gas is usually pure nitrogen or argon or, as in the case of infant formulae, a mixture of nitrogen and carbon dioxide. The final oxygen content in the sealed package should be well below 1%, but this low a concentration is difficult to achieve in modern high-speed packaging lines. "Difficult" in this sense means "costly" – always the Achilles heel in food production!

4. *Addition of n-3 PUFA as microencapsulated powders to food products*

Preformed microcapsules containing fish oils rich in long-chain *n*-3 PUFA can be added to food products as a free-flowing powder. This technology, called microencapsulation, is a fairly old process, especially in the pharmaceuticals industry, and was first used to encapsulate printing inks for typewriter ribbons (National Cash Register Company, 1930).

Microencapsulation is essentially the packaging of an active ingredient into extremely small parcels with a diameter of 10–500 μm and containing 10–60% (w/w) or more of active component. With a higher core loading the microcapsules would become unstable. The active material is either in the form of a single micro-droplet surrounded by the shell (which forms the boundary between the fish oil droplet and the surrounding air), or many minute sub-droplets dispersed within a honeycomb of matrix material which also forms the external barrier. In the case of tuna fish oil, the sub-droplets are 1–5 μm in diameter, and are dispersed individually within innumerable tiny chambers within the honeycomb. The shell of the microcapsule is designed to completely contain the active component until it is to be released.

The matrix material forming the microcapsule into which the oil droplets are embedded has to provide a number of different functions if it is to work satisfactorily. It must:

- mask the odors and off-flavors of the contents;
- protect the contents against oxidation or chemical decomposition;
- be easy to use in food processing (e.g. as a free-flowing, dust-free powder);
- be of neutral color, to blend into the food product;
- be of neutral or designated flavor to integrate with the end product;
- survive processing when used as an additive;
- possess some controlled-release property to release the active component when required;
- contain only materials approved by current food regulations pertaining at the point of sale;
- enhance and extend shelf life;
- allow wider product application;
- be affordable or cost neutral to the consumer.

The final cost of an acceptable, stable, active ingredient – however desirable – is very important to the food manufacturer. Only a few microencapsulation processes available to the food industry combine a relatively low production cost with good shelf life stability and ease of incorporation.

There are many ways to microencapsulate active components, such as spray drying, film coating, coacervation, carrageenan entrapment, molecular encapsulation using β-cyclodextrin, double emulsions, liposomes, and microemulsions (Vilstrup, 2001). Of these, spray drying is currently the best technology available to the food industry to produce stable, cost-effective, microencapsulated ingredients or products. Spray dryers, first constructed in 1878 (Hayashi, 1989), are now widely used in the dairy industry and, in a modified form, by infant formulae manufacturers.

Commercially available spray dryers in combination with a highly efficient two-stage homogenizer can be used to microencapsulate edible oils containing long-chain *n*-3 PUFA. The spray drier for this purpose consists of feed pump, rotary atomizer, air heater, air disperser, drying chamber, butterfly valves, and systems for exhaust air cleaning and powder recovery. Associated equipment includes tanks, powder-dispersing systems, and a two-stage homogenizer with a nozzle pressure of up to 250 bar. All equipment directly in contact with ingredients or product must be made of high-quality stainless steel.

In the case of heat-sensitive materials, air flow in the spray drier is co-current; i.e. the drying air and particles move through the drying chamber in the same direction. The rotary atomizer is capable of moving at more than 12 000 rpm and can be adjusted to produce spherical droplets of a specific size.

Water-soluble ingredients are dispersed in hot water and contacted in a high-speed premixer with the oil at the same temperature as the water. The coarse emulsion is then pumped to the two-stage homogenizer and from there to the rotary atomizer. The average oil droplet diameter in the emulsion should be around 2 μm. The particle size of the powder is controlled, to a certain extent, through the wheel speed of the atomizer. Balancing the inlet and outlet temperatures of the drying chamber controls the final moisture content of the powder. The two-stage homogenizer can be replaced by a more advanced microfluidizer designed to work under constant pressure. This produces a more uniform-sized microemulsion.

The water-soluble matrix material usually consists of a food-grade material which is a good emulsifier or which assists emulsification (e.g. hydrolysed starches, gums, or proteins), or blends of such ingredients and a film-former like maltodextrin (corn syrup). Antioxidants and a chelator to bind free metal ions are also added to the water phase at this stage. A second food-grade emulsifier can be added to the oil phase, which should be stabilized with antioxidant(s) and, in the case of fish oil, stored under nitrogen before processing.

Microencapsulation is an added cost in food production, and spray drying technology has proven to be the most cost-competitive approach.

H. Products based on *n*-3 PUFA in the marketplace

Foods enriched in *n*-3 PUFA, especially the long-chain *n*-3 PUFA, include various infant formulae now on sale in Europe, Asia, and Australia. Sources of the PUFA are fish oils, egg lipids, or microbial oils. Most products contain microencapsulated PUFA (Clover Corp, 2001). The level of DHA as a percentage of total fatty acids in the infant formulae are in the range of 0.2–0.5% (Clover Corp, 2001). In Japan and Korea, a range of foods containing DHA have been produced, including baby cereals, baby meals, milk products, cheeses, mayonnaise, low-fat spreads, yoghurt, baked goods (biscuits, bread, cakes), hamburgers, sausages, desserts, ice cream, health beverages, juices, and confectionery. The DHA content of these foods is in the range of 1–220 mg/100 g product. Other products produced worldwide include (Clover Corp, 2001):

- bovine milk (1.2% DHA as a percentage of total fatty acids) in Australia;
- semi-skimmed milk (60 mg EPA+DHA/100 ml) in Italy;
- eggs (>200 mg DHA/100 g) in various countries;
- cheese (130 mg EPA+DHA/100 g) in Canada;
- bread (39 mg long-chain *n*-3 PUFA/100 g) in Australia;
- margarine (800 mg EPA+DHA/100 g) in Australia.

I. Bioavailability of *n*-3 fatty acids from functional foods

A number of studies have shown that *n*-3-enriched eggs, milk, and processed foods containing 290–600 mg long-chain *n*-3 PUFA per serving significantly improve plasma and blood cell long-chain *n*-3 PUFA levels (Jiang and Simm, 1993; Roche and Gibney, 1994; Lovegrove *et al.*, 1997; Sorensen *et al.*, 1998; Farrell, 1998; Visioli *et al.*, 2000). A recent study showed that the daily consumption of 1.3 g of stable fish oil incorporated into 93 g bread for 4 weeks significantly reduced plasma TAG levels and significantly increased plasma high-density lipoprotein (HDL) cholesterol levels in hyperlipidemic subjects, without increasing lipid oxidation (Liu *et al.*, 2001b). Another study showed that there was no difference in the bioavailability of long-chain *n*-3 PUFA from capsules compared with microencapsulated products (Wallace *et al.*, 2000). Lower doses of long-chain *n*-3 PUFA (98 mg/day) in bread, present in microencapsulated form, have also been shown to significantly improve plasma *n*-3 PUFA status (Yep *et al.*, 2002).

References

Abedin, L, Lien, EL, Vingrys, AJ and Sinclair, AJ (1999) The effects of dietary α-linolenic acid compared with docosahexaenoic acid on brain, retina, liver, and heart in the guinea pig. *Lipids*, 34, 475–482.

Abeywardena, MY, McLennan, PL and Charnock, JS (1991) Differential effects of dietary fish oil on myocardial prostaglandin I_2 and thromboxane A_2 production. *Am. J. Physiol.*, 260, 379–385.

Abril, R and Barclay, W (1998) Production of docosahexaenoic acid-enriched poultry eggs and meat using an algae-based feed ingredient. *World Rev. Nutr. Diet.*, 83, 77–88.

Adam, O (1995) Anti-inflammatory diet in rheumatic diseases. *Eur. J. Clin. Nutr.*, 49, 703–717.

Adams, PB, Lawson, S, Sanigorski, A and Sinclair, AJ (1996) Arachidonic acid to eicosapentaenoic acid ratio in blood correlates positively with clinical symptoms of depression. *Lipids*, 31, S157–S161.

Ando, H, Ryu, A, Hashimoto, A, Oka, M and Ichihashi, M (1998) Linoleic acid and alpha-linolenic acid lightens ultraviolet-induced hyperpigmentation of the skin. *Arch. Dermatol. Res.*, 290, 375–381.

Assies, J, Lieverse, R, Vreken, P, Wanders, RJ, Dingemans, PM and Linszen, DH (2001) Significantly reduced docosahexaenoic and docosapentaenoic acid concentrations in erythrocyte membranes from schizophrenic patients compared with a carefully matched control group. *Biol. Psych.*, 49, 510–522.

Bao, DQ, Mori, TA, Burke, V, Puddey, IB and Beilin, LJ (1998) Effects of dietary fish and weight reduction on ambulatory blood pressure in overweight hypertensives. *Hypertension*, 32, 710–717.

Barclay, W, Abril, R, Abril, P, Weaver, C and Ashford, A (1998) Production of docosahexaenoic acid from microalgae and its benefits for use in animal feeds. *World Rev. Nutr. Diet.*, 83, 61–76.

Belluzzi, A, Brignola, C, Campieri, M, Pera, A, Boschi, S and Miglioli, M (1996) Effect of an enteric-coated fish oil preparation on relapses in Crohn's disease. *N. Engl. J. Med.*, 334, 1557–1616.

Berger, A, Crozier, G, Bisogno, T, Cavaliere, P, Innis, S and DiMarzo, V (2001) Anandamide and diet: inclusion of dietary arachidonate and docosahexaenoate leads to increased brain levels of the corresponding *N*-acylethanolamines in piglets. *Proc. Natl Acad. Sci. USA*, 98, 6402–6406.

Bjerve, S (1987) Alpha-linolenic acid deficiency in adult women. *Nutr. Rev.*, 45, 15–19.

Bjerve, KS, Fischer, S and Alme, K (1987) Alpha-linolenic acid deficiency in man: effect of ethyl linolenate on plasma and erythrocyte fatty acid composition and biosynthesis of prostanoids. *Am. J. Clin. Nutr.*, 46, 570–576.

Billman, GE, Kang, JX and Leaf, A (1999) Prevention of sudden cardiac death by dietary pure omega-3 polyunsaturated fatty acids in dogs. *Circulation*, 99, 2452–2457.

Bowen, RA and Clandinin, MT (2000) High dietary 18:3*n*-3 increases the 18:3*n*-3 but not the 22:6*n*-3 content in the whole body, brain, skin, epididymal fat pads, and muscles of suckling rat pups. *Lipids*, 35, 389–394.

British Nutrition Foundation (1992) *Unsaturated Fatty Acids: Nutritional & Physiological Significance. The Report of the British Nutrition Foundation's Task Force*, British Nutrition Foundation, London.

Budowski, P, Hawkey, CM and Crawford, MA (1980) Protective effect of alpha-linolenic acid in encephalomalacia in chickens. *Ann. Nutr. Aliment.*, 34, 389–399.

Burr, ML, Fehily, AM, Gilbert, JF, Rogers, S, Holliday, RM, Sweetnam, PM, Elwood, PC and Deadman, NM (1989) Effect of changes in fat, fish, and fibre intakes on death from myocardial infarction: Diet and reinfarction trial (DART). *Lancet*, 334, 757–761.

Calder, PC and Zurier, RB (2001) Polyunsaturated fatty acids and rheumatoid arthritis. *Curr. Opin. Clin. Nutr. Metab. Care*, 4, 115–121.

Cave, WT, Jr (1997) Omega-3 polyunsaturated fatty acids in rodent models of breast cancers. *Breast Cancer Res. Treat.*, 46, 239–246.

Cho, HP, Nakamura, M and Clarke, SD (1999a) Cloning, expression, and fatty acid regulation of the human delta-5 desaturase. *J. Biol. Chem.*, 274, 37335–37339.

Cho, HP, Nakamura, MT and Clarke, SD (1999b) Cloning, expression, and nutritional regulation of the mammalian delta-6 desaturase. *J. Biol. Chem.*, 274, 471–477.

Christensen, JH, Skou, HA, Madsen, T, Torring, I and Schmidt, EB (2001) Heart rate variability and *n*-3 polyunsaturated fatty acids in patients with diabetes mellitus. *J. Intern. Med.*, 249, 545–552.

Christie, WW (1981) The composition, structure and function of lipids in the tissues of ruminant animals. In: *Lipid Metabolism in Ruminant Animals* (WW Christie, ed.), Pergamon Press, Oxford, pp.95–192.

Clarke, SD and Jump, DB (1994) Dietary polyunsaturated fatty acid regulation of gene transcription. *Annu. Rev. Nutr.*, 14, 83–98.

Clover Corp (2001) *Technical Commercial Files*, Clover Corp, Sydney, Australia.

Collett, ED, Davidson, LA, Fan, YY, Lupton, JR and Chapkin, RS (2001) *n*-6 and *n*-3 polyunsaturated fatty acids differentially modulate oncogenic Ras activation in colonocytes. *Am. J. Physiol. Cell Physiol.*, 280, 1066–1075.

Crozier, GL, Fleith, M, Traitler, H and Finot, PA (1989) Black currant seed oil feeding and fatty acids in liver lipid classes of guinea pigs. *Lipids*, 24, 460–466.

Curtis, CL, Hughes, CE, Flannery, CR, Little, CB, Harwood, JL and Caterson, B (2000) *n*-3 fatty acids specifically modulate catabolic factors involved in articular cartilage degradation. *J. Biol. Chem.*, 275, 721–724.

Darlington, LG and Stone, TW (2001) Antioxidants and fatty acids in the amelioration of rheumatoid arthritis and related disorders. *Br. J. Nutr.*, 85, 251–269.

De Caterina, R, Endres, S, Kristensen, SD and Schmidt, EB (1993) *n*-3 Fatty Acids and Vascular Disease, Springer-Verlag, London.

De Deckere, EAM, Korver, O, Verschuren, PM and Katan, M (1998) Health aspects of fish and *n*-3 polyunsaturated fatty acids from plant and marine origin. *Eur. J. Clin. Nutr.*, 52, 749–753.

De Lorgeril, M, Salen, P, Martin, J-L, Moniaud, I, Delaye, I and Mamelle, N (1999) Mediterranean diet, traditional risk factors and rate of cardiovascular complications after myocardial infarction: Final report of the Lyon Diet Heart Study. *Circulation*, 99, 779–785.

Diboune, M, Ferard, G, Ingenbleek, Y, Tulasne, PA, Calon, B, Hasselmann, M, Saude, P, Spielmann, D and Metais, P (1992) Composition of phospholipid fatty acids in red blood cell membranes of patients in intensive care units: effects of different intakes of soybean oil, medium-chain triglycerides, and black-currant seed oil. *J. Parenter. Enter. Nutr.*, 16, 136–141.

Djousse, L, Pankow, JS, Eckfeldt, JH, Folsom, AR, Hopkins, PN, Province, MA, Hong, Y, and Ellison, RC (2001) Relation between dietary linolenic acid and coronary artery disease in the National Heart, Lung and Blood Institute Family Heart Study. *Am. J.Clin. Nutr.*, 74, 612–619.

Donadio, JV, Bergstrahl, EJ, Offord, KP, Spencer, DC and Holley, KE (1994) A controlled trial of fish oil in IgA nephropathy. *N. Engl. J. Med.*, 331, 1194–1199.

Dyerberg, J (1986) Linolenate derived polyunsaturated fatty acids and prevention of atherosclerosis. *Nutr. Rev.*, 44, 125–134.

Edwards, RH, and Peet, M (1999) Essential fatty acid intake in relation to depression. In: *Phospholipid Spectrum Disorder in Psychiatry* (M Peet, I Glenn and DF Horrobin, eds), Marius Press, Lancashire, pp.211–221.

Estabrook, R, Chacos, N, Martin-Wixstron, C and Capdevila, J (1982) Cytochrome P-450: A versatile vehicle of variable veracity. In: *Oxygenases and Oxygen Metabolism* (M Nozake, ed.), Academic Press, New York, pp.371–387.

Ezaki, O, Takahashi, M, Shigematsu, T, Shimamura, K, Kimura, J, Ezaki, H and Gotoh, T (1999) Long-term effects of dietary alpha-linolenic acid from perilla oil on serum fatty acid composition and on the risk factors of coronary heart disease in Japanese elderly subjects. *J. Nutr. Sci. Vitaminol.* (Tokyo), 45, 759–762.

Fantone, JC, Marasco, WA, Elgas, LJ and Ward, PA (1983) Anti-inflammatory effects of prostaglandin E1: in vivo modulation of the formyl peptide chemotactic receptor on the rat neutrophil. *J. Immunol.*, 130, 1495–1497.

Farrell, DJ (1998) Enrichment of hen eggs with *n*-3 long-chain fatty acids and evaluation of enriched eggs in humans. *Am. J. Clin. Nutr.*, 68, 538–544.

Ferretti, A, Nelson, GJ, Schmidt, PC, Bartolini, G, Kelley, DS and Flanagan, VP (1998) Dietary docosahexaenoic acid reduces the thromboxane/prostacyclin synthetic ratio in humans. *J. Nutr. Biochem.*, 9, 88–92.

Fiennes, RNTW, Sinclair, AJ and Crawford, MA (1973) Essential fatty acid studies in primates. Linolenic acid requirements of Capuchins. *J. Med. Prim.*, 2, 155–169.

Fitzpatrick, FA and Murphy, RC (eds) (1989) *Cytochrome P-450, Structure Metabolism and Biochemistry*, Plenum Press, New York.

Frankel, EN (1998) *Lipid Oxidation*, The Oily Press, Dundee.

Frontini, MG, Srinivasan, SR, Elkasabany, A and Berenson, GS (2001) Distribution and cardiovascular risk correlates of serum triglycerides in young adults from a biracial community: the Boga lusa Heart Study. *Atherosclerosis*, 155, 201–209.

Fu, Z and Sinclair, AJ (2000a) Increased alpha-linolenic acid intake increases tissue alpha-linolenic acid content and apparent oxidation with little effect on tissue docosahexaenoic acid in the guinea pig. *Lipids*, 35, 395–400.

Fu, Z and Sinclair, AJ (2000b) Novel pathway of metabolism of alpha-linolenic acid in the guinea pig. *Pediatr. Res.*, 47, 414–417.

Furse, RK, Rossetti, RG and Zurier, RB (2001) Gammalinolenic acid, an unsaturated fatty acid with anti-inflammatory properties, blocks amplification of IL-1β production by human monocytes. *J. Immunol.*, 167, 490–496.

Gallerani, M, Manfredini, R, Caracciolo, S, Scapoli, C, Molinari, S and Fersini, C (1995) Serum cholesterol concentrations in parasuicide. *Br. Med. J.*, 310, 1632–1636.

Gerster, H (1998) Can adults adequately convert a-linolenic acid to eicosapentaenoic acid and docosahexaenoic acid? *Int. J. Vit. Nutr. Res.*, 68, 159–173.

GISSI-Prevenzione Investigators (1999) Dietary supplementation with *n*-3 polyunsaturtated fatty acids and vitamin E after myocardial infarction: results of the GISSI-Prevenzione trial. *Lancet*, 354, 447–455.

Greiner, RC, Winter, J, Nathanielsz, PW and Brenna, JT (1997) Brain docosahexaenoate accretion in fetal baboons: bioequivalence of dietary alpha-linolenic and docosahexaenoic acids. *Pediatr. Res.*, 42, 826–834.

Greiner, RS, Moriguchi, T, Hutton, A, Slotnick, BM and Salem, N (1999) Rats with low levels of brain docosahexaenoic acid show impaired performance in olfactory-based and spatial learning tasks. *Lipids*, 34, S239–S243.

Gunstone, FD, Harwood, JL and Padley, FB (eds) (1994) *The Lipid Handbook* (2nd edn), Chapman & Hall, London.

Hamberg, M, Svensson, J and Samuelsson, B (1975) Thromboxanes: A new group of biologically active compounds derived from prostaglandin endoperoxides. *Proc. Natl Acad. Sci. USA*, 72, 2994–2998.

Hammarstrom, S (1982) Leukotriene formation by mastocytoma and basophilic leukemia cells. *Prog. Lipid Res.*, 20, 89–95.

Hansen, HS and Jensen, B (1985) Essential function of linoleic acid esterified in acylglucosylceramide and acylceramide in maintaining the epidermal water permeability barrier. Evidence from feeding studies with oleate, linoleate, arachidonate, columbinate and α-linolenate. *Biochim. Biophys. Acta*, 834, 357–363.

Harder, DR, Campbell, WB and Roman, RJ (1995) Role of cytochrome P-450 enzymes and metabolites of arachidonic acid in the control of vascular tone. *J. Vasc. Res.*, 32, 79–92.

Hase, T, Mizuno, T, Onizawa, K, Kawasaki, K, Nakagiri, H, Komine, Y, Murase, T, Meguro, S, Tokimitsu, I, Shimasaki, H and Itakura, H (2001) Effects of α-linolenic acid-rich diacylglycerol on diet-induced obesity in mice. *J. Oleo. Sci.*, 50, 701–710.

Hayashi, H (1989) Drying technologies of foods – history and future. *Drying Technol.*, 7, 315–369.

Hegsted, DM, McGandy, RB, Myers, ML and Stare, FL (1965) Quantitative effects of dietary fat on serum cholesterol in man. *Am. J. Clin. Nutr.*, 17, 281–295.

Hibbeln, JR and Salem, N, Jr (1995) Dietary polyunsaturated fatty acids and depression: when cholesterol does not satisfy. *Am. J. Clin. Nutr.*, 62, 1–9.

Hill, DJ, Griffiths, DH and Rowley, AF (1999) Trout thrombocytes contain 12- but not 5-lipoxygenase activity. *Biochim. Biophys. Acta*, 1437, 63–70.

Hillard, CJ and Campbell, WB (1997) Biochemistry and pharmacology of arachidonyl-ethanolamide, a putative endogenous cannabinoid. *J. Lipid Res.*, 38, 2383–2398.

Holman, RT (1968) Biological activities of and requirements for polyunsaturated fatty acids. *Prog. Chem. Fats Other Lipids*, 9, 611–680.

Horrobin, DF, Huang, YS, Cunnane, SC and Manku, MS (1984) Essential fatty acids in plasma, red blood cells and liver phospholipids in common laboratory animals as compared to humans. *Lipids*, 19, 806–811.

Hu, FB, Stampfer, MJ, Manson, JE, Rimm, EB, Wolk, A, Colditz, GA, Hennekens, CH and Willett, WC (1999) Dietary intake of alpha-linolenic acid and risk of fatal ischemic heart disease among women. *Am. J. Clin. Nutr.*, 69, 890–897.

Igarashi, M and Miyazawa, T (2000) Newly recognized cytotoxic effect of conjugated trienoic fatty acids on cultured human tumor cells. *Cancer Lett.*, 148, 173–179.

Ikeda, I, Cha, JY, Yanagita, T, Nakatani, N, Oogami, K, Imaizumi, K and Yazawa, K (1998a) Effects of dietary alpha-linolenic, eicosapentaenoic and docosahexaenoic acids on hepatic lipogenesis and beta-oxidation in rats. *Biosci. Biotechnol. Biochem.*, 62, 675–680.

Ikeda, I, Yoshida, H, Tomooka, M, Yosef, A, Imaizumi, K, Tsuji, H and Seto, A (1998b) Effects of long-term feeding of marine oils with different positional distribution of eicosapentaenoic and docosahexaenoic acids on lipid metabolism, eicosanoids production, and platelet aggregation in hypercholesterolemic rats. *Lipids*, 33, 897–904.

Iso, H, Rexrode, KM, Stampfer, MJ, Manson, JE, Colditz, GA, Speizer, FE, Hennekens, CH and Willett, WC (2001) Intake of fish and omega-3 fatty acids and risk of stroke in women. *JAMA*, 285, 304–312.

James, MJ and Cleland, LG (1997) Dietary n-3 fatty acids and therapy for rheumatoid arthritis. *Sem. Arthr. Rheum.*, 27, 85–97.

Jarvinen, R, Knekt, P, Hakulinen, T, Rissanen, H and Heliovaara, M (2001) Dietary fat, cholesterol and colorectal cancer in a prospective study. *Br. J. Cancer*, 85, 357–361.

Jiang, Z and Sim, JS (1993) Consumption of n-3 polyunsaturated fatty acid-enriched eggs and changes in plasma lipids of human subjects. *Nutrition*, 9, 513–518.

Johansson, A (1999) *Availability of Seed Oils from Finnish Berries* (PhD thesis), University of Turku, Finland.

Kang, JX and Leaf, A (2000) Prevention of fatal cardiac arrythmias by polyunsaturated fatty acids. *Am. J. Clin. Nutr.*, 71, 202S–207S.

Katsuragi, Y, Takeda, Y, Abe, C, Mpri, K, Toi, T, Takei, A, Shimasaki, H and Itakura, H (2001) Effect of dietary α-linolenic acid-rich diacylglycerol on body fat in man (2): Effects on resting metabolism and fat metabolism. *J. Oleo. Sci.*, 50, 747–752.

Keys, A, Anderson, JT and Grande, F (1965) Serum cholesterol response to changes in the diet. IV Particular saturated fatty acids in diet. *Metabolism*, 14, 776–787.

Kidd, PM (2000) Attention deficit/hyperactivity disorder (ADHD) in children: rationale for its integrative management. *Altern. Med. Rev.*, 5, 402–428.

Kinsella, JE, Lokesh, B and Stone, RA (1990) Dietary n-3 polyunsaturated fatty acids and amelioration of cardiovascular disease: possible mechanisms. *Am. J. Clin. Nutr.*, 52, 1–28.

Kiss, L, Bieniek, E, Weissmann, N, Schutte, H, Sibelius, U, Gunther, A, Bier, J, Mayer, K, Henneking, K, Padberg, W, Grimm, H, Seeger, W and Grimminger, F (1998) Simultaneous analysis of 4- and 5-series lipoxygenase and cytochrome P450 products from different biological sources by reversed-phase high-performance liquid chromatographic technique. *Anal. Biochem.*, 261, 16–28.

Kris-Etherton, PM, Taylor, DS, Yu, S, Moriarty, K, Fishell, V, Morgan, R, Zhao, G, Heimbach, JT and Etherton, TD (1998). Present estimates of omega 3 fatty acid intakes in the US. *World Rev. Nutr. Dietet.*, 83, 228.

Kurlack, LO and Stephenson, TJ (1999) Plausible explanations for effects of long chain polyunsaturated fatty acids on neonates. *Arch. Dis. Child.: Fetal Neonatal Edn*, 80, 148–154.

Kyle, D (2001) The large-scale production and use of a single-cell oil highly enriched in docosahexaenoic acid. In: *Omega 3 Fatty Acids: Chemistry, Nutrition and Health Effects* (F Shahidi and JW Finley, eds), American Chemical Society, Washington, pp.92–107.

Lands, WEM (1986) *Fish and Human Health*, Academic Press, Orlando.

Lee, JH, Sugano, M and Ide, T (1988) Effects of various combinations of omega 3 and omega 6 polyunsaturated fats with saturated fat on serum lipid levels and eicosanoid production in rats. *J. Nutr. Sci. Vitaminol. (Tokyo)*, 34, 117–129.

Leyton, J, Drury, PJ and Crawford, MA (1987) Differential oxidation of saturated and unsaturated fatty acids in vivo in the rat. *Br. J. Nutr.*, 57, 383–393.

Li, D (1998) *The Influence of Diet on Atherosclerotic and Thrombotic Risk Factors in Healthy Men* (PhD thesis), RMIT University, Melbourne, Australia.

Li, D, Ng, A, Mann, N and Sinclair, AJ (1998) Meat fat can make a significant contribution to dietary arachidonic acid. *Lipids*, 33, 437–440.

Li, D, Sinclair, A, Wilson, A, Nakkote, S, Kelly, F, Abedin, L, Mann, N and Turner, A (1999a) Effect of dietary alpha-linolenic acid on thrombotic risk factors in vegetarian men. *Am. J. Clin. Nutr.*, 69, 872–882.

Li, D, Sinclair, A, Mann, N, Turner, A, Ball, M, Kelly, F, Abedin, L and Wilson, A (1999b) The association of diet and thrombotic risk factors in healthy male vegetarians and meat-eaters. *Eur. J. Clin. Nutr.*, 53, 612–619.

Li, D, Mansor, M, Zhuo, SSR, Woon, T, Anthony, MA and Sinclair, AJ (2002) Omega-3 polyunsaturated fatty acid content of canned meats commonly available in Australia. *Food Austr.*, 54, 311–315.

Liu, G, Bibus, DM, Bode, AM, Ma, WY, Holman, RT and Dong, Z (2001a) Omega 3 but not omega 6 fatty acids inhibit AP-1 activity and cell transformation in JB6 cells. *Proc. Natl Acad. Sci. USA*, 98, 7510–7515.

Liu, M, Wallin, T and Saldeen, T (2001b) Effect of bread containing stable fish oil on plasma phospholipid fatty acids, triglycerides, HDL-cholesterol, and malondialdehyde in subjects with hyperlipidemia. *Nutr. Res.*, 21, 1403–1410.

Löliger, J (1983) Natural Antioxidants. In: *Rancidity of Foods* (JC Allen and RJ Hamilton, eds), Elsevier, London, pp.89–107.

Lovegrove, JA, Brooks, CN, Murphy, MC, Gould, BJ and Williams, CM. (1997) Use of manufactured foods enriched with fish oils as a means of increasing long-chain *n*-3 polyunsaturated fatty acid intake. *Br. J. Nutr.*, 78, 223–236.

Maes, M, Christophe, A, Delanghe, J, Altamura, C, Neels, H and Meltzer, HY (1999) Lowered omega3 polyunsaturated fatty acids in serum phospholipids and cholesteryl esters of depressed patients. *Psychiatr. Res.*, 85, 275–291.

Mann, NJ, Sinclair, AJ, Pille, M, Johnson, LG, Warrick, GE, Reder, E and Lorenz, R (1997) The effect of short-term diets rich in fish, red meat, or white meat on thromboxane and prostacyclin synthesis in humans. *Lipids*, 32, 635–644.

Mantzioris, E, James, MJ, Gibson, RA and Cleland, LG (1994) Dietary substitution with an α-linolenic acid-rich vegetable oil increases eicosapentaenoic acid concentrations in tissues. *Am. J. Clin. Nutr.*, 59, 1304–1309.

Martinez, M, Vazquez, E, Garcia-Silva, MT, Manzanares, J, Bertran, JM, Castello, F and Mougan, I (2000) Therapeutic effects of docosahexaenoic acid ethyl ester in patients with generalised peroxisomal disorders. *Am. J. Clin. Nutr.*, 71, 376S–385S.

Mayatepek, E and Hoffmann, GF (1995) Leukotrienes: biosynthesis, metabolism, and pathophysiologic significance. *Pediatr. Res.*, 37, 1–9.

McDonald, BE, Gerrard, JM, Bruce, VM and Corner, EJ (1989) Comparison of the effect of canola oil and sunflower oil on plasma lipids and lipoproteins and on in vivo thromboxane A2 and prostacyclin production in healthy young men. *Am. J. Clin. Nutr.*, 50, 1382–1388.

McMahon, B, Mitchell, S, Brady, HR and Godson, C (2001) Lipoxins: revelations on resolution. *Trends Pharmacol. Sci.*, 22, 391–395.

Menard, CR, Goodman, KJ, Corso, TN, Brenna, JT and Cunnane, SC (1998) Recycling of carbon into lipids synthesized de novo is a quantitatively important pathway of alpha-[U-^{13}C]linolenate utilization in the developing rat brain. *J. Neurochem.*, 71, 2151–2180.

Mest, HJ, Beitz I, Block, HU and Forster, W (1983) The influence of linseed oil diet on fatty acid pattern in phospholipids and thromboxane formation in platelets in man. *Klin. Wochenschr.*, 61, 187–191.

Mitchell, DC, Gawrisch, K, Litman, BJ and Salem, N, Jr (1998) Why is docosahexaenoic acid essential for nervous system function? *Biochem. Soc. Trans.*, 26, 365–370.

Moncada, S and Vane, JR (1979) Arachidonic acid metabolites and the interactions between platelets and blood vessel walls. *N. Engl. J. Med.*, 300, 1142–1148.

Moore, SA, Hurt, E, Yoder, E, Sprecher, H and Spector, AA (1995) Docosahexaenoic acid synthesis in human skin fibroblasts involves peroxisomal retroconversion of tetracosahexaenoic acid. *J. Lipid Res.*, 36, 2433–2443.

Mori, TA, Bao, DQ, Burke, V, Puddey, IB and Beilin, LJ (1999) Docosahexaenoic acid but not eicosapentaenoic acid lowers ambulatory blood pressure and heart rate in humans. *Hypertension*, 34, 253–260.

Morris, MC, Sack, F and Rosner, B (1993) Does fish oil lower blood presure? A meta-analysis of controlled trials. *Circulation*, 88, 523–533.

Mufti, RM, Balon, R and Arfken, CL (1998) Low cholesterol and violence. *Psychiatr. Serv.*, 49, 221–214.

Nagao, T, Watanabe, H, Goto, N, Onizawa, K, Taguchi, H, Matsuo, N, Yasukawa, T, Tsushima, R, Shimasaki, H and Itakura, H (2000) Dietary diacylglycerol suppresses accumulation of body fat compared to triacylglycerol in men in a double-blind controlled trial. *J. Nutr.*, 130, 792–797.

Napier, JA, Michaelson, LV and Stobart, AK (1999) Plant desaturases: harvesting the fat of the land. *Curr. Opin. Plant Biol.*, 2, 123–127.

National Cash Register Company, Inc (1930) US Patent

Newton, IS (1998) Global food fortification perspectives of long-chain omega 3 fatty acids. *World Rev. Nutr. Dietet.*, 83, 199–209.

Pan, DA and Storlien, LH (1993) Dietary lipid profile is a determinant of tissue phospholipid fatty acid composition and rate of weight gain in rats. *J. Nutr.*, 123, 512–519.

Parker-Barnes, J, Das, T, Bobik, E, Leonard, AE, Thurmond, JM, Chaung, LT, Huang, YS and Mukerji, P (2000) Identification and characterisation of an enzyme involved in the elongation of *n*-6 and *n*-3 polyunsaturated fatty acids. *Proc. Natl Acad. Sci. USA*, 97, 8284–8289.

Partonen, T, Haukka, J, Virtamo, J, Taylor, PR and Lonnqvist, J (1999) Association of low serum total cholesterol with major depression and suicide. *Br. J. Psych.*, 175, 259–262.

Pawlosky, R, Barnes, A and Salem, N (1994) Essential fatty acid metabolism in the feline: relationship between liver and brain production of long-chain polyunsaturated fatty acids. *J. Lipid Res.*, 35, 2032–2040.

Pegorier, JP (1998) Regulation of gene expression by fatty acids. *Curr. Opin.Clin. Nutr. Metab. Care*, 1, 329–334.

Pereira, C, Li, D and Sinclair, AJ (2001) The alpha-linolenic acid content of green vegetables commonly available in Australia. *Int. J. Vitamin Nutr. Res.*, 71, 223–228.

Petrik, MBH, McEntree, MF, Johnson, BT, Obukowicz, MG and Whelan, J (2000) Highly unsaturated *n*-3 fatty acids, but not α-linolenic acid, conjugated linoleic or γ-linolenic acids, reduce tumorigenesis in Apc^{minl+} mice. *J. Nutr.*, 130, 2434–2443.

Phinney, DS, Odin, RS, Johnson, SB and Holman, RT (1990) Reduced arachidonate in serum phospholipids and cholesteryl esters associated with vegetarian diets in humans. *Am. J. Clin. Nutr.*, 51, 385–392.

Price, PT, Nelson, CM and Clarke, SD (2000) Omega-3 polyunsaturated fatty acid regulation of gene expression. *Curr. Opin. Lipidol.*, 11, 3–7.

Raz, A, Minkes, MS and Needlemen, P (1977) Endoperoxides and thromboxanes: structural determinants for platelet aggregation and vasoconstriction. *Biochim. Biophys. Acta*, 488, 305–311.

Reisbick, S, Neuringer, M and Connor, WE (1992) Postnatal deficiency of omega-3 fatty acids in monkeys: fluid intake and urine concentration. *Physiol. Behav.*, 51, 473–479.

Ren, B, Thelen, AP, Peters, JM, Gonzalez, FJ and Jump, DB (1997) Polyunsaturated fatty acid suppression of hepatic fatty acid synthase and S14 gene expression does not require peroxisome proliferator-activated receptor alpha. *J. Biol. Chem.*, 272, 26827–26832.

Richardson, AJ and Puri, BK (2000) The potential role of fatty acids in attention-deficit/hyperactivity disorder. *Prostaglandins Leukotrienes Essent. Fatty Acids*, 63, 79–87.

Richardson, AJ, Easton, T and Puri, BK (2000) Red cell and plasma fatty acid changes accompanying symptom remission in a patient with schizophrenia treated with eicosapentaenoic acid. *Eur. Neuropsychopharmacol.*, 10, 189–193.

Roche, H and Gibney, MJ (1994) The effect of consumption of fish oil-enriched spreadable fats on platelet phospholipid fatty acid composition in human volunteers. *Int. J. Vitamin Nutr. Res.*, 64, 237–242.

Rokkones, T (1953) A dietary factor essential for hair growth in rats. *Int. Z. Vitaminforsch.*, 25, 86–98.

Rothman, D, DeLuca, P and Zurier, RB (1995) Botanical lipids: effects on inflammation, immune responses, and rheumatoid arthritis. *Semin. Arthritis Rheum.*, 25, 87–96.

Salem, N and Ward, GR (1993) Are omega-3 fatty acids essential nutrients for mammals? World *Rev. Nutr. Dietet.*, 72, 128–147.

Salem, N, Wegher, B, Mena, P and Uauy, R (1996) Arachidonic and docosahexaenoic acids are biosynthesised from their 18-carbon precursors in human infants. *Proc. Natl Acad. Sci. USA*, 93, 49–54.

Salem, N, Litman, B, Kim, HK and Gawrisch, K (2001) Mechanisms of action of docosahexaenoic acid in the nervous system. *Lipids*, 36, 945–959.

Serhan, CN and Oliw, E (2001) Unorthodox routes to prostanoid formation: new twists in cyclooxygenase-initiated pathways. *J. Clin. Invest.*, 107, 1481–1489.

Sheaff Greiner, RC, Zhang, Q, Goodman, KJ, Giussani, DA, Nathanielsz, PW and Brenna, JT (1996) Linoleate, alpha-linolenate, and docosahexaenoate recycling into saturated and monounsaturated fatty acids is a major pathway in pregnant or lactating adults and fetal or infant rhesus monkeys. *J. Lipid Res.*, 37, 2675–2686.

Simopoulos, AP, Leaf, A and Salem, N (2000) Workshop statement on the essentiality of and recommended dietary intakes for omega 6 and omega 3 fatty acids. *Prostaglandins Leukotrienes Essent. Fatty Acids*, 63, 119–121.

Sinclair, AJ (1975) Incorporation of radioactive polyunsaturated fatty acids into liver and brain of the developing rat. *Lipids*, 10, 175–184.

Sinclair, AJ (2000) Commentary on the Workshop Statement. *Prostaglandins Leukotrienes Essent. Fatty Acids*, 63, 135–137.

Sinclair, AJ and Vingrys, AJ (1998) Estimation of long chain *n*-3 polyunsaturated fatty acid status of Australians. *Proc. Nutr. Soc. Austr.*, 22, 196.

Sinclair, AJ, Dunstan, GA, Naughton, JM, Sanigorski, AJ and O'Dea, K (1992) The lipid content and fatty acid composition of commercial marine and freshwater fish and mollusks from temperate Australian waters. *Austr. J. Nutr. Diet.*, 49, 77–83.

Sinclair, AJ, Oon, KS, Lim, L, Li, D and Mann, N (1998) The omega-3 fatty acid content of canned, smoked and fresh fish in Australia. *Austr. J. Nutr. Diet.*, 55, 116–120.

Singh, RB, Niaz, MA, Sharma, JP, Kumar, R, Rastogi, V and Moshiri, M (1997) Randomized, double-blind, placebo-controlled trial of marine omega-3 oil and mustard oil in patients with suspected acute myocardial infarction: the Indian Experiment of Infarct Survival-4. *Cardiovasc. Drugs Therap.*, 11, 485–491.

Sorensen, NS, Marckmann, P, Hoy, C, Van Duyvenvoorde, W and Princen, HMG (1998) Effect of fish-oil-enriched margarine on plasma lipids, low-density-lipoprotein particle composition, size, and susceptibility to oxidation. *Am. J. Clin. Nutr.*, 68, 235–241.

Sprecher, H, Chen, Q, Mohammed, BS and Luthria, DL (1998) The role played by microsomes and peroxisomes in regulating unsaturated fatty acid biosynthesis. In: *Essential Fatty Acids and Eicosanoids, Invited Papers from the Fourth International Congress* (RA Riemersma, RW Kelly and R Wilson, eds), AOCS Press, Champaign, pp.1–6.

Stoll, AL, Severus, WE, Freeman, MP, Rueter, S, Zboyan, HA, Diamond, E, Cress, KK and Marangell, LB (1999) Omega 3 fatty acids in bipolar disorder: a preliminary double-blind, placebo-controlled trial. *Arch. Gen. Psychiatr.*, 56, 407–412.

Svaneborg, N, Moller, JM, Schmidt, EB, Varming, K, Lervang, HH and Dyerberg, J (1994) The acute effects of a single very high dose of n-3 fatty acids on plasma lipids and lipoproteins in healthy subjects. *Lipids*, 29, 145–147.

Taguchi, H, Watanabe, H, Onizawa, K, Nagao, T, Gotoh, N, Yasukawa, T, Tsushima, R, Shimasaki, H and Itakura, H (2000) Double-blind controlled study on the effects of dietary diacylglycerol on postprandial serum and chylomicron triacylglycerol responses in healthy humans. *J. Am. Coll. Nutr.*, 19, 789–796.

Takei, A, Katsuragi, Y, Abe, C, Mori, K, Takeda, Y, Seo, Y, Takase, H, Takahashi, H, Tohata, A, Chikama, A, Fumoto, S, Meguro, S, Komine, Y, Nagao, T, Hase, T, Tokimitsu, I, Shimasaki, H and Itakura, H (2001) Effect of dietary α-linolenic acid-rich diacylglycerol on body fat in man (1): Lowering effect on body fat. *J. Oleo. Sci.*, 50, 735–746.

Terry, P, Lichtenstein, P, Feychting, M, Ahlbom, A and Wolk, A (2001) Fatty fish consumption and risk of prostate cancer. *Lancet*, 357, 1764–1766.

UK Department of Health (1991) *Dietary Reference Values for Food Energy and Nutrients for the United Kingdom. Committee on Medical Aspects of Food Policy Report on Health and Social Subjects, No 41*, Her Majesty's Stationery Office, London.

Vermunt, SH, Mensink, RP, Simonis, MM and Hornstra, G (2000) Effects of dietary alpha-linolenic acid on the conversion and oxidation of ^{13}C-alpha-linolenic acid. *Lipids*, 35, 137–142.

Vilstrup, P (2001) *Microencapsulation of Food Ingredients* (1st edn), Leatherhead Publishing, Leatherhead.

Visioli, F, Rise, P, Plasmati, E, Pazzucconi, F, Sirtori, CR and Galli, C. (2000) Very low intakes of n-3 fatty acids incorporated into bovine milk reduce triacylglycerol and increase HDL-cholesterol concentrations in healthy subjects. *Pharmacol. Res.*, 41, 571–576.

von Schacky, C, Siess, W, Fischer, S and Weber, PC (1985) A comparative study of eicosapentaenoic acid metabolism by human platelets in vivo and in vitro. *J. Lipid Res.*, 26, 457–464.

von Schacky, C, Angerer, P, Kothny, W, Theisen, K, and Mudra, H (1999) The effect of dietary ω-3 fatty acids on coronary atherosclerosis. A randomized, double-blind, placebo-controlled trial. *Ann. Intern. Med.*, 130, 554–562.

Voss, AM, Reinhart, S, Sankarappa, S and Sprecher, H (1991) Metabolism of 22:5n-3 to 22:6n-3 in rat liver is independent of 4-desaturase. *J. Biol. Chem.*, 266, 19995–20000.

Wallace, JM, Mc Cabe, AJ, Robson, PJ, Keogh, MK, Murray, CA, Kelly, PA, Marquez-Ruiz, G, McGlynn, H, Gilmore, WS and Strain, JJ (2000) Bioavailability of n-3 polyunsaturated fatty acids in foods enriched with microencapsulated fish oil. *Ann. Nutr. Metab.*, 44, 157–162.

Warner, K and Eskin, NAM (1995) *Methods to Assess the Quality and Stability of Oils and Fat-Containing Foods*, AOCS Press, Champaign.

Weisinger, HS, Armitage, JA, Sinclair, AJ, Vingrys, AJ, Burns, PL and Weisinger, RS (2001) Perinatal omega-3 fatty acid deficiency affects blood pressure later in life. *Nature Med.*, 7, 258–259.

Weisinger, HS, Vingrys, AJ and Sinclair, AJ (1996) The effect of docosahexaenoic acid on the electroretinogram of the guinea pig. *Lipids*, 31, 65–70.

Weisinger, HS, Vingrys, AJ, Bui, BV and Sinclair, AJ (1999) Effects of dietary *n*-3 fatty acid deficiency and repletion in the guinea pig retina. *Invest. Ophthalmol. Vis. Sci.*, 40, 327–338.

Woods J, Ward, G and Salem, N (1996) Is docosahexaenoic acid necessary in infant formula? Evaluation of high linolenate diets in the neonatal rat. *Pediatr. Res.*, 40, 687–694.

Yamazaki, K, Fujikawa, M, Hamazaka, T, Yano, S and Shono, T (1992) Comparison of the conversion rates of alpha-linolenic acid (18:3*n*-3) and stearidonic acid (18:4*n*-3) to longer polyunsaturated fatty acids in rats. *Biochim. Biophys. Acta*, 1123, 18–26.

Yep, YL, Li, D, Mann, NJ, Bode, O and Sinclair, AJ (2002) Bread enriched with microencapsulated tuna oil increases plasma omega 3 fatty acids in humans. *Asia Pacific J. Clin. Nutr.*, 11(4).

Zampelas, A, Peel, AS, Gould, BJ, Wright, J and Williams, CM (1994) Polyunsaturated fatty acids of the *n*-6 and *n*-3 series: effects on postprandial lipid and apolipoprotein levels in healthy men. *Eur. J. Clin. Nutr.*, 48, 842–848.

Zeller, S, Barclay, W and Ruben, A (2001) Production of docosahexaenoic acid from microalgae. In: *Omega 3 Fatty Acids: Chemistry, Nutrition and Health Effects* (F Shahidi and JW Finley, eds), American Chemical Society, Washington, pp.108–124.

Ziboh, VA, Miller, CC and Cho, Y (2000) Metabolism of polyunsaturated fatty acids by skin epidermal enzymes: generation of antiinflammatory and antiproliferative metabolites. *Am. J. Clin. Nutr.*, 71, S361–S366.

CHAPTER 9

Oils containing oleic, palmitoleic, γ-linolenic and stearidonic acids

BAORU YANG[1], FRANK D. GUNSTONE[2] and HEIKKI KALLIO[1]

[1]*University of Turku, Turku, Finland;* [2]*Scottish Crop Research Institute, Invergowrie, Scotland, UK*

A. High-oleic oils

Vegetable oils with high levels (>80%)* of oleic acid (18:1*n*-9) must necessarily be low in saturated acids and in polyunsaturated fatty acids (PUFA) (especially 18:3) and this represents an improved composition for frying oils (and for lubricants and biofuels). Saturated acids are avoided on health grounds and because they promote the crystallization considered undesirable in both frying oils and lubricants. PUFA, especially α-linolenic acid (18:3*n*-3) and other *n*-3 acids, though desirable on health grounds, promote oxidative deterioration in both food and non-food products, and food processors prefer oils which meet nutritional requirements and, at the same time, allow the production of foods with a long shelf life (Gunstone, 2001a, 2001b).

Olive oil has long been recognized as an oil rich in oleic acid and its use is actively promoted as a component of the Mediterranean lifestyle. How far this healthy lifestyle is due to olive oil as opposed to other factors and how far the benefits of olive oil are related to its high level of oleic acid rather than the many minor phenolic components present in this oil remains uncertain (Luchetti, 2001). In addition to olive oil and the high-oleic oils to be discussed later, almond (61% oleic acid), and macadamia (50% oleic acid, together with 22% palmitoleic acid (see Section B below)) are also oleic-rich oils.

As shown in Table 1, rapeseed/canola, sunflower, soybean, and safflower can all be modified by traditional breeding methods or by genetic modification to produce high-oleic varieties containing >70–90% oleic acid. The content of saturated acids is generally <10% and is lowest in the modified canola oils.

*Unless otherwise indicated, all fatty acid concentrations are given as weight percentages of total fatty acids in the sample.

Table 1. High-oleic oils (≥74% oleic acid) derived from rapeseed/canola, sunflower, soybean, and safflower oils[a]

Oil	Fatty acid content (wt% of total fatty acids) Palmitic (16:0)	Stearic (18:0)	Oleic (18:1*n*-9)	Linoleic (18:2*n*-6)	α-Linolenic (18:3*n*-3)	References
Rapeseed/ canola	4	1	84	5	3	Gunstone (1998)
	3	3	82	7	4	Abidi *et al.* (1999)
	3	3	78	8	5	Abidi *et al.* (1999)
	4	2	76	13	3	Daun (2000)
	4	2	75	13	2	Daun (2000)
	7[b]		75	8	6	Petukhov *et al.* (1999a, 1999b)
	7[b]		75	11	7	Loh and Lampert (1998)
	7[b]		75	13	5	Loh and Lampert (1998)
Sunflower	3	2	92	2	<0.5	Reske *et al.* (1997)
	3	4	88	3	<0.5	Daun (2000)
	2	4	87	3	<0.5	Marquez-Ruiz *et al.* (1999)
	10[b]		80	9[c]		Optimum Quality Grains (1998)
	7[b]		80	12	<0.5	Loh and Lampert (1998)
	5	11	79	2	<0.5	Reske *et al.* (1997)
	3	4	75	15	<0.5	Daun (2000)
mid-oleic	8–9[b]		60–65	25–30	<1	Gupta (1998, 2000)
Soybean	6	3	86	2	2	Liu (1999)
	11[b]		81	4	4	Loh and Lampert (1998)
	10[b]		>80	7[c]		USDA (2000)
	9	3	79	3	6	Wilson (1999)
Safflower	6	2	74	16	<0.5	Gunstone (1998)

[a]Adapted from Gunstone (2001c). [b]Combined total of 16:0 plus 18:0. [c]Combined total of 18:2*n*-6 plus 18:3*n*-3.

Linoleic acid (18:2*n*-6) contents are mainly in the ranges of 3–5% or 13–16%. α-Linolenic acid is still present at low levels (2–7%) in oils based on rapeseed or soybeans. Much effort is devoted to modifying soybean or canola plants to furnish oils with reduced α-linolenic acid levels which can be used for frying without brush hydrogenation.

DuPont commercialized a genetically engineered high-oleic soybean oil with >80% oleic acid and predicted that close to 1 million acres will be planted with this line in the USA in the early years of this decade. As expected the oil has high oxidative stability (Kinney, 1998).

Table 2. Fatty acid compositions of traditional and high-oleic varieties of safflower and sunflower oils[a]

	Fatty acid content (wt% of total fatty acids)				
Oil	**Palmitic (16:0)**	**Stearic (18:0)**	**Oleic (18:1*n*-9)**	**Linoleic (18:2*n*-6)**	**α-Linolenic (18:3*n*-3)**
Safflower					
Traditional	4–6	1–2	12–16	75–79	0
High-oleic	4.5	1.5	75–80	16–20	0
Sunflower					
Traditional	7	4	20	68	<1
High-oleic	6	4	83	5	<1
'90+' type	3.5–4.0	1.0–1.5	90.5–91.0	2.8–3.0	0.5–0.8
'NuSun'	<4–5	3–4	50–75	20–30	<1

[a]Data taken from Smith (1996); Gupta (2000); Anonymous (2001).

The soybean oil described by Wilson (1999) with almost 80% oleic acid is a non-genetically modified oil resulting from the 'Better Bean Initiative' and was expected to be ready for sowing during the 2000 season.

Australian investigators (Xu *et al.,* 1999, 2000) have reported a high-oleic canola oil (71–74% oleic, 13–14% linoleic, and 5.5–7.0% α-linolenic acids) and a high-oleic, low-α-linolenic canola oil (69–74% oleic, 16–21% linoleic, and 2.0–2.7% α-linolenic acids). The latter oil has been designated 'monola'. It is advertised as a superior frying oil and has been available in limited quantities in Australia since 1999 (Anon., 2000).

High-oleic varieties of sunflower and of safflower were produced by conventional seed breeding techniques over 20 years ago and have been available in the marketplace for some time (Table 2). In general the oils contain ~80% oleic acid but levels up to ~90% are also available. About 300 000 tonnes of high-oleic sunflower oil is produced annually, most of it in the USA and in France. In 1999, about 500 tonnes were the 90%-oleic variety. These high-oleic oils have potential as a starting material for chemical manipulation but they are also useful in food products such as spraying oils for snacks and crackers, etc., as frying oils, in preparing special foods for infants and the elderly, and for margarines, spreads, mayonnaise, and salad dressings (Anon., 2001).

An interesting new oil is a mid-oleic sunflower oil known as 'NuSun'. This contains 7–9% saturated acids, 50–75% oleic acid, 20–30% linoleic acid, and <1% α-linolenic acid. In its level of oleic acid, it falls between those of commodity sunflower oil (~20%) and high-oleic varieties (~70–90%). This non-genetically modified material was developed in the USA and was grown on 40 000 hectares in 1998, producing 20 000–25 000 tonnes of oil. The planted area rose to ~160 000 hectares in 1999 and ~400 000 hectares in 2000. It is hoped that in 3–5 years most of the sunflower grown in the USA on 1.2–1.5 million hectares will be of this variety and that it will become the commodity sunflower oil of that country and possibly elsewhere also. In the USA NuSun is expected to substitute for cottonseed, sunflower, soybean, and canola oils and to provide a low-cost alternative to palm olein and to mixtures of fully hydrogenated and partially hydrogenated soybean oil which contain *trans* acids (Gupta, 1998, 2000). This product meets the requirements for a new oil: versatility in application (cooking, frying, baking, salad dressing), abundant supply, and affordable price. More research is being directed towards further reducing saturated fatty acid levels and modifying the tocopherol content and composition to improve oxidative stability.

Following the use of the name 'canola' to designate low-erucic acid rapeseed oil, similar terms have been coined for high-oleic sunflower ('sunola'), safflower ('saffola'), soybean ('soyola'), and canola ('monola') oils.

B. Palmitoleic acid

1. Occurrence

Palmitoleic acid (16:1*n*-7), also referred to as zoomaric acid, is synthesized *de novo* in microorganisms, plants, animals, and humans by Δ^9-desaturation of palmitic acid (16:0). It can be converted to *cis*-vaccenic acid (18:1*n*-7) through two-carbon chain elongation.

Palmitoleic acid is commonly found at low levels in animal and plant tissues (Gunstone, 1994). It is present in most fish oils at levels of around 10% (w/w) of total fatty acids. In animal depot fat, the level of palmitoleic acid varies between 3% and 10%, depending on location and animal species. In human and animal milk, the proportion of palmitoleic acid, together with other 16:1 isomers, is typically 2–6%, with exceptionally high levels found in grey seal milk (16%) and camel colostrum (14%) and milk (10%) (Gorban and Izzeldin, 2001).

Palmitoleic acid represents <1% of total fatty acids in most vegetable oils (Padley *et al.*, 1994), with higher levels in a few plant sources of which sea buckthorn and macadamia are the best known.

Sea buckthorn (*Hippophaë rhamnoides*) is a berry-bearing, hardy bush of the family Elaeagnaceae, naturally distributed in Asia and Europe and cultivated also in North America. The species is classified into nine subspecies, of which three (*H. rhamnoides* subsp. *sinensis*, subsp. *rhamnoides* and subsp. *mongolica*) are most commonly used for commercial purposes. In the soft parts, the berries contain an oil (20–30% dry weight) having 20–50% palmitoleic acid, depending on the origins of the berries (Yang, 2001).

Macadamia is a large evergreen tree indigenous to the coastal rain forest of

Australia. Ten species of the genus *Macadamia* have been described, and two (*M. integrifolia* and *M. tetraphylla*) are grown commercially in Hawaii, Australia, and New Zealand. The lipid content in macadamia nut ranges from 70% to 80% (Kaijser *et al.*, 2000) and the oil from *M. integrifolia* contains about 30% palmitoleic acid (Gummeson *et al.*, 2000). Palmitoleic acid attains levels of 17–34% in oils from cultivars of *M. tetraphylla* grown in New Zealand (Kaijser *et al.*, 2000). Its high content of oleic acid (40–60%) and palmitoleic acid taken together make macadamia nut oil the most monounsaturated natural edible oil. This oil is widely used in cosmetic products, and its application as a food supplement has been increasing with the knowledge of the beneficial effects of monounsaturated fatty acids (MUFA).

The seed oil of cat's claw (*Doxantha unguis-cati* L.) contains 64% palmitoleic acid, and the sum of palmitoleic acid and *cis*-vaccenic acid is ~80% (w/w) of total fatty acids (Cahoon *et al.*, 1998). The seeds of *Kermadecia sinuata*, native to New Caledonia, contain 42% oil with palmitoleic acid (33%) and other 16:1 isomers (36%) as the major fatty acids (Vickery, 1969).

Lipids from *Saccharomyces* species contain a high proportion of palmitoleic acid, the level in *S. cerevisiae* being up to 50% of total fatty acids (Padley *et al.*, 1994).

2. *Levels of palmitoleic acid in human and animal tissues and health conditions*

The level of palmitoleic acid varies among different tissues in humans. In plasma lipids this fatty acid normally represents <4% of total fatty acids (Messner *et al.*, 1998). Higher levels are found in Greenland Inuit (9%), probably due to their high dietary intake of meat of marine origin (Dyerberg *et al.*, 1975). Site-specific differences exist in the level of the fatty acid in human adipose tissues.

Alterations in the level of palmitoleic acid have been found in tissues under certain physiological and disease conditions, including essential fatty acid deficiency (EFAD), cancer, cardiovascular disease, and diabetes mellitus. These changes are often associated with changes in the relative levels of *n*-6 and *n*-3 PUFA in the tissues.

EFAD is a physiological, sometimes pathological, condition caused by abnormalities in dietary intake, absorption, and/or metabolism of *n*-6 and *n*-3 PUFA. A typical feature of EFAD is a decrease in the levels of linoleic acid and arachidonic acid (20:4*n*-6), an increase in the levels of palmitoleic, oleic, and eicosatrienoic (20:3*n*-9) acids, and increases in the eicosatrienoic:arachidonic, palmitoleic:linoleic, and oleic:linoleic acid ratios in plasma phospholipids (Chardon *et al.*, 1985). In EFAD the relative proportion of palmitoleic acid is also increased in other tissues in both human and animal models (Berghaus *et al.*, 1975; Alam *et al.*, 1994; Gómez-Tubío *et al.*, 1999). Pregnancy and high alcohol intake induce or exacerbate the biochemical changes characteristic of EFAD, including an increased proportion of palmitoleic acid (Sanjurjo *et al.*, 1993; Gómez-Tubío *et al.*, 1999). A higher proportion of palmitoleic acid in liver and subcutaneous fat is associated with a higher degree of fat infiltration in the liver of alcoholics (Berghaus *et al.*, 1975).

The levels of palmitoleic and myristoleic (14:1*n*-5) acids in gluteal fat showed an inverse association with breast cancer risk in a multi-centre investigation performed in five European countries (Simonsen *et al.*, 1998). On the other hand, increased levels of palmitoleic acid and decreased levels of PUFA have been reported in cultured cancer cells (Reynier *et al.*, 1991) and in malignant tissues (Engelbrecht *et al.*, 1998).

Associations have been reported between the levels of palmitoleic acid in certain tissues and cardiovascular disease and related risk factors. The percentage of palmitoleic acid in plasma phosphatidylcholine is inversely correlated with the plasma low-density lipoprotein (LDL) cholesterol concentration in subjects with primary hyperlipemia (Žák *et al.*, 1998). A significant inverse association was reported between the proportion of palmitoleic acid in erythrocyte phospholipids and the plasma apolipoprotein B level among 45–64 year-old men randomly selected from the urban community of Lille, France. However, its level did not correlate significantly with the levels of plasma total or LDL cholesterol (Theret *et al.*, 1993). The proportion of palmitoleic acid in percutaneous fat was inversely associated with the recurrence of stenosis (narrowing of blood vessels) (Foley *et al.*, 1992). A multiple logistic regression analysis of the lipids and fatty acids in plasma suggested that the relative content of palmitoleic acid in percutaneous fat at the time of angioplasty was the best predictor of restenosis (Foley *et al.*, 1992). There is evidence suggesting that a higher proportion of palmitoleic acid in plasma cholesterol esters may be associated with increased incidence of hypertension (Zheng *et al.*, 1999; Simon *et al.*, 1996) and coronary artery disease (Siguel and Lerman, 1994).

Epidemiological studies have revealed a positive correlation between the proportions of palmitoleic acid, γ-linolenic acid (GLA; 18:3*n*-6), and dihomo-γ-linolenic acid (DGLA; 20:3*n*-6) in serum cholesterol esters and the incidence of glucose intolerance and non-insulin-dependent diabetes mellitus (Vessby *et al.*, 1994). There is also a correlation between the level of palmitoleic acid in serum and the risk of developing non-insulin-dependent diabetes mellitus. This may be attributable to dietary factors but there is a strong possibility that genetically determined differences in the activities of enzymes involved in the synthesis and metabolism of fatty acids in the body may contribute to the correlation. This occurs through the effects of the enzymes on lipid composition and on insulin sensitivity (Borkman *et al.*, 1993).

Palmitoleic acid can be synthesized endogenously as well as acquired directly from the diet, but a clear distinction between dietary intake and internal synthesis as the source of observed tissue concentrations is not possible. The mechanism responsible for the changes in tissue palmitoleic acid levels reviewed above is not clear. Considering the low content of palmitoleic acid in dietary sources, these changes are unlikely to be due to an alteration in dietary content. Rather, they may be the result of an up- or down-regulation of lipogenesis and of the activity of Δ^9-desaturase induced by dietary factors such as low essential fatty acid intake and/or disease conditions. Different mechanisms may prevail depending on the tissues and physiological and disease conditions.

An increase in the level of palmitoleic acid helps to maintain the unsaturation index of membrane phospholipids and partially compensates for the loss of fluidity and function of cell membranes resulting from EFAD. An increased level of palmitoleic acid and a suppressed level of PUFA make the cell membrane less susceptible to oxidative stress. Elevated levels of palmitoleic acid observed in certain tumours may have been a result of an increased uptake of the fatty acid by tumour cells (Sauer and Dauchy, 1992) or of its preferential transference by the hosts as a salvage response to the malignant situations (Simonsen *et al.*, 1998). Enhanced utilization of essential fatty acids, following up-regulated activities of the enzymes responsible for the formation of leukotrienes and prostaglandins, may also contribute to the increase in the relative content of palmitoleic acid in tumour cells (Reynier *et al.*, 1991). Therefore the increased level of palmitoleic acid in these tissues does not necessarily reflect the effect of a dietary supplementation with the fatty acid. The low incidence of coronary heart disease and certain cancers in the Mediterranean population aroused interest in the possible positive role of MUFA in human nutrition. A diet rich in MUFA is considered to protect against coronary heart disease and certain hormone-related cancers, but most of the published data relating to MUFA have been obtained from studies with oleic acid. Physiological effects and bioactivities of fatty acids are influenced not only by the degree of unsaturation and the position and configuration of double bonds, but also by the chain length. Therefore, physiological effects demonstrated by oleic acid cannot be directly extrapolated to palmitoleic acid. Unfortunately, experimental data obtained directly from the latter are limited.

3. *Effects of palmitoleic acid on cancer*

Palmitoleic acid and its C_{18} homologue *cis*-vaccenic acid (both *n*-7 monoene acids) have been reported to be inhibitors of mutagenesis and carcinogenesis (Hayatsu *et al.*, 1988). Oral treatment with palmitoleic acid markedly prolonged the survival time of mice with Ehrlich ascites tumours, and the anti-tumour activity of palmitoleic acid is stronger than that of oleic acid (Ito *et al.*, 1982). When palmitoleic acid-treated tumour cells were compared with control tumour cells, the total lipids and phospholipids contents were significantly lower and the fatty acid compositions of phosphatidylcholine, cholesterol esters, and triacylglycerols were considerably different (Ito *et al.*, 1982).

Results obtained from human prostate cell cultures and cell-free systems demonstrated inhibitory effects of palmitoleic, oleic, and myristoleic acids on the activity of human and rat microsomal 5α-reductase, which converts testosterone into 5α-dihydrotestosterone (Liang and Liao, 1992). The latter binds to androgen receptors and functions in the nucleus to regulate the expression of specific genes. 5α-Dihydrotestosterone promotes benign prostatic hyperplasia and female hirsutism. Certain 4-azasteroids are potent competitive inhibitors of 5α-dihydrotestosterone and have shown therapeutic value for treating prostate cancer. The inhibitory effect of palmitoleic acid on microsomal 5α-reductase activity may represent a mechanism related to its anti-tumour effects.

An anti-tumour formulation containing sodium palmitoleate, lactose, fatty acid sucrose esters, and magnesium stearate has been patented and is claimed to be effective against Ehrlich ascites tumours in mice (Toyo Jozo Co Ltd, 1984).

4. Effects of palmitoleic acid on hyperlipemia, hypertension and diabetes mellitus

A macadamia nut-enriched diet (42% energy as fat, 20% energy from macadamia nuts) for 4 weeks lowered serum total and LDL cholesterol levels compared with the baseline values (on a pre-entry diet containing 37.1% of total energy as fat) (Colquhoun *et al.*, 1996). Hypocholesterolemic and hypoglyceridemic activities of ethyl palmitoleate were reported to be comparable with those of ethyl linoleate and ethyl linolenate (Budijanto *et al.*, 1992). Feeding rats with a palmitoleic acid-enriched diet provided by macadamia nut oil decreased LDL cholesterol and increased high-density lipoprotein (HDL) cholesterol levels in plasma, accompanied by a decrease in the fat deposit in liver (Brillhart and Maurer, 1992). A diet containing 1% (w/w of total diet) palmitoleic acid significantly reduced the incidence of stroke and improved the survival rate in stroke-prone spontaneously hypertensive rats. The preventive effect was ascribed to a possible direct metabolic improvement of vascular smooth muscle cells (Yamori *et al.*, 1986).

Glucotoxicity and lipotoxicity contribute to the impaired β-cell function observed in non-insulin-dependent diabetes mellitus. Exposure of rat pancreatic islets to palmitic acid in a culture medium containing glucose markedly increased β-cell DNA fragmentation, resulting in increased apoptosis rate coupled with reduced proliferation capacity of β-cells and impaired insulin secretion (Maedler *et al.*, 2001). In contrast, palmitoleic acid, alone or in combination with palmitic acid, did not affect DNA fragmentation. Palmitoleic acid improved β-cell secretory function that had been reduced by palmitic acid and enhanced β-cell proliferation in a medium containing glucose (Maedler *et al.*, 2001).

Table 3. Fatty acid composition of oil from the soft part of sea buckthorn (*Hippophaë rhamnoides*) berries of different subspecies[a]

	Fatty acid content (wt% of total fatty acids)						
Subspecies	**Palmitic (16:0)**	**Palmitoleic (16:1*n*-7)**	**Stearic (18:0)**	**Oleic (18:1*n*-9)**	**Vaccenic (18:1*n*-7)**	**Linoleic (18:2*n*-6)**	**α-Linolenic (18:3*n*-3)**
mongolica	30–40	40–50	1–3	4–7	5–7	5–8	1–2
rhamnoides	20–30	30–40	1–2	13–20	8–10	6–15	1–6
sinensis	15–30	12–35	1–2	12–25	6–10	3–6	1–10

[a]Data taken from Yang (2001).

5. *Sea buckthorn (Hippophaë rhamnoides) oils*

Sea buckthorn berries have been used traditionally in China as both food and medicine. Two oils with different fatty acid compositions can be isolated from the berries. The seed oil is highly unsaturated, with linoleic and α-linolenic acids together constituting up to 70% of the total fatty acids, but the oil from the soft parts is rich in palmitoleic and palmitic acids. The level of palmitoleic acid in oil from the soft part varies with the origin of the berries. The highest level is found in berries of *H. rhamnoides* subsp. *mongolica,* native to the Altai regions, and the lowest level in berries of subsp. *sinensis,* which grows naturally in China. Table 3 presents a summary of the fatty acid compositions of oils from the soft parts of berries of different origins. 'Ruet', 'Luchezarnaya', 'Dar Katuni', 'Vitaminaya' and 'Maslichnaya' are commercial cultivars selected from the natural sea buckthorn population in the Altai region. The majority (60–80%) of triacylglycerols in the oil from the soft part of these berries has palmitoleic acid in the *sn*-2 position. Commercial oils, isolated by supercritical CO_2 extraction or by freon extraction from dried soft parts of berries from different sources, are available. Centrifugation is commonly used for isolating oil from berry juice.

Oil from the soft parts of sea buckthorn berry is already used as a raw material for functional foods and cosmetics in China, Russia, Europe, and North America. Animal experiments and clinical trials have shown protective, tissue-regenerative, and anti-inflammatory effects of the oil on skin and mucosa (Yang, 2001). There is evidence that the oil may regulate immune function, inhibit oxidation, suppress the growth of cancer cells, and reduce risk factors for cardiovascular disease. The physiological effects of sea buckthorn oils have recently been reviewed (Yang, 2001).

Sea buckthorn (soft part) oil is also rich in carotenoids (up to 1%, w/w of oil), tocopherols (0.3–0.5%), and phytosterols (2–3%) (Yang, 2001) and the benefits observed with sea buckthorn oil are probably the result of a synergy between fatty acids and other bioactive components.

6. *Enrichment of palmitoleic acid*

The aerobic culture *S. cerevisiae* (AKU4109) has been used to produce palmitoleic acid (Nobuhiro and Yasu, 1992). The Δ^9-desaturase gene from this source has been expressed in eggplant (*Solanum melongena* L.) (Xing and Chin, 2000) and in tomato (*Lycopersicon esculentum*) (Wang *et al.*, 1998), resulting in a marked increase in the level of palmitoleic acid in leaves and other tissues. Increasing the production of this fatty acid in plants has been suggested as a possible approach to enhance their disease resistance, due to its cytotoxic effect against certain phytopathogenic fungi (Xing and Chin, 2000; Wang *et al.*, 1998). Transgenic oil plants with the Δ^9-desaturase gene from *S. cerevisiae* may be a possible source of oils with a high content of palmitoleic acid.

cDNA for acyl-ACP (acyl carrier protein) desaturase, isolated from developing seeds of cat's claw, has been successfully expressed and a peptide with Δ^9-desaturase activity specific for palmitoyl-ACP produced in *Escherichia coli* (Cahoon *et al.*, 1998). Efforts are now under way to express the cDNA of acyl-ACP desaturase isolated from developing macadamia seeds in *E. coli* (Gummeson *et al.*, 2000). Expression of Δ^9-desaturase genes of plant origin in microorganisms offers a promising alternative source of products enriched in palmitoleic acid.

7. *Summary*

Alterations in tissue levels of palmitoleic acid are often found under certain physiological and disease conditions, frequently in association with changes in the relative levels *n*-6 and *n*-3 PUFA. Neither the mechanism nor the physiological/pathological significance of these changes is well understood. Palmitoleic acid showed potentially protective effects against cardiovascular disease and certain cancers in studies carried out *in vivo* and *in vitro* but further investigations are needed to verify these effects due to the limited data available.

Oils from macadamia nut and the soft parts of sea buckthorn berries are both rich sources of palmitoleic acid. A wide range of positive health effects of sea buckthorn oil are indicated by the long history of application of the berries in China, and by clinical trials and animal experiments carried out in several countries. In addition to the fatty acids, several unsaponifiable components in sea buckthorn oil (phytosterols, tocopherols, and carotenoids) probably play an important role in the health effects of the oil. Isolation of palmitoyl-ACP-specific Δ^9-desaturases and expression of the corresponding genes in plants and micro-organisms provides promising alternative enriched sources of palmitoleic acid.

C. γ-Linolenic acid (GLA)

GLA is an important member of the *n*-6 family of PUFA and there are several circumstances in which it might be desirable to add it to the diet. GLA is available in seed oils of evening primrose, borage, and blackcurrant, and methods of raising the concentration of this acid in these sources are described below. The nutritional and medical uses of GLA are also outlined. Useful reviews on the physical and chemical properties (Gunstone) and the nutritional and medical properties (Horrobin) of GLA were published in 1992.

1. *Metabolic pathway of GLA*

GLA is the first intermediate in the bioconversion of linoleic acid to arachidonic acid (Figure 1). This change involves the introduction of a third double bond at the Δ^6-position under the catalytic influence of the Δ^6-desaturase enzyme, and is believed to be the rate-limiting step in the metabolic pathway.

For humans in good health this presents no problem, but conditions such as ageing, smoking, diabetes, high alcohol intake, stress-related hormones, and viral infections inhibit this rate-determining step. In these cases it is considered wise to supplement the normal dietary intake of GLA. Nutritional factors can also influence this metabolic pathway.

2. *Occurrence of GLA in seed oils and microorganisms*

As GLA is an intermediate in the conversion of linoleic acid to arachidonic acid, it is not surprising that GLA is found at low levels in many animal fats. It is present in both cow's milk fat (~0.1%, w/w of milk) and human milk fat (0.35–1.0%, w/w of milk fat) and its presence in the latter has often been cited as an indication of its importance in human nutrition. However, richer sources of GLA are found in the seed oils of some less common plants and are also produced by some microorganisms.

At present there are three commercial sources of GLA from seed oils: evening primrose (*Oenothera biennis* and *O. lamarckiana*), borage (*Borago officinalis*), and blackcurrant (*Ribes nigrum*). Typical fatty acid profiles for these three oils are given in Table 4. Ucciani (1995) found GLA in 164 higher plant species belonging to 10 botanical families and Aitzetmüller (2001) considers that there are probably 120–150 individual botanical species each containing >5% GLA in their seed oils. Guil-Guerrero *et al.* (2001a) described 49 plant species which are potential sources of GLA.

Evening primrose oil (extracted from seeds containing 20–24%, w/w oil) was the first to be developed commercially and used as a source of GLA. It contains 6–14% (generally 9–10%) GLA accompanied by high levels of linoleic acid. Borage (starflower) is a richer source of GLA both in terms of the oil content of the seeds (25–36%) and the level of GLA in the oil (~22%). It has less linoleic acid than evening primrose oil and contains a series of *n*-9 higher monoene acids (20:1, 22:1, and 24:1) at a combined level of about 8%.

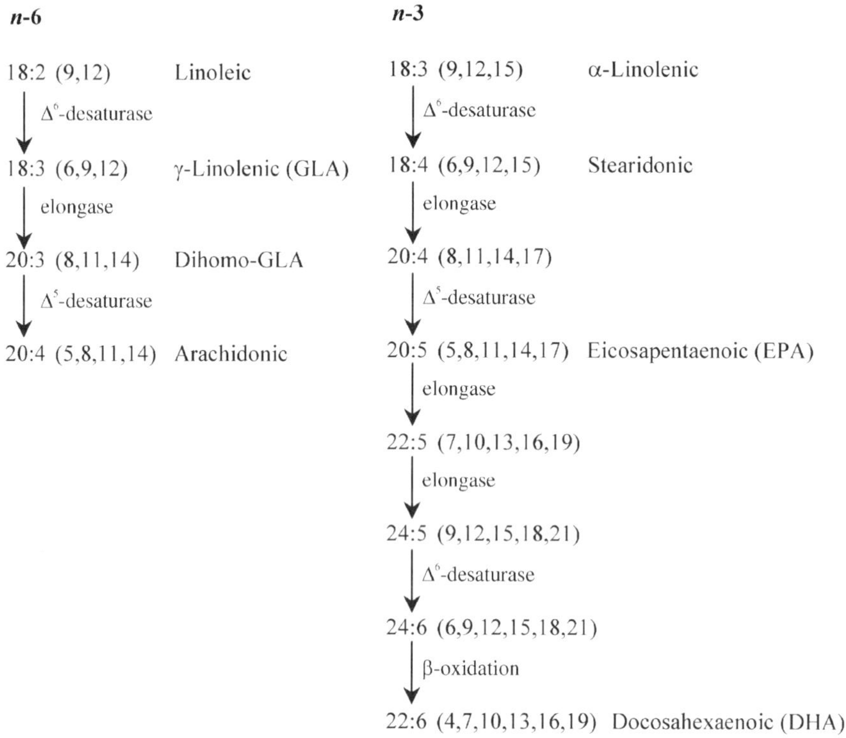

Figure 1. There are two important families of PUFA: the *n*-6 family based on linoleic acid, and the *n*-3 family based on α-linolenic acid. Since animals – including humans – cannot produce linoleic or α-linolenic acids themselves, these acids are essential parts of the diet, obtainable only from vegetable sources or from animals which have already derived these acids from a plant source. Once ingested, the two C_{18} acids can be metabolized to important C_{20} and C_{22} acids through a series of changes involving desaturation and elongation. The same enzymes are required for both families of acids and there is competition for these. The most important acids in these sequences are arachidonic acid in the *n*-6 series, and eicosapentaenoic acid (EPA) and docosahexaenoic acid (DHA) in the *n*-3 series. All these acids are important components of phospholipid membranes, and the two C_{20} acids are also precursors of prostaglandins and other eicosanoids. The locations of the double bonds are indicated in brackets after the abbreviated structure of each fatty acid.

Blackcurrant oil is intermediate between the other two sources in GLA content but also contains α-linolenic acid and stearidonic acid (see Section D below).

Evening primrose oil grows in many countries in Europe and also in New Zealand but for commercial reasons the crop is now grown mainly in China (75–80%) and Eastern Europe. Seed is produced at a level of 1000–1200 kg/ hectare. Annual production is estimated to be about 12 000 tonnes of seed

Table 4. Fatty acid composition of evening primrose oil, borage oil, and blackcurrant oil and of oils from *Mortierella isabelina* and *Mucor javanicus*[a]

	Evening primrose	**Borage**[b]	**Blackcurrant seed**	***Mortierella isabelina***	***Mucor javanicus***
Oil content (wt% of seed)	16–26	27–35	22–26		
Fatty acids (wt% of oil):					
Palmitic (16:0)	7–10	10	6–7	27	22–25
Stearic (18:0)	1.5–3.5	4	1.5	6	5–8
Oleic (18:1*n*-9)	6–11	18	9–11	44	38–41
Linoleic (18:2*n*-6)	65–80	37	46–49	12	10–12
GLA (18:3*n*-6)	6–14[c]	22	14–16	8	15–18
α-Linolenic (18:3*n*-3)	ND	ND	13–14	ND	ND
Stearidonic (18:4*n*-3)	ND	ND	2.5–3	ND	ND

[a]Data taken from Gunstone (1992). [b]Also contains 20:1 (4%), 22:1 (2.5%), 24:1 (1%). [c]Generally 9–10%. ND, not detected.

yielding 1500–1800 tonnes of oil. Over the 10 years to 1999 the price of the oil fell from US$45/kg to US$17.5/kg (Clough, 2001a, 2001b).

The UK is the main source of borage (starflower) but it is also grown in the Netherlands, Canada, New Zealand, and Poland. Annual production of the seed is estimated at 1500–2000 tonnes, which provides about 500–700 tonnes of oil. Borage seed costs US$3000–4000/tonne. The price of the oil fell from US$60/kg to US$35/kg over the 10 years to 1999 (Clough, 2001a, 2001b).

Blackcurrant seed oil is a by-product from the production of blackcurrant juice and jelly, and production is estimated to be ~50–100 tonnes/year.

These oils are most often supplied in capsules, with added tocopherol and ascorbyl palmitate to serve as antioxidants, and have a shelf life of about 3 years.

There are several reports of plants such as rapeseed/canola being genetically modified to produce GLA but commercial products are not yet available. One example cited by Palombo *et al.* (2000) has the fatty acid composition detailed in Table 5. Another group (Liu *et al.*, 2001) introduced Δ^6-desaturase and Δ^{12}-desaturase, both from the fungus *Mortierella alpina*, into canola plant cells. The modified canola plants produced an oil rich in GLA (Table 5). The major triacylglycerols in this modified oil were GGL (23 mol%), GLO (20 mol%), GGO (12 mol%), GGG (11 mol%), and GLL (8 mol%), where G represents GLA, L represents linoleic acid, O represents oleic acid, and the symbols indicate all 6 possible stereoisomers with the indicated fatty acid composition.

The distribution of these fatty acyl chains between the *sn*-1, 2, and 3 positions has been examined in a number of ways. Results for GLA are summarized in Table 6.

Because of the high level of linoleic acid in evening primrose oil its major

Table 5. Fatty acid composition of canola oil before and after genetic modification to increase the GLA content

	Fatty acid content (wt% of total fatty acids)						
	Palmitic (16:0)	**Stearic (18:0)**	**Oleic (18:1*n*-9)**	**Linoleic (18:2*n*-6)**	**GLA (18:3*n*-6)**	**Other**	**References**
Before	4.2	2.7	71.5	19.9	ND	1.7	Liu *et al.* (2001)
After (1)	5.3	3.0	20.0	22.4	43.0	6.3	Liu *et al.* (2001)
After (2)	4.2	3.7	25.0	26.0	37.0	4.1	Palombo *et al.* (2000)

ND, not detected.

Table 6. Distribution of GLA between the *sn*-1, 2, and 3 positions of TAG in evening primrose, borage, and blackcurrant seed oils

	GLA content (wt% of total fatty acids)				
	TAG	***sn*-1**	***sn*-2**	***sn*-3**	**References**
Evening primrose (1)	9.6	7.2	10.7	10.9	Laakso and Christie (1990)
Evening primrose (2)	9.3	3.6	10.7	13.5	Lawson and Hughes (1988)
Borage	24.8	4.0	40.4	30.1	Lawson and Hughes (1988)
Blackcurrant	15.9	4.1	17.4	25.8	Lawson and Hughes (1988)

triacylglycerols contain two (or three) linoleic acid chains: LLL (54.3 mol%), LLG (17.6 mol%), LLO (13.7 mol%), and LLP (7.9 mol%), where the symbols are as described above and P represents palmitic acid. The triacylglycerol composition of borage oil is more complex with 10 molecular species each accounting for >4 mol% of all triacylglycerols: LLG (15.4 mol%), OLG (12.3 mol%), LLL (10.1 mol%), LGG (9.8 mol%), PLG (9.5 mol%), PLL/OOG (8.9 mol%), OLL (8.2 mol%), SLG/POG (5.0 mol%), LLE (4.1 mol%), OOL (4.1 mol%), where the symbols are as described above and E represents 20:1 and S represents stearic acid.

Moulds belonging to the Mucorales family produce GLA as their only 18:3 acid. Commercial processes for GLA production have been developed in Japan based on *Mortierella isabelina* and in the UK based on *Mucor javanicus*. The UK production is now discontinued following a change in ownership of the producer company. These products differ from the seed oils in their higher levels of saturated and monoene acids and their lower level of linoleic acid (Table 4). This makes the mould-derived oils more suitable as a source of pure

GLA or of upgraded triacylglycerols since the necessary separations are easier to effect.

The blue–green *Spirulina* algae *S. platensis* (18–21% GLA, w/w of total fatty acids) and *S. maxima* (12%, w/w GLA) are also potential sources of GLA. However, the GLA is present in various galactosyldiacylglycerols or phosphatidic acids rather than in triacylglycerols.

3. Sources with higher GLA levels

It is possible to raise the level of GLA in seed oils by selective enzymatic reactions. Evening primrose and borage oils with twice their normal GLA levels have been produced and offered for sale. Although mainly triacylglycerols, these products may contain up to 20% (w/w of oil) diacylglycerols. Procedures for increasing the GLA content of the oils are based on the fact that some lipases discriminate against acids having a double bond close to the acyl function, and so GLA with its Δ^6-unsaturation can be distinguished from the more common unsaturated acids with Δ^9-unsaturation (oleic and linoleic). This differentiation is apparent in hydrolysis, esterification, acidolysis, and alcoholysis reactions promoted by appropriate lipases. The following examples are typical.

Borage oil (22% GLA) is hydrolysed completely (*Pseudomonas* species, 35°C, 24 h) and the borage acids are selectively esterified with lauric alcohol in the presence of *Rhizopus delemar* lipase at 30°C for 20 h. The latter enzyme discriminates against GLA and this concentrates in the unesterified acids fraction, which finally contains 70% GLA (74% recovery). If the esterification is repeated (using *Candida rugosa* lipase, 35°C, 15 h), 94% GLA is obtained with 68% recovery (Shimada *et al.*, 1997).

Selective hydrolysis of borage oil with *Candida rugosa* lipase (35°C, 15 h) gave glycerol esters containing a maximum of 46% GLA. After removal of free acids, a second partial hydrolysis raised the GLA concentration to 54% with 76% recovery. On a pilot-plant scale, 7 kg of borage oil gave 1.5 kg of upgraded oil (56% GLA) after removal of free acids by molecular distillation. Starting with an already upgraded borage oil (10 kg, 45% GLA), high-quality GLA (2.1 kg, 98% pure, 49% recovery) was obtained by hydrolysis and two separate partial esterifications (Shimada *et al.*, 1998).

Clough (2001a) reports "The limited number of commercial products available are based on either borage oil or evening primrose oil, and typically the reactions are run to achieve a concentration approximately 2.5 times the original concentration. A yield of 1 tonne of finished product from 3–4 tonnes of oil input would be normal. The resulting blends of triacylglycerols and diacylglycerols are typically diluted with further amounts of the original raw material to produce a standardized product containing 20% GLA if based on evening primrose oil or 40% if based on borage oil. A finished product containing 80–85% (w/w) triacylglycerols and 15–20% (w/w) diacylglycerols would be normal."

Structured triacylglycerols (see Chapter 6) of the type CGC (where C represents caprylic acid, and G represents GLA) have been prepared in a two-step process. In the first stage the triacylglycerol GGG is prepared from glycerol, GLA, and *Candida antarctica* lipase. This is then subjected to acidolysis with caprylic acid or interesterification with ethyl caprylate in the presence of *Rhizopus delemar* lipase. A typical reaction product contains 53% caprylic acid and 47% GLA with CCG (58%) and CGG (29%) as the major triacylglycerols (Kawashima *et al.*, 2001). This particular work was carried out only on a laboratory scale but there are many reports of this type of enzymatic process being carried out on a kilogram scale. At present such compounds are probably too expensive to be incorporated into functional foods and are more likely to be used in pharmaceutical preparations.

4. *Nutritional and medical uses of GLA*

The major applications for GLA-rich oils are in the area of health food supplements, but markets have also been developed in infant nutrition products, pet foods, and cosmetics. The total market value for these oils is around US$50 million/year.

Available evidence indicates that only around 5–10% of the daily intake of linoleic acid can be converted to GLA and beyond. For a 60 kg adult with a dietary intake of 5–20 g/day of linoleic acid, the endogenous rate of formation of GLA will be 250–1000 mg/day or around 4–17 mg/kg body weight/day. Human breast milk contains 100–400 mg/l of GLA+DGLA. A 5 kg baby consuming 1 litre of milk per day receives 20–80 mg/kg body weight/day of these two acids combined (Horrobin, 1992).

DGLA is converted to two main eicosanoids: prostaglandin E_1 (PGE_1) and 15-hydroxy-DGLA. PGE_1 is anti-inflammatory, anti-thrombotic, a vasodilator, and stimulates the formation of cAMP, which inhibits phospholipase A_2; 15-hydroxy-DGLA is a potent natural anti-inflammatory agent which inhibits the conversion of arachidonic acid to its 5- and 12-lipoxygenase metabolites. GLA and DGLA are considered to be useful in the treatment of diabetes, atopic eczema, inflammation, stress, cardiovascular disease, and cancer (Horrobin, 1992).

Soft gelatin capsules containing evening primrose oil have been sold as nutritional supplements for more than 30 years, and borage oil (also marketed as starflower oil) capsules have been sold for more than 15 years. Samples may be blended with other oils (e.g. fish, linseed) or have added vitamins, minerals, or herbal extracts. Nutritional supplements probably account for 80% of the total usage of GLA oils.

GLA has been used to treat a wide range of conditions, including atopic eczema, dermatitis and other inflammatory skin conditions, diabetic neuropathy, breast pain, premenstrual syndrome, high blood pressure, and cancer. Not all the claims made for these treatments have been confirmed. Several of these issues are elaborated in two books devoted to current research work on GLA edited by Huang and colleagues (1996, 2001).

Doses have not been fully defined, but the following levels have been recommended for adults (levels refer to GLA itself, not GLA-containing oils): for nutritional purposes, 25–50 mg/day; for therapeutic use, 100–500 mg/day; for pharmacological effects, 500–2000 mg/day (Horrobin, 1992).

Two products have been licensed as pharmaceuticals: 'Efamast' for relief of cyclical mastalgia and 'Epogam' for treatment of atopic eczema. There is also a range of pre-term and full-term infant formulations on the market (particularly in Europe) containing GLA from various sources at levels up to 0.9% (w/w of formula).

Many skin care products (creams, lotions, soaps) contain GLA oils at levels between 0.1% and 2.0% (w/w of product).

5. *Functional foods containing GLA*

As already indicated, GLA-containing oils are used mainly in encapsulated form as nutritional supplements. They are also added to some infant formulas and some enteral nutrition products (Clough, 2001a). A few pet food companies add GLA oils to cat and dog speciality products.

Beyond this there is only anecdotal evidence of the use of GLA in functional foods. The authors are indebted to Peter Clough for much of the following information.

These include:

- A fortified bread, targeted at women, containing soy isoflavones and evening primrose oil.
- 'Honey Royale' is honey to which lemon juice and evening primrose oil have been added.
- Some "complete-meal-in-a-drink" powders available in the USA contain GLA.
- Drinks available in Japan such as 'Honey Lemon' and 'Linofresh' contain GLA.
- Trials have been conducted on spreads in which some sunflower oil (2–3%) has been replaced by borage oil, but there is no evidence that these products have been launched.
- By incorporating GLA oils into chicken rations it is possible to produce eggs with a modified fatty acid composition. This is a widely used method of incorporating particular fatty acids, such as omega-3 (n-3) acids and conjugated linoleic acid, into eggs.

D. Stearidonic acid

1. Dietary sources

Vertebrates, including humans, lack the Δ^{12}- and Δ^{15}-desaturases responsible for the production of linoleic and α-linolenic acids, the two essential fatty acids in human nutrition. Stearidonic acid (18:4*n*-3) has an important intermediary position in the synthesis of other polyunsaturated *n*-3 fatty acids (Figure 1) and has a significant role in lipid metabolism in humans. It holds a similar position in the metabolism of *n*-3 acids (see also Chapter 8) to that of GLA in the *n*-6 series (Figure 1). The content of this fatty acid is low in all human tissues and organs and also in most foodstuffs. It is synthesized by Δ^{6}-desaturase from α-linolenic acid, and this enzyme is commonly agreed to be the rate-limiting factor in the desaturation–elongation cascade of *n*-3 PUFA in humans. In many cultures and diets, fish and other seafoods are common sources of stearidonic acid, which makes up 1–3% of the total fatty acids in many fish species (Robles-Medina *et al.*, 1998).

Several plant species in the three families Boraginaceae, Primulaceae, and Grossulariaceae accumulate stearidonic acid and/or GLA in their seed oils (Gunstone, 1992). Stearidonic acid does not exist in abundance in common edible plant crops and its dietary availability is low. When present in higher plants, it is not elongated by plant tissues. Its important metabolites, such as eicosapentaenoic acid (EPA; 20:5*n*-3) and docosahexaenoic acid (DHA; 22:6*n*-3), have not yet been recognized in seed oils. Analogously the *n*-6 counterpart of stearidonic acid – GLA – does not produce arachidonic acid in plants. However, primitive gymnosperm seeds represent a biological exception. They synthesize both arachidonic acid and EPA (Wolff *et al.*, 1999). Transgenic plants containing arachidonic acid or EPA do not yet exist due to lack of recombinant enzymes involved in the elongation of PUFA. It is likely, however, that transgenic plants with higher amounts of stearidonic acid or GLA will be available in the near future (see Section C.2 above).

There are some plant sources of stearidonic acid but the oils have not yet been properly tested as food ingredients. Seed oils of many species of the Boraginaceae family (e.g. *Echium asperrimum*, *E. creticum*, *E. sabulicola*, *E. boissieri*, *E. plantagineum*, *E. fastuosum*, *E. vulgare*, *Buglosoides arvensis*, *Asperugo procumbens*, *Lithospermum arvense*, *Li. officinale*, *Lappula squarrosa*, *L. myosotis*, *L. intermedia*, *L. granulata* and *Myosotis caespitosa*) contain stearidonic acid at levels between 10% and 20% (Tsevegsüren and Aizetmüller, 1996; Velasco and Goffman, 1999; Guil-Guerrero *et al.*, 2000; Cisòwski *et al.*, 2001; Guil-Guerrero *et al.*, 2001). The highest level (21.4%) was found in the seed oil of *Hackelia deflexa* by Tsevegsüren and Aizetmüller (1996). The oil contents of the seeds are typically between 10% and 25%.

Within the family Grossulariaceae, seed oils of various *Ribes* species

(*R. alpinum*, *R. montigenum*, *R. nigrum*, *R. rubrum*, *R. spicatum*, *R. uva-crispa*) contain stearidonic acid at levels between 2% and 6% (Wolf *et al.*, 1983; Traitler *et al.*, 1984, 1988; Olsson *et al.*, 1990; Ucciani *et al.*, 1995; Johansson *et al.*, 1997). Ucciani *et al.* (1995) reported stearidonic acid in *Grossularia burejens* seed oil (6%). The oil contents of these currant seeds vary from 15% to 30%.

In the family Primulaceae, seed oils of some species of the genus *Primula* (*P. florindae*, *P. sikkimensis*, *P. alpicola*) contain GLA and stearidonic acid, the latter at levels of 10–15% (Aizetmüller and Werner, 1991).

In addition to seeds, other plant sources of stearidonic acid have been examined. Leaves of common borage contain 6–9 mg stearidonic acid per gram of leaf dry weight (Sewón and Tyysjärvi, 1993). Earlier studies had shown that leaves of Caryophyllaceae species contain stearidonic acid (Jamieson and Reid, 1969, 1971), and small amounts of this fatty acid is found in blackcurrant leaf lipids together with 16:3*n*-3. Stearidonic acid was most abundant in the phosphatidylcholine fraction of blackcurrant leaves (Dobson, 2000). However, the content of stearidonic acid in the vegetative parts of plants is generally too low to be used as a profitable source of the fatty acid.

Some marine algae contain stearidonic acid in high quantities, often together with 16:4*n*-3 and EPA (Takagi *et al.*, 1985). The algae *Ulva pertusa* and *Undaria pinnatifida* contain a combined total of up to 40% stearidonic acid plus 16:4*n*-3. *Undaria pinnatifida* also contains trace amounts of GLA (Ishihara *et al.*, 2000).

In addition to the fatty acid content, the triacylglycerol composition and the distribution of the fatty acids between the *sn*-1, *sn*-2 and *sn*-3 positions may affect the nutritional value of the oils. In blackcurrant seed oil, stearidonic acid is mainly located in the *sn*-1 or *sn*-3 position (Kallio *et al.*, 1997), usually the *sn*-3 position (Lawson and Hughes, 1988). The proportion of stearidonic acid is about 8% in *Onosmodium hispidissimum* (false cromwell) and about 90% of it is equally distributed between the *sn*-2 and *sn*-3 positions (MacKenzie *et al.*, 1993).

2. *Preparation of stearidonic acid oils*

Speciality oils may be produced by all the common methods of edible oil isolation and purification. In many cases the oil content is low and combined pressing and solvent extraction methods have been applied. Extraction by supercritical CO_2 also has potential because of its efficiency in extracting low-oil seeds combined with continuously reducing production costs in modern factories. Furthermore, some degree of fractionation according to unsaturation and chain length is possible. Supercritical fluid extraction is recommended for the production of speciality oils because it proceeds at a low temperature, in a low-oxygen atmosphere, is solvent-free, and provides the possibility of aseptic extraction (Manninen *et al.*, 1997).

Whatever the method of isolation and purification, natural edible seed oils with a sufficient content of stearidonic acid can be used as food ingredients or nutritional supplements. Stearidonic acid in more concentrated form, however, is needed for clinical nutrition and research purposes.

Stearidonic acid can be concentrated by multi-step processing of the fatty acids liberated by saponification or lipolysis. PUFA can be separated from the less unsaturated fatty acids by urea fractionation to achieve a 5 fold concentration of stearidonic acid from both fish oil and blackcurrant seed oil (Traitler *et al.*, 1988; Wille *et al.*, 1991). The process will partially separate EPA and DHA from mixtures of GLA and stearidonic acid. Thereafter, chromatographic methods such as reversed-phase or argentation chromatography are needed for further purification of the stearidonic acid (Traitler *et al.*, 1988; Guil-Guerrero and Belarbi, 2001). As a result, products containing ~90% stearidonic acid (w/w of total fatty acids) from fish oil (Wille *et al.*, 1991; Robles-Medina *et al.*, 1995), blackcurrant seed oil (Wille *et al.*, 1991), and marine microalgae *Isochrysis galbana* oil (Robles-Medina *et al.*, 1995) have been described.

Selective enzymatic removal of unwanted fatty acids as dodecyl esters has been used to concentrate stearidonic acid from the oils of two edible marine algae, *Undaria pinnatifida* and *Ulva pertusa*. After removal of the lauryl esters, unreacted stearidonic acid was concentrated to 95 wt% purity by reversed-phase chromatography (Ishihara *et al.*, 2000).

Purification of stearidonic acid and its derivatives with transition metals capable of forming π-complexes has been demonstrated. Fatty acids of seed oil of *Echium plantagineum* containing 12.2 wt% stearidonic acid were converted to ethyl esters and treated with silver nitrate solution. After various extraction procedures an ethyl ester phase containing 94.6 wt% stearidonic acid was obtained (Coupland and Newbould, 1997).

These processes needs to be scaled up and improved to reach the level of industrial applications.

3. *Metabolism of stearidonic acid*

Of the long-chain *n*-3 fatty acids, DHA is important in maintaining normal cerebral cortex and other neuro-related tissues and is essential for functions of the retina. EPA is an eicosanoid precursor, yielding products which reduce platelet aggregation and the risk of arterial thrombosis, among other functions. Stearidonic acid is an essential, but less investigated, intermediary metabolite in the biosynthesis of EPA and DHA in humans.

The Δ^6- and Δ^5-desaturases of fatty acids are of special importance in human liver microsomes in the perinatal period during the active development of the nervous system. Δ^6-desaturase converts α-linolenic acid to stearidonic acid and linoleic acid to GLA, and Δ^5-desaturase converts 20:4*n*-3 to EPA and DGLA to arachidonic acid (Figure 1). The enzyme activities are lower in fetuses than in adults (Poisson *et al.*, 1993) and are regulated by several hormones (Spielmann *et al.*, 1989). In humans Δ^6-desaturase is commonly the rate-limiting factor in the long-chain PUFA synthesis sequence, and a high content of linoleic acid has long been known to be a potent inhibitor of this enzyme. The two essential fatty acids, linoleic acid and α linolenic acid, compete with each other for the Δ^6-desaturase. Several other factors, including hormonal control, regulate the synthesis and conversion of stearidonic acid. Blackcurrant seed oil, with its *n*-6:*n*-3 ratio of 5 and good ratio of GLA to stearidonic acid seems to cover the requirements of an optimal oil for maintaining health (Johansson *et al.*, 1997).

In animal studies stearidonic acid is more effective than α-linolenic acid for the formation of long-chain *n*-3 PUFA (Huang *et al.*, 1991; Yamazaki *et al.*, 1992; Kabakibi *et al.*, 2001), so dietary stearidonic acid can compensate for low activity of human Δ^6-desaturase in α-linolenic acid-based diets.

4. *Nutritional and health effects of stearidonic acid*

Many unsaturated fatty acids, including stearidonic acid, are claimed to have anti-cancer properties at concentrations not harmful to normal cells. This has been shown, for example, in $APC^{Min/+}$ and nude mice (Kabakibi *et al.*, 2001). When the incorporation and metabolism of α-linolenic and stearidonic acids were compared in non-malignant NIH-3T3 cells, α-linolenic acid slightly stimulated growth but stearidonic acid showed a clear inhibitory effect (Cantrill *et al.*, 1993).

Stearidonic acid is a significant factor in controlling inflammation (Kabakibi *et al.*, 2001). It inhibits mouse and human epidermal phospholipase A_2 in a dose-dependent manner (Finnen and Horrobin, 1991). The concentration required to inhibit the enzyme activity by 50% was also reported. When phospholipase A_2 is inhibited by stearidonic acid, less arachidonic acid is liberated to be a substrate of cyclooxygenase and of the 5-, 12- and 15-lipoxygenases, which may produce several inflammation-promoting eicosanoids (Finnen and Horrobin, 1991). Stearidonic acid seems also to inhibit 5-lipoxygenase and to reduce the formation of leukotrienes from arachidonic acid *in vitro*. This indicates suppression of inflammation (Guichardant *et al.*, 1993). The inhibitory effect of stearidonic acid was dose-dependent.

Blackcurrant seed oil supplementation increases EPA levels slightly in human plasma lipids (Spielmann *et al.*, 1989). An 8-week supplementation with blackcurrant seed oil significantly decreased plasma total cholesterol and LDL cholesterol levels and increased plasma HDL cholesterol levels in hyperlipidemic patients (Spielmann *et al.*, 1989). Blackcurrant oil also reduces the tendency of human platelets to aggregate (Kockmann *et al.*, 1989).

Atopic eczema is related to a decreased content of PUFA beyond the Δ^6-desaturase stage. Oral administration of CO_2-extracted alpine currant seed oil, containing 4% stearidonic acid and 18% α-linolenic acid, relieved the symptoms of atopic dermatitis in children (Johansson *et al.*, 1999).

Animal and human studies have shown that PUFA significantly lower blood pressure. In hypertensive rats, blackcurrant seed oil effected a greater reduction of blood pressure than did borage oil or evening primrose oil in a 7-week study (Engler, 1993). Whether this was due to the stearidonic acid remains open. Blackcurrant oil decreased significantly the resting diastolic blood pressure in mildly hypertensive men when compared with safflower oil in a double-blind, parallel study (Deferne and Leeds, 1996).

5. *Stearidonic acid and functional foods*

When referring to functional foods, the non-defined status and requirements (outside of Japan) have to be taken into account. Present novel-food legislation sets special demands on the history and properties of the products, and increasing research and development is necessary to support these requirements for stearidonic acid.

Stearidonic acid metabolism seems to be a regulatory centre in lipid metabolism. The rate of synthesis of stearidonic acid from α-linolenic acid is limited in humans by the relatively low activity of Δ^6-desaturase. This could be overcome by using dietary sources of stearidonic acid. Supplementation is also proposed for persons intolerant to fish or other seafoods. Several investigations have shown the potential of stearidonic acid in reducing the risk of cancer, controlling inflammation, reducing the risk of cardiovascular disease, and relieving various skin symptoms, but none of these claims can yet be regarded as scientifically proven.

The "functional food concept" based on stearidonic acid is relevant due to the common use of blackcurrant seed oil (*Ribes nigrum*) and some other food-quality seed oils. Novel plant-breeding and development techniques should make it possible to produce oils with selected composition such as a canola oil with a high content of selected fatty acids (Kabakibi *et al.*, 2001), but only when we know what we want.

References

Abidi, SL, List, GR and Rennick, KA (1999) Effect of genetic modification on the distribution of minor constituents in canola oil. *J. Am. Oil Chem. Soc.*, 76, 463–467.

Aitzetmüller, K (2001) Database will show plant species with GLA. *Lipid Technol.*, 13, 27.

Aitzetmüller, K and Werner, G (1991) Stearidonic acid (18:4*n*-3) in *Primula florindae*. *Phytochemistry*, 30, 4011–4013.

Alam, SQ, Henderson, M and Alam, BS (1994) Effects of essential fatty acid deficiency on the fatty acid composition and arachidonic acid levels in rat maxillae and mandibles. *Calcified Tissue Int.*, 55, 169–172.

Anonymous (2000) Australians using high-oleic canola oil for frying. *INFORM,* 11, 1061.

Anonymous (2001) High-oleic oils. http://www.high-oleic.com

Berghaus, G, Kurz, A and Dotzauer, G (1975) Vergleich des Fettsäuremusters zwischen menschlichem Fett- und Lebergewebe bei Nichtalkoholikern und Alkoholikern, *Dtsch. Med. Wschr.*, 100, 1233–1238.

Borkman, M, Storlien, LJ, Pan, DA, Jenkin, AB, Chisholm, DJ and Campbell, LB (1993) The relationship between insulin sensitivity and the fatty acid composition of skeletal-muscle phospholipids. *N. Engl. J. Med.*, 328, 238–244.

Brillhart, DD and Maurer, GL (1992) Food and pharmaceutical compositions containing short chain monounsaturated fatty acids. World Patent WO 92/01450 A1.

Budijanto, S, Ito, M, Furukawa, Y and Kimura, S (1992) Effect of various dietary fatty acid ethyl esters on plasma cholesterol and lipoprotein metabolism in rats. *J. Clin. Biochem. Nutr.*, 13, 13–22.

Cahoon, EB, Shah, S, Shanklin, J and Browse, J (1998) A determinant of substrate specificity predicted from the acyl-acyl carrier protein desaturase of developing cat's claw seeds. *Plant Physiol.*, 117, 593–598.

Cantrill, RC, Huang, YS, Ells, GW and Horrobin, DF (1993) Comparison of the metabolism of α-linolenic acid and its Δ^6 desaturation product, stearidonic acid, in cultured NIH-3T3 cells. *Lipids*, 28, 163–166.

Chardon, P, Crastes de Paulet, P, Lemat, AM, Crastes de Paulet, A, Kienlen, J and du Cailar, J (1985) Exclusive enteral feeding with Enteronutril® and its effects on serum lipids. *Ann. Fr. Anesth. Réanim.*, 4, 333–338.

Cisòwski, W, Zielinska-Stasiek, M, Stolyhwo, A, Migas, P and Kamieniec, A (2001) Gas–liquid chromatographic analysis of fatty acids obtained from the seeds of some *Boraginaceae* plants. *Acta Chromatographica*, 11, 215–223.

Clough, P (2001a) Specialty vegetable oils containing γ-linolenic acid and stearidonic acid. In: *Structured and Modified Lipids* (FD Gunstone, ed.), Marcel Dekker, New York, pp.75–117.

Clough, P (2001b) Sources and production of speciality oils containing GLA and stearidonic acid. *Lipid Technol.*, 13, 9–12.

Colquhoun, DM, Humphries, JA, Moores, D and Somerset, SM (1996) Effects of a macadamia nut enriched diet on serum lipids and lipoproteins compared with a low fat diet. *Food Austr.*, 48, 216–222.

Coupland, K and Newbould, C (1997) Treatment of oils. World Patent WO 97/46649.

Daun, JK (2000) Canadian Grain Commission, Winnipeg, Canada, personal communication.

Deferne, JL and Leeds, AR (1996) Resting blood pressure and cardiovascular reactivity to mental arithmetic in mild hypertensive males supplemented with blackcurrant seed oil. *J. Human Hypertension*, 10, 531–537.

Dobson, G (2000) Leaf lipids of *Ribes nigrum*: a plant containing 16:3, α-18:3, γ-18:3 and 18:4 fatty acids. *Biochem. Soc. Trans.*, 28, 583–586.

Dyerberg, J, Bang, HO and Hjorne, N (1975) Fatty acid composition of the plasma lipids in Greenland Eskimos. *Am. J. Clin. Nutr.*, 28, 958–966.

Engelbrecht, AM, Louw, L and Cloete, F (1998) Comparison of the fatty acid composition in intraepithelial and infiltrating lesions of the cervix: part II, free fatty acid profiles. *Prostaglandins Leukotr. Essent. Fatty Acids*, 59, 253–257.

Engler, MM (1993) Comparative study of diets enriched with evening primrose, blackcurrant, borage or fungal oils on blood pressure and pressor responses in spontaneously hypertensive rats. *Prostaglandins Leukotr. Essent. Fatty Acids*, 49, 809–814.

Finnen, MJ and Horrobin, DF (1991) Stearidonic acid and/or eicosatatraenoic acid for the treatment of inflammations. European Patent 0460484 A1.

Foley, JB, Tounger, K, Foley, D, Kinsella, A, Molloy, M, Crean, PA, Gearty, G, Gibney, M and Walsh, MJ (1992) Lipids and fatty acids and their relationship to restenosis. *Catheter. Cardio. Diag.*, 25, 25–30.

Gómez-Tubío, A, Carreras, O, Tavares, E and Delgado, MJ (1999) Effects of chronic ethanol consumption on fatty acid profile of heart tissue in rats. *Alcohol. Clin. Exp. Res.*, 23, 404–407.

Gorban, AMS and Izzeldin, OM (2001) Fatty acids and lipids of camel milk and colostrum. *Int. J. Food Sci. Nutr.*, 52, 283–287.

Guichardant, M, Traitler, H, Spielmann, D, Sprecher, H and Finot, PA (1993) Stearidonic acid, an inhibitor of 5-lipoxygenase pathway. A comparison with timnodonic and dihomogammalinolenic acid. *Lipids*, 28, 321–324.

Guil-Guerrero, JL and Belarbi, EH (2001) Purification process for cod liver oil polyunsaturated fatty acids. *J. Am. Oil Chem. Soc.*, 78, 477–484.

Guil-Guerrero, JL, Gómez-Mercado, FF, García-Morato, P and Campra-Madrid, P (2000) Occurrence and characterization of oils rich in γ-linolenic acid Part I: *Echium* seeds from Macronesia. *Phytochemistry*, 53, 451–456.

Guil-Guerrero, JL, García Maroto, FF and Giménez-Giménez, A (2001) Fatty acid profiles from forty-nine plant species that are potential new sources of γ-linolenic acid. *J. Am. Oil Chem. Soc.*, 78, 677–684.

Gummeson, PO, Lenman, M, Lee, M, Singh, S and Stymne, S (2000) Characterisation of acyl-ACP desaturases from *Macadamia integrifolia* Maiden & Betche and *Nerium oleander* L. *Plant Sci.*, 154, 53–60.

Gunstone, FD (1992) Gamma linolenic acid – Occurrence and physical and chemical properties. *Progr. Lipid Res.*, 31, 145–161.

Gunstone, FD (1994) Fatty acid structure. In: *The Lipid Handbook* (2nd edn) (FD Gunstone, JL Harwood and FB Padley, eds), Chapman and Hall, London, pp.1–19.

Gunstone, FD (1998) Movements towards tailor-made fats. *Progr. Lipid Res.*, 37, 277–305.

Gunstone, FD (2001a) Why are structured lipids and new lipid sources required? In: *Structured and Modified Lipids* (FD Gunstone, ed.), Marcel Dekker, New York, pp.1–10.

Gunstone, FD (2001b) Procedures used for lipid modification. In: *Structured and Modified Lipids* (FD Gunstone, ed.), Marcel Dekker, New York, pp.11–36.

Gunstone, FD (2001c) Oilseed crops with modified fatty acid composition. *J. Oleo. Sci.*, 50, 269–279.

Gupta, MK (1998) NuSun – the future generation of oils. *INFORM*, 9, 1150–1154.

Gupta, MK (2000) Nu-Sun – healthy oil at a commodity price. *Lipid Technol.*, 12, 29–33.

Hayatsu, H, Arimoto, S and Negishi, T (1988) Dietary inhibitors of mutagenesis and carcinogenesis. *Mutat. Res.*, 202, 429–446.

Horrobin, DF (1992) Nutritional and medical importance of gamma linolenic acid. *Progr. Lipid Res.*, 31, 163–194.

Huang, YS and Mills, DE (eds) (1996) *γ-Linolenic Acid: Metabolism and its Roles in Nutrition and Medicine*, AOCS Press, Champaign.

Huang, YS and Ziboh, A (eds) (2001) *Recent Advances in Biotechnology and Clinical Applications of Gamma Linolenic Acid*, AOCS Press, Champaign.

Huang, YS, Smith, RS, Redden, PR, Cantrill, RC and Horrobin, DF (1991) Modification of liver fatty acid metabolism in mice by *n*-3 and *n*-6 delta-6-desaturase substrates and products. *Biochim. Biophys. Acta*, 1082, 319–327.

Ishihara, K, Murata, M, Kaneniwa, M, Saito, H, Komatsu, W and Shinohara, K (2000) Purification of stearidonic acid (18:4(*n*-3)) and hexadecatetraenoic acid (16:4(*n*-3)) from algal fatty acid with lipase and medium pressure liquid chromatography. *Biosci. Biotechnol. Biochem.*, 64, 2454–2457.

Ito, H, Kasawa, K, Naruse, S and Shimura, K (1982) Anti-tumour effect of palmitoleic acid on Ehrlich ascites tumour. *Cancer Lett.*, 17, 197–203.

Jamieson, GR and Reid, EH (1969) Leaf lipids of some members of the *Boraginaceae* family. *Phytochemistry*, 8, 1489–1494.

Jamieson, GR and Reid, EH (1971) Leaf lipids of some members of *Caryophyllaceae*. *Phytochemistry*, 10, 1575–1577.

Johansson, A, Laakso, P and Kallio, H (1997) Characterization of seed oils of wild, edible Finnish berries. *Z. Lebensm. Unters. Forsch.*, 204, 300–307.

Johansson, A, Isolauri, E, Salminen, S, Laakso, P, Turjanmaa, K, Katajisto, J and Kallio, H (1999) Alpine currant seed oil as a source of polyunsaturated fatty acids in the treatment of atopic eczema. In: *Functional Foods – A New Challenge for the Food Chemists* (R Lásztity, W Pfannhauser, L Simon-Sarkadi and S Tömösközi, eds), Publishing company of TUB (Technical University of Budapest), Budapest, pp.530–536.

Kabakibi, A, Hummert, SL, Whelan, J, McEntee, FM, Ursin, V and Obukowicz, MG (2001) Metabolic, anti-cancer, and anti-inflammatory actions of stearidonic acid. 92nd AOCS Annual Meeting & Expo, Minneapolis, Minnesota, May 13–16, 2001. *INFORM*, Suppl., S62.

Kaijser, A, Dutta, P and Savage, G (2000) Oxidative stability and lipid composition of macadamia nuts grown in New Zealand. *Food Chem.*, 71, 67–70.

Kallio, H, Currie, G, Gibson, R and Kallio, S (1997) Mass spectrometry of food lipids: Negative ion chemical ionization / collision induced dissociation analysis of oils containing γ-linolenic acids as an example. *Ann. Chim.*, 87, 187–198.

Kawashima, A, Shimada, Y, Yamamoto, M, Sugihara, A, Nagoa, T, Komemushi, S and Tominaga, Y (2001) Enzymatic synthesis of high-purity structured lipids with caprylic acid at 1,3-positions and polyunsaturated fatty acid at 2-position. *J. Am. Oil Chem. Soc.*, 78, 611–616.

Kinney, AJ (1998) Plants as industrial chemical factories – new oils from genetically engineered soybeans. *Fett/Lipid*, 100, 173–176.

Kockmann, V, Spielmann, D, Traitler, H and Lagarde, M (1989) Inhibitory effect of stearidonic acid (18:4*n*-3) on platelet aggregation and arachidonate oxygenation. *Lipids*, 24, 1004–1007.

Laakso, P and Christie, WW (1990) Chromatographic resolution of chiral diacylglycerol derivatives: potential in the stereospecific analysis of triacyl-*sn*-glycerols. *Lipids,* 25, 349–353.

Lawson, LD and Hughes, BG (1988) Triacylglycerol structure of plant and fungal oils containing γ-linolenic acid. *Lipids*, 23, 313–317.

Liang, T and Liao, S (1992). Inhibition of steroid 5α-reductase by specific aliphatic unsaturated fatty acids. *Biochem. J.*, 285, 557–562.

Liu, KS (1999) Soy oil modification: products, applications. *INFORM*, 10, 868–878.

Liu, JW, DeMichele, S, Bergana, M, Bobik, E, Hastilow, C, Chuang, LT, Mukerji, P and Huang, YS (2001) Characterisation of oil exhibiting high γ-linolenic acid from a genetically transformed canola strain. *J. Am. Oil Chem. Soc.*, 78, 489–493.

Loh, W and Lampert, D (1998) Reducing or eliminating trans fatty acids in shortenings. *Cargill Foods Res. Dept Tech. Bull.*, 20(11), 1–6.

Luchetti, F (2001) Olive oil: facts and figures. *Lipid Technol.*, 13, 32–36.

MacKenzie, SL, Giblin, EM and Mazza, G (1993) Stereospecific analysis of *Onosmodium hispidissimum* Mack. seed oil triglycerides. *J. Am. Oil Chem. Soc.,* 70, 629–631.

Maedler, K, Spinas, GA, Dyntar, D, Moritz, W, Kaiser, N and Donath, MY (2001) Distinct effects of saturated and monounsaturated fatty acids on β-cell turnover and function. *Diabetes*, 50, 69–76.

Manninen, P, Häivälä, E, Sarimo, S and Kallio, H (1997) Distribution of microbes in supercritical CO_2 extraction of sea buckthorn (*Hippophaë rhamnoides*) oils. *Z. Lebensm. Unters. Forsch.*, 204, 202–205.

Marquez-Ruiz, G, Garces, R, Leon-Camacho, M and Mancha, M (1999) Thermoxidative stability of triacylglycerols from mutant sunflower seeds. *J. Am. Oil Chem. Soc.*, 76, 1169–1174.

Messner, T, Sihm, H and Vessby, B (1998) Fatty acid composition in serum among males 4–16 years after myocardial infarction. *Int. J. Circumpolar Health*, 57, 22–31.

Nobuhiro, A and Yasu, M (1992) Manufacture of POA with yeast. Japanese Patent JP 04108394 A2.

Olsson, U, Kaufmann, P and Herslöf, PG (1990) Multivariate optimization of a gas–liquid chromatographic analysis of fatty acid methyl esters of blackcurrant seed oil. *J. Chromatogr.*, 550, 385–394.

Optimum Quality Grains (1998) http://www.itsoptimum.com

Padley, FB, Gunstone, FD and Harwood, JL (1994) Occurrence and characteristics of oils and fats. In: *The Lipid Handbook* (FD Gunstone, JL Harwood and FB Padley, eds), Chapman & Hall, London, pp.47–183.

Palombo, JD, DeMichelle, SJ, Liu, JW, Bistrian, BR and Huang, YS (2000) Comparison of growth and fatty acid metabolism in rats fed diets containing equal levels of γ-linolenic acid from high γ-linolenic acid canola oil or borage oil. *Lipids,* 35, 975–981.

Petukhov, I, Malcolmson, LJ, Przybylski, R and Armstrong, L (1999a) Frying performance of genetically modified canola oils. *J. Am. Oil Chem. Soc.,* 76, 627–632.

Petukhov, L, Malcolmson, LJ, Przbylski, R and Armstong, L (1999b) Storage stability of potato chips fried in genetically modified canola oils. *J. Am. Oil Chem. Soc.,* 76, 889–896.

Poisson, JP, Dupuy, RP, Sarda, P, Descomps, B, Narce, M, Rieu, D and de Paulet, AC (1993) Evidence that liver microsomes of human neonates desaturate essential fatty acids. *Biochim. Biophys. Acta*, 1167, 109–113.

Reske, J, Siebrecht, J and Hazebroek, J (1997) Triacylglycerol composition and structure in genetically modified sunflower and soybean oils. *J. Am. Oil Chem. Soc.*, 74, 989–998.

Reynier, M, Sari, H, d'Anglebermes, M, Ah Kye, E and Pasero, L (1991) Differences in lipid characteristics of undifferentiated and enterocytic differentiated HT human colonic cells. *Cancer Res.*, 51, 1270–1277.

Robles-Medina, A, Giménez-Giménez, A, García-Camacho, F, Sánchez-Pérez, JA, Molina-Grima, E and Contreras-Gómez, A (1995) Concentration and purification of stearidonic, eicosapentaenoic, and docosahexaenoic acids from cod liver oil and the marine microalga *Isochrysis galbana*. *J. Am. Oil Chem. Soc.*, 72, 575–583.

Robles-Medina, A, Molina-Grima, E, Giménez-Giménez, A and González, MJI (1998) Downstream processing of algal polyunsaturated fatty acids. *Biotechnol. Adv.*, 16, 517–580.

Sanjurjo, P, Matorras, R, Ingunza, N, Alonso, M, Rodríguez-Alarcón, J and Perteagudo, L (1993) Coss-sectional study of percental changes in total plasmatic fatty acids during pregnancy. *Horm. Metab. Res.*, 25, 590–592.

Sauer, LA and Dauchy, RT (1992) Uptake of plasma lipids by tissue-isolated hepatomas 7288CTC and 7777 in vivo. *Br. J. Cancer*, 66, 290–196.

Sewón, P and Tyysjärvi, E (1993) Stearidonic and γ-linolenic acid contents of common borage leaves. *Phytochemistry*, 33, 1029–1032.

Shimada, Y, Sugihara, A, Shibahiraki, M, Fujita, H, Nakano, H, Nagao, T, Terai, T and Tominaga, Y (1997) Purification of γ-linolenic acid from borage oil by a two-step enzymatic method. *J. Am. Oil Chem. Soc.*, 74, 1465–1470.

Shimada, Y, Sakai, N, Sugihara, A, Fujita, H, Honda, Y and Tominaga, Y (1998) Large-scale putification of γ-linolenic acid by selective esterification using *Rhizopus delemar* lipase. *J. Am. Oil Chem. Soc.*, 75, 1539–1543.

Siguel, EN and Lerman, RH (1994) Altered fatty acid metabolism in patients with angiographically documented coronary artery disease. *Metabolism*, 43, 982–993.

Simon, AJ, Fong, J and Bernert, JT (1996) Serum fatty acids and blood pressure. *Hypertension*, 27, 303–307.

Simonsen, NR, Navajas, JFC, Martin-Moreno, JM, Huttunen, JK, Martin, BC, Thamm, M,

Kardinaal, AFM, van't Veer, P, Kok, FJ and Kohlmeier, L (1998) Tissue stores of individual MUFA and breast cancer: the EURAMIC study. *Am. J. Clin. Nutr.*, 68, 134–141.

Smith, JR (1996) Oleic safflower. In: *Safflower* (JR Smith, ed.), AOCS Press, Champaign, pp.357–369.

Spielmann, D, Traitler, H, Crozier, G, Fleith, M, Bracco, U, Finot, PA, Berger, M and Holman, RT (1989) Biochemical and bioclinical aspects of blackcurrant seed oil ω3–ω6 balanced oil. *NATO ASI Ser. A*, 171, 309–321.

Takagi, T, Asahi, M and Itabashi, Y (1985) Fatty acid composition of twelve algae from Japanese waters. *Yukagaku*, 34, 1008–1012.

Theret, N, Bard, JM, Nuttens, MC, Lecerf, JM, Delbart, C, Romon, M, Salomez, JL and Fruchart, JC (1993) The relationship between the phospholipid fatty acid composition of red blood cells, plasma lipids, and apolipoproteins. *Metabolism,* 42, 562–568.

Toyo Jozo Co Ltd (1984) Antitumour formulation containing POA. Japanese Patent JP 59062523 A2.

Traitler, H, Winter, H, Richli, U and Ingenbleek, Y (1984) Characterization of γ-linolenic acid in *Ribes* seed. *Lipids*, 19, 923–928.

Traitler, H, Wille, HJ and Studer, A (1988) Fractionation of blackcurrant seed oil. *J. Am. Oil Chem. Soc.*, 65, 755–760.

Tsevegsüren, N and Aitzetmüller, K (1996) γ-Linolenic and stearidonic acids in Mongolian Boraginaceae. *J. Am. Oil Chem. Soc.*, 73, 1681–1684.

Ucciani, E (1995) Sources potentielles d'acide gamma-linolenique: une revue. *Oleag. Corps Gras Lipides,* 2, 319–322.

Ucciani, E, Costagliola, G and Gruber, M (1995) Composition en acides gras d'huile de *Saxifragaceae*. *Oleag. Corps Gras Lipides,* 2, 491–493.

US Department of Agriculture (2000) Notice of release of Soyola soybean. http://www.ars.usda.gov/is/pr/2000/000327.htm

Velasco, L and Goffman, FD (1999) Chemotaxonomic significance of fatty acids and tocopherols in Boraginaceae. *Phytochemistry*, 52, 423–426.

Vessby, B, Aro, A, Skarfors, E, Berglund, L, Salminen, I and Lithell, H (1994) The risk to develop NIDDM is related to the fatty acid composition of the serum cholesterol esters. *Diabetes*, 43, 1353–1357.

Vickery, JR (1969) Composition of the seed oil of *Kermadecia sinuata*. *Austr. J. Sci.*, 31, 334.

Wang, CL, Chin, CK and Chen, A (1998) Expression of the yeast delta-nine desaturase gene in tomato enhances its resistance to powdery mildew. *Physiol. Mol. Plant Pathol.*, 52, 371–383.

Wille, H-J, Traitler, H and Lagarde, M (1991) Verfahren zur Anreicherung von Stearidonsäure. *Fat Sci. Technol.*, 93, 363–368.

Wilson, RF (1999) Alternatives to genetically-modified soybeans – the Better Bean initiative. *Lipid Technol.*, 11, 107–110.

Wolf, RB, Kleiman, R and England, RE (1983) New sources of γ-linolenic acid. *J. Am. Oil Chem. Soc.*, 65, 1858–1860.

Wolff, RL, Christie, WW, Pédrono, F and Marpeau, AM (1999) Arachidonic, eicosapentaenoic, and biosynthetically related fatty acids in the seed lipids from a primitive gymnosperm, *Agathis robusta*. *Lipids,* 34, 1083–1095.

Xing, JS and Chin, CK (2000) Modification of fatty acids in eggplant affects its resistance to *Verticillium dahliae*. *Physiol. Mol. Plant Pathol.*, 56, 217–225.

Xu, XQ, Tran, VH, Palmer, MV, White, K and Salisbury, P (1999) Chemical and physical analyses and sensory evaluation of six deep-frying oils. *J. Am. Oil Chem. Soc.,* 76, 1091–1099.

Xu, XQ, Tran, VH, Palmer, MV, White, K and Salisbury, P (2000) Chemical, physical, and sensory properties of Monola oil, palm olein and their blends in deep frying trials. *Food Austr.*, 52, 77–82.

Yamori, Y, Nara, Y, Tsubouchi, T, Sogawa, Y, Ikeda, K and Horie, R (1986) Dietary prevention of stroke and its mechanism in stroke-prone spontaneously hypertensive rats, Preventive effect of dietary fiber and palmitoleic acid. *J. Hypertens.*, 4(Suppl. 3), 449–452.

Yamazaki, K, Fujikawa, M, Hamazaki, T, Yano, S and Shono, T (1992) Comparison of the conversion rates of α-linolenic acid (18:3(*n*-3)) and stearidonic acid (18:4(*n*-3)) to longer polyunsaturated fatty acids in rats. *Biochim. Biophys. Acta*, 1123, 18–26.

Yang, B (2001) *Lipophilic Components of Sea Buckthorn (Hippophaë rhamnoides) Seeds and Berries and Physiological Effects of Sea Buckthorn Oils* (PhD thesis), University of Turku, Turku, Finland (ISBN 951-29-2221-5).

Žák A, Zeman, M and Tvrzická, E (1998) Fatty acid composition of plasma phosphatidyl choline and lipid and lipoprotein concentrations in hyperlipoproteinaemia. II. Relationship with B lipoproteins. *Čas. Lék. Čes.*, 137, 400–403.

Zheng, JZ, Folsom, AR, Ma, J, Arnett, DK, McGovern, PG and Eckfeldt, JH (1999) Plasma fatty acid composition and 6-year incidence of hypertension in middle-aged adults: the Atherosclerosis Risk in Communities (ARIC) Study. *Am. J. Epidemiol.*, 150, 492–500.

CHAPTER 10

Conjugated linoleic acid

CLAIRE E. FERNIE

Scottish Crop Research Institute, Invergowrie, Scotland, UK

A. Introduction

Conjugated linoleic acid (CLA) is the collective name given to a group of C_{18} positional and geometric isomers containing two conjugated double bonds, rather than the more common methylene-interrupted double bond sequence. An example of one such isomer is shown in Figure 1.

These compounds have been recognized in milk for over 50 years but in the past decade they have provoked new interest as they have shown potential in animal and cell systems for the treatment of a number of major medical conditions, including cancer, obesity, and diabetes. Synthetic CLA has been available for a number of years as a dietary supplement but the focus of new research into CLA is now turning towards incorporation into functional foods, for both human nutrition and animal husbandry.

B. Potential health benefits

Many claims for health benefits have been made for CLA in both animals and humans. These have been reviewed in detail elsewhere (Yurawecz *et al.*, 1999a) and will only be discussed briefly here. The first indication that CLA might be biologically active was reported in 1979 when an extract from grilled beef was found to inhibit mutagenesis in a rat liver cell fraction (Pariza *et al.*, 1979). Studies on the potentially anti-carcinogenic properties of CLA, both *in vitro* and *in vivo* (Pariza *et al.*, 2001), have demonstrated that the isomer with an active effect on cancer cells is probably the 9-*cis,* 11-*trans* isomer. *In vivo* studies in rodents showed that dietary CLA supplementation decreased the number and size of mammary tumours, regardless of other lipid intake (Ip *et al.*, 1996), and that topically applied CLA reduced the size and number of skin

C.E. Fernie is funded through European Union grant number QLK1-CT99-00076.

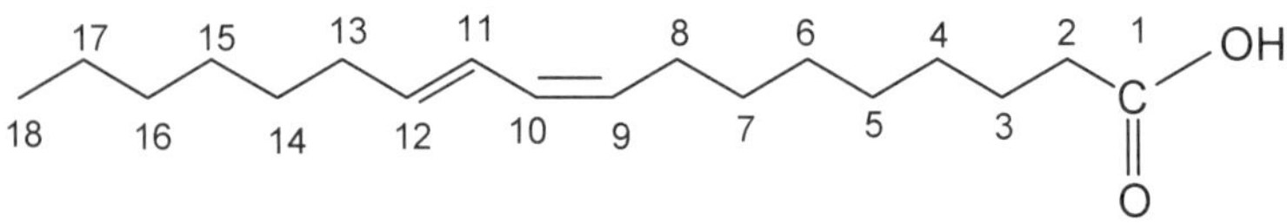

Figure 1. Structure of 9-*cis*, 11-*trans* conjugated linoleic acid (CLA).

tumours (Belury *et al.*, 1996). Although some work has been done on prostate, forestomach, and colon cancers, the focus of most studies has been on breast cancer. A potential link has been demonstrated in humans between increased levels of CLA in breast tissue and a decreased risk of breast cancer (Lavillonnière and Bougnoux, 1999). As CLA is active in tumours at levels from 0.1% (w/w) in rats (Ip *et al.*, 1994), dietary products enriched with CLA may have a role to play in the prevention or limitation of tumour growth, particularly in older people. This is difficult to demonstrate conclusively as it requires long-term follow-up studies. The use of CLA as a potential cancer treatment, rather than as a preventative measure, is more likely via a pharmaceutical rather than a functional food approach.

One of the more promising areas for the use of CLA in functional foods or animal feeds is in the alteration of body composition. Several animal studies have demonstrated that CLA reduces body fat, increases lean mass, and improves feed efficiency. These effects have been attributed to the 10-*trans,* 12-*cis* isomer, and are discussed in detail below. Trials in humans, with mixtures of isomers, have so far been inconclusive. Variable amounts of CLA (1.7–7 g/day) and different categories of subjects (males, females, students, bodybuilders) have been examined (Jahreis *et al.*, 2000; Blankson *et al.*, 2000). Slight changes were observed in some cases but few consistent effects on body composition have been demonstrated. In some of the studies, however, the concurrent effects of exercise and dietary lipid content may not have been properly considered. Studies have been conducted on adults with good nutritional status and health rather than on developing juveniles or individuals with medical conditions. All of these factors may help to explain the differences between results from animal and human trials. As more becomes known about the biochemical mechanisms behind the effects associated with CLA, particularly those of specific isomers, it should become possible to target CLA-enriched functional foods towards the sections of the population that would most benefit from them. Currently it looks unlikely that CLA will become the ultimate off-the-shelf weight-loss aid for the general population as many had hoped.

In rodents and chicks, CLA has also shown effects on immune function, by altering the fatty acid composition of the phospholipids and thus affecting eicosanoid and cytokine production (Cook *et al.*, 1999). In these same animal groups, CLA has a direct effect on bone remodelling, also via eicosanoid production (Watkins *et al.*, 1999). Although these effects have not been reproduced in humans, the few trials so far reported were conducted with healthy individuals (Kelley *et al.*, 2000). A functional food containing CLA may have a role in maintaining remission in people with immune-associated disorders such as asthma, rheumatoid arthritis, and inflammatory bowel disease, and in delaying, if not preventing, their onset in population groups with a genetic predisposition to these diseases. It would not, however, be appropriate as a treatment for active disease. Although human studies have yet to be conducted, the effectiveness of CLA on estrogen-responsive human mammary cancer cells suggests a possible role for a CLA-containing functional food in the prevention of osteoporosis, which has a hormonal element to its etiology. CLA has been shown to affect atherosclerosis in rabbits and hamsters (Kritchevsky, 1999), but these effects have not yet been reproduced in humans. *In vitro* trials with the 9-*cis,* 11-*trans* isomer have shown that it can bind to and activate peroxisome proliferator-activated receptors and so may have a role to play in preventing insulin resistance and type II diabetes, but *in vivo* data are scarce.

C. CLA availability and functional foods

Humans may have a limited capacity to produce CLA endogenously from vaccenic acid (11-*trans* 18:1) (Salminen *et al.*, 1998) but the majority of CLA in human tissues is of dietary origin. Other monogastric animals such as pigs, poultry, and fish also require CLA to be included in their diet in order to accumulate it in their tissues. Conversely, ruminants such as cows, goats, and sheep have enzymes that allow them to synthesize CLA and its intermediate precursor, vaccenic acid, in small but significant quantities. These compounds then accumulate in the milk and meat of ruminant animals. CLA is therefore available at low levels in a number of natural products (Table 1), mainly of ruminant origin and containing mainly the 9-*cis,* 11-*trans* isomer (also known as rumenic acid).

Two of the items in Table 1 merit further discussion. The highest levels of CLA are seen in products that have been subjected to some degree of processing, i.e. cheese and ghee. The high level of CLA in cheeses is dependent on the CLA level of the raw milk, the microflora present, the length of ageing, the processing technology used (including heat treatment), and the addition of proteins as whey or caseinates (Prandini *et al.*, 2001; Lin *et al.*, 1998). Ghee is the name given to a variety of products, which are essentially clarified butter fat (i.e. butter with the non-fat solids and moisture removed). These materials are

Table 1. CLA concentrations in a number of foods

Food	CLA(mg/g total fat)	References
Lamb and beef	4.0–5.0	Hamilton (2001)
Butter	12.3–14.2	O'Shea *et al.* (2000)
Live yoghurt	3.3–4.6	O'Shea *et al.* (2000)
Cheeses	1.5–19.9	O'Shea *et al.* (2000); Prandini *et al.* (2001)
Milk	3.0–6.0	Prandini *et al.* (2001)
Ghee	10.0–25.0	Sserunjogi *et al.* (1998)
Safflower oil	0.7	Precht and Molkentin (2000)

common across Asia, the Middle East, and Africa. The high concentration of CLA is thought to be due to isomerization of linoleic acid during the clarification process, which involves heating to 110–120ºC (Sserunjogi *et al.*, 1998).

The levels of CLA in these foods fall well short of the amounts (extrapolated from animal trials) likely to be active in humans, estimated at 1.5–3.5 g/day (Banni and Martin, 1998). Therefore, any attempt to produce an animal-derived product containing CLA for functional food use must focus on increasing the natural CLA content. The literature values for CLA contents in foods cannot be considered totally reliable, as analytical procedures used in the past have led to inaccuracies. Only recently have enhanced techniques using high-performance liquid chromatography (HPLC), gas chromatography (GC), and gas chromatography – mass spectrometry (GC–MS) allowed the concentration of CLA components to be accurately assessed along with conclusive identification of individual isomer profiles (Christie *et al.*, 2001).

There are two distinct functional food approaches to using CLA. The first of these is to produce a food product with an active health benefit due to its CLA content (e.g. by altering the feeding regime of cows to give heightened CLA levels in milk). The second approach involves using CLA in animal feed to alter the body composition of the animal, leading to the production of a functional food such as extra-lean ham or low-fat milk but where the functionality is not due to the CLA content. There are currently very few products containing enhanced CLA concentrations on the market. One such product is 'OmegaSmart' cheese (manufactured by PurNutra Inc, Canada). This became available in June 2001 and is one of the first products to indicate the level of CLA on its label (44 mg/serving.) However, as its name indicates, this product is marketed on the basis of its high level of *n*-3 fatty acids, not on its CLA content, and makes no specific health claims. The increased level of CLA is due to the feeding of increased levels of docosahexaenoic acid (DHA; 22:6*n*-3) and other *n*-3 fatty acids to the cows producing the raw milk. A link between DHA feeding and CLA content has been reported elsewhere and is discussed below.

Commercial methods are also available to produce bulk quantities of individual CLA isomers and mixtures for incorporation into food. The current scarcity of food products containing CLA is in part due to the inconclusive nature of the health claims in humans and the lack of specific toxicity trials and safety data.

D. CLA and animal husbandry

There are two distinct types of animal model used with respect to CLA and food. Ruminant and non-ruminant animals have different lipid metabolisms and therefore each group will be discussed individually.

1. Ruminant CLA production

Ruminants (i.e. cows, sheep, goats, camels) are able to produce CLA naturally through the biohydrogenation of unsaturated fatty acids in the rumen. This process involves the action of the rumen bacterium *Butyrivibrio fibrisolvens* on dietary linoleic acid (via linoleic acid isomerase) to give CLA as an intermediate in the formation of vaccenic acid and subsequently 18:0 (stearic acid). CLA is also formed in tissues such as the mammary gland by the action of Δ^9-stearoyl coenzyme A desaturase on vaccenic acid. Both CLA and its precursor can accumulate in sufficient amounts that they are absorbed and incorporated into the muscle tissue and milk of the animal. The amounts of CLA produced depend on the microbial population of the rumen and can be altered significantly by dietary regimes. The feeding of CLA is not a suitable strategy for increasing the content in ruminant tissues, as it is rapidly biohydrogenated in the rumen to stearic acid. Increases in CLA level in ruminants can only be brought about by manipulation of the rumen processes leading to CLA production. Exceptionally, CLA that is fed can be protected in some way from rumen reactions to allow subsequent absorption in the duodenum or abomasum.

The sequence of reactions leading to the formation of CLA in the rumen is shown in Figure 2.

2. CLA manipulation in ruminant-derived products

Factors affecting the fatty acid profile and CLA level of cow's milk include breed, diet, and stage of lactation (early-lactation milk generally has a lower solid fat content, i.e. a lower proportion of saturated fat). Even among animals with similar characteristics, CLA levels can differ considerably, probably due to differences in rumen microbial populations or levels of Δ^9-desaturase activity. Unsaturated lipids can be partially protected from rumen hydrogenation by feeding whole seeds, encapsulated fatty acids, or fatty acids as their calcium salts.

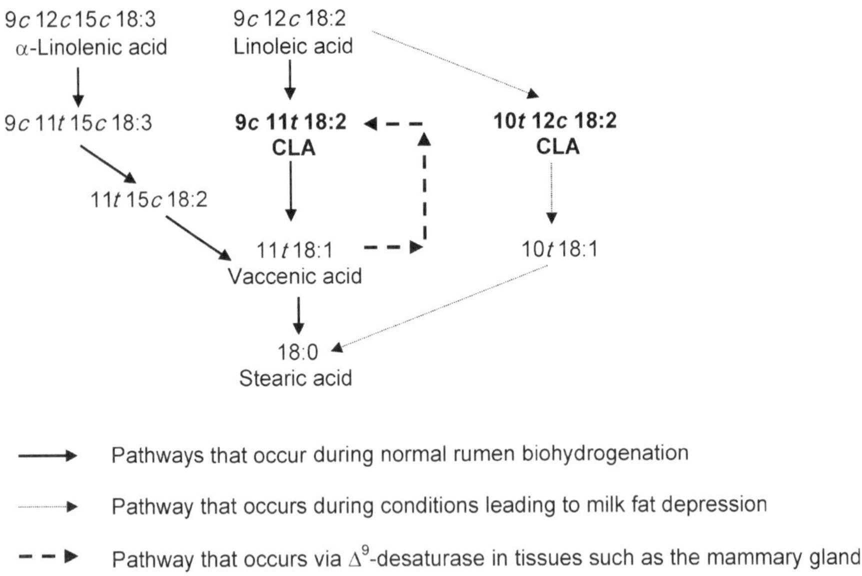

Figure 2. Biohydrogenation pathways of unsaturated fatty acids in the rumen and mammary gland. CLA isomers are highlighted in bold type.

The majority of the work done to increase CLA levels in milk has focused on dietary manipulation. Cows fed pasture have higher CLA levels than those fed silage or concentrates, mainly due to the increased levels of linoleic and α-linolenic acids in fresh grass. The amounts of these fatty acids vary seasonally, with the first spring growth having the highest levels. This is partly related to the ratio of shoot to stem over the growing year. The fatty acid profiles of grasses are distinct for each species, when the plants receive exactly the same treatment, but even the time and frequency of cutting can lead to significant differences. It may be possible to breed grasses for high lipid content, but management regimes are equally important (Dewhurst *et al.*, 2001). The carbohydrate content of feed could affect the rate of microbial fermentation in the rumen. Thus, the decline in the levels of certain sugars and soluble fibre during processes to preserve forage may also help to explain the differences in CLA production between animals fed on pasture (which is high in starch, sugars, and soluble fibre) and those fed on silage (Kelly *et al.*, 1998). A number of studies have looked at the effects of feeding different oils on the level of CLA in milk fat (Dhiman *et al.*, 2000; Baer *et al.*, 2001; Donovan *et al.*, 2000; Chilliard *et al.*, 2000, 2001). Feeding soybean oil (mainly linoleic acid) led to greater increases in CLA levels than feeding linseed oil (mainly α-linolenic acid). Both treatments led to a reduction in medium-chain fatty acids. At the highest levels of feeding, milk fat levels were depressed. Similar results have been seen with other linoleate-rich oils (Fearon, 2001).

Feeding menhaden (fish) oil at 2% of the dry matter intake of the diet also lowered milk fat concentrations. The level of CLA was increased by approximately 3.6 fold, but there was also a concurrent 4–5 fold rise in the level of vaccenic acid. Increasing the oil intake to 3% had no additional effects. The level of *n*-3 fatty acids in the milk was increased, mainly due to increased levels of eicosapentaenoic acid (EPA; 20:5*n*-3), as was the level of total *trans* monoenes. The effects of pasteurization and oxidation on the raw milk were examined but no significant changes were observed and the CLA isomer profile was unchanged. There were no significant flavour differences found between the milk (and butter made from the milk) of cows fed 2% menhaden oil and control samples. The higher polyunsaturated fat content of the butter meant that it was softer at 4°C and 20°C, but the acid value and peroxide value of the butter were similar to those of control samples even after 3 months. A consumer evaluation of milk from cows fed a fish oil diet found no difference in acceptability compared with the control milk (Ramaswamy *et al.*, 2001).

The mechanisms for the formation of CLA from fish oil are unknown at this time. It has been suggested that the high level of *n*-3 polyunsaturated fatty acids (PUFA) alters the microbial fermentation in the rumen in some way and perhaps inhibits the hydrogenation of 18:1 fatty acids to stearic acid. The large increase in vaccenic acid on feeding menhaden oil to increase CLA content is an important factor. CLA levels are highly correlated with the total *trans* level in milk fat, and higher levels of *trans* fat are associated with reduced production of milk fat (referred to as "milk fat depression"). However, higher vaccenic acid levels in the rumen lead to higher CLA levels in milk fat by Δ^9-desaturation in the mammary glands. Although vaccenic acid doesn't inhibit milk fat synthesis to the same extent as other *trans* fats, it may have an effect on *de novo* fatty acid synthesis.

Trans monoenes are generally regarded as harmful to human health, with detrimental effects on factors affecting cardiovascular disease, but most of the work has been done using partially hydrogenated vegetable oils high in the 9-*trans* and 10-*trans* 18:1 isomers. Little work has been done specifically with the 11-*trans* isomer. A study of nurses in the USA compared diets where the *trans* fat content over the ten preceding years was either from a vegetable source (mainly 9- and 10-*trans*) or from ruminant sources (mainly 11-*trans*). The former group had a significantly increased risk of cardiovascular disease, at the highest level of intake, compared with the latter (Willett *et al.*, 1993). The CLA content of milk fat in cows can be increased relatively easily by dietary alteration; the greater challenge may be to reduce the levels of 9-*trans* and 10-*trans* 18:1, while still retaining the levels of 11-*trans* 18:1, CLA, and other PUFA. One possible strategy would be feeding a diet of unsaturated fat, but with a high fibre content (Voigt and Hagemeister, 2001).

As detailed above, many of the diets used to increase CLA content cause a concurrent decrease in milk fat content. Milk fat depression has been recognized in cows for over a century. In the past, when milk was marketed on the basis of its fat content, this was a serious cause for concern, but in recent times the emphasis has moved more towards milk with high protein content, and lower levels of fat are regarded as more healthy. This effect is seen in cows and in other lactating animals fed CLA supplements. Baumgard *et al.* (2001) induced milk fat depression by abomasomally infusing 10-*trans,* 12-*cis* CLA and concluded that there was a requirement for a 10-*trans* double bond. An extensive review by Bauman and Griinari (2001) led the authors to postulate a unifying theory for all of the dietary factors involved. They suggest that there is a requirement both for an alteration in rumen microbial processes (e.g. by feeding a low-fibre diet) and for a source of unsaturated fatty acids. Under these conditions, there is an alteration in the normal biohydrogenation processes, leading to the production of 10-*trans* 18:1, via 10-*trans,* 12-*cis* CLA, and that it is the 10-*trans* 18:1 that is responsible for the suppression of milk fat synthesis. However, this has yet to be confirmed. There is little information available on the factors affecting CLA content in beef, but the levels in meat fat are thought to mirror those found in milk.

Goat's milk is used widely as a starting material for cheeses. It is also an alternative to cow's milk for those who have a sensitivity or allergy to cow's milk, although the allergic reaction itself is generally considered to be triggered by non-lipid components. Feeding canola oil as a replacement for barley in the diet has been used to increase the levels of CLA in goat's milk (Mir *et al.*, 1999). As the percentage of canola oil was increased from 0% to 4%, the CLA content of the milk also increased, from 1.05% to 3.20% of total fat, which is a larger increase than that observed in cattle. In contrast to cattle, however, the milk fat content increased by >1% at the highest level of canola feeding (6%), with no effect on milk yield, protein content, or lactose content. The fatty acid composition of goat's milk differs from cow's milk in that it contains higher levels of linoleic acid and α-linolenic acid and more short- to medium-chain fatty acids (mainly due to a higher content of 10:1). This may reflect differences in fatty acid metabolism, which could account for the increase in milk lipid content. Another study reported a 10 fold increase in the CLA content of goat's milk by including encapsulated CLA in the diet (Fearon, 2001).

As has been mentioned, supplementing ruminant diets with CLA is not an effective way of increasing tissue CLA levels, as the dietary supplement is rapidly hydrogenated in the rumen. One possible way of avoiding this, investigated in lambs, is by feeding CLA to pre-ruminant animals. At this stage of development, fluid meals pass directly to the abomasum for absorption, avoiding the rumen. Four-week-old cross-bred lambs were fed 0.33 g CLA dissolved in olive oil in 50 ml of milk replacer for 21 days, before weaning onto pelleted feed (Mir *et al.*, 2000). The CLA-fed lambs had less adipose fat but the

CLA content of the tissues was not affected. Another study on lambs of a similar age given milk from grass-fed ewes showed an increase in CLA content in both liver and adipose tissues, but this may have been related to the high levels of linoleic and α-linolenic acids in the milk, rather than to the level of CLA (Banni *et al.*, 1996). In the former study, some of the lambs that had been fed the control diet pre-weaning were switched post-weaning to a diet that contained 6% safflower oil by dry weight, for 80 days. The rate of food intake was reduced post-weaning in the lambs that had been fed CLA pre-weaning and also in those control animals that subsequently received the safflower oil diet. Only the CLA diet lowered the growth rate during this period. The safflower oil diet increased adipose fat and significantly increased the levels of tissue CLA. Both diets altered the fatty acid profiles of the tissue lipids. The researchers considered that the safflower oil regime would lead to an increase in the level of CLA in off-the-shelf lamb meat cuts such that an intake of 250 g lamb per day would give a CLA intake of 446 mg. Oil feeding in sheep is a strategy that has also been used to decrease the levels of ciliate protozoa in the rumen (Ivan *et al.*, 2001). These protozoa decrease protein availability and increase methane production. Feeding sunflower seed oil at 6% of the dry matter intake to male sheep over 1 year old reduced the protozoa in the rumen to a negligible level and led to a concurrent increase in the CLA content of both muscle tissues and subcutaneous fat. The fatty acid profiles of the tissues were altered such that the concentrations of myristic and palmitic acids decreased, and those of stearic and linoleic acids increased. This would be regarded as a more favourable composition from the point of view of cardiovascular health.

3. *Poultry and egg production*

Increased consumption of chicken (and eggs) at the expense of red meat, as a healthier and lower-fat option, has made poultry and poultry products a potentially significant target in the campaign to increase the CLA content of foods. Variations in CLA concentration due to different dietary regimes and the subsequent effects on the quality of the resulting food products have been examined. In ruminants, tissue CLA is generally assumed to consist mainly of the 9-*cis,* 11-*trans* isomer. However, in poultry trials, since it is accepted that poultry cannot synthesize CLA endogenously, the composition of the CLA isomer mixtures that are fed is generally stated. In the majority of the recent studies discussed here, the CLA used contained equivalent (or near equivalent) amounts of the 9-*cis,* 11-*trans* and 10-*trans,* 12-*cis* isomers, with minor (<2%) amounts of other positional or geometric isomers.

Aydin *et al* (1999) showed that 0.5% (w/w of total diet) CLA is enough to induce 100% chick mortality, but this result was not confirmed by other studies (Latour *et al.*, 2000). Further studies relating to CLA and eggs are concerned with the effects on the lipid composition of eggs as a food source, and on egg hatchability and therefore the efficiency of poultry meat production. Adverse effects on chick hatchability have been prevented by feeding a diet containing 5% CLA plus 10% olive oil (Aydin *et al.*, 2001). A 5 fold increase in yolk CLA content was observed (which was substantially lower than that achieved by feeding CLA alone) but hatchability was improved to >85% on the diet with added olive oil, compared with 100% mortality on the pure CLA diet. This appears to be due to a relationship between the monoene/saturate ratio and embryonic mortality. By reducing 18:1 and increasing 18:0 levels, CLA appears to have a detrimental effect on the lipid balance required for chick survival. [Interestingly, a patent has been granted for the use of this property of CLA in controlling the population of birds, such as pigeons, regarded as pests in urban areas (Cook *et al.*, 1996).] Studies aimed at increasing the CLA levels of eggs as a food source have investigated the feeding of 0–5% CLA (w/w of diet) using soybean oil as the control (Chamruspollert and Sell, 1999; Ahn *et al.*, 1999; Du *et al.*, 1999). The highest levels of CLA led to a lower rate of egg production. But, despite the lower feed intake and body weight gain, egg weight was not influenced. Apart from EPA, the levels of all of the other main fatty acids in yolk were altered, with an increase in saturates and a decrease in unsaturated fatty acids. These reflected alterations in the dietary lipid profiles, but may have also been due to an inhibition of Δ^9-desaturase enzymes by CLA. The decrease in long-chain PUFA could have resulted either from competition with CLA for Δ^6-desaturase and elongase enzymes, or from a lack of dietary substrates for their formation. The total CLA contents of eggs increased linearly with dietary CLA level, and feeding CLA at a level of 5% (w/w) of diet led to an estimated CLA content of 310–365 mg per egg. Some isomeric preferences were demonstrated (such as higher levels of 9-*cis,* 11-*trans* CLA in the triacylglycerol fraction compared with the phospholipids) and the total content of CLA isomers in the triacylglycerol fraction of yolk was higher than in the phospholipid fraction. The triacylglycerol fraction also contained 16:2 fatty acids (presumably conjugated), which may suggest a limitation to the β-oxidation cycle for CLA.

Both increased dietary CLA and increased storage time led to an increased percentage of yolk and a decrease in percentage of albumin, due partly to redistribution of water between the two compartments. The decrease in percentage of yolk lipid over time was not affected by the concentration of dietary CLA. The water content and pH of yolks were increased on CLA feeding, and may be due to CLA incorporation into the yolk membrane leading to increased permeability. Some reports mention an increased reddish colouring to the yolks of eggs with increased CLA levels. The partitioning of minerals between the yolk and albumin may influence this, but the data are not consistent. However, the yolks of eggs with increased CLA levels have repeatedly been described as more firm than controls (perhaps due to the increased saturated fat content), to the point of being “rubbery” and “elastic”.

This textural difference also increased over time on refrigeration. It is possible to alleviate some of these changes in fatty acid profile by feeding CLA together with a source of unsaturated fat. Feeding CLA with soy oil or flax oil (Du *et al.*, 2000) increased the levels of linoleic acid and α-linolenic acid, respectively, but each treatment still led to an overall decrease in unsaturated fatty acids and an increase in palmitic and stearic acids, compared with feeding the oils without added CLA. The long-chain metabolites of linoleic and α-linolenic acids were increased in the phospholipid fraction, but no CLA metabolites were observed. CLA combined with flax oil increased the levels of DHA compared with flax oil alone, suggesting that CLA may promote the synthesis of *n*-3 PUFA, or at least affect DHA metabolism. This would be a desirable outcome from the point of view of both human and chick health as DHA is an important component of brain tissue.

A few trials have investigated the effects of feeding CLA on chicken meat composition and on sensory profiles. Broiler chickens fed an equivalent mixture of 9-*cis,* 11-*trans* and 10-*trans,* 12-*cis* CLA isomers over 42 days showed an increase in saturates and decrease in monoenes as was demonstrated in egg yolks (Simon *et al.*, 2000). It was suggested that the increase in saturates may improve the oxidative stability of the meat. Feed intake and body weight were not significantly affected, but body composition was altered such that there was an increase in liver and leg protein, with a concurrent decrease in the fat contents of liver, leg, and breast. The CLA concentration increased to 10% of the total fatty acids, with the 9-*cis,* 11-*trans* isomer incorporated at higher levels than the 10-*trans,* 12-*cis* isomer. The highest level of CLA was found in the abdominal fat and the lowest in the liver. CLA also reduced plasma and LDL cholesterol levels. Another study using a mixture of four CLA isomers (8-*trans,* 10-*cis*; 9-*cis,* 11-*trans*; 10-*trans,* 12-*cis* and 11-*cis,* 13-*trans*) gave similar results (Szymczyk *et al.*, 2001). The CLA was substituted for sunflower oil at levels of up to 1.5% (w/w of total diet). The 9-*cis,* 11-*trans* and 11-*trans,* 13-*cis* isomers were incorporated to a greater degree than the other isomers. No differentiation between tissues was reported. At the highest level of CLA feeding, the researchers estimated that 300 g of leg meat could provide 1.5 g of CLA to the human diet. Patties of chicken meat from birds fed a mixture of 9-*cis,* 11-*trans* CLA and 10-*trans,* 12-*cis* CLA substituted for soybean oil at levels of up to 1.5% were analysed for volatiles and oxidative stability after irradiation and/or vacuum packing (Du *et al.*, 2001). The fatty acid composition of the meat varied from controls as reported previously, but there were no significant differences in the amount of fat or the moisture content of the meat patties. Vacuum packing reduced oxidation significantly, as measured by volatile production. Dietary CLA had a small, but distinct, effect on reducing volatiles, but did not prevent lipid oxidation in aerobic conditions and so did not act as an antioxidant. Instead, the increase in saturates induced by CLA feeding and the replacement of linoleic acid by CLA reduced the main precursor to hexanal and pentanal formation, and this was thought to be responsible for the slight increase in oxidative stability. CLA had no effect on odour as assessed by a trained sensory panel.

4. *Pig breeding*

Experiments with pigs have focused less on incorporating CLA into muscle tissues (although this is regarded as an added benefit) and more towards exploiting the ability of CLA to increase protein and reduce carcass fat in certain animal models (Dugan *et al.*, 1997). High levels of polyunsaturates are generally regarded as detrimental to meat quality. Not only do they decrease the oxidative stability of the meat but they also make it less firm and more difficult to carve. Feeding CLA (containing 25% of 9-*cis,* 11-*trans* and 35% of 10-*trans,* 12-*cis* as the only significant CLA isomers) to grower pigs at levels up to 1.2% (w/w of diet) increased the saturated fatty acids and decreased monounsaturates and polyunsaturates in adipose tissue and latissimus muscle. Both CLA isomers were incorporated into these tissues (Ramsay *et al.*, 2001). The weights of the heart, liver, and kidneys were not affected and there was no reduction in total lipid content in response to the CLA. This may have been due to the slower rate of fat deposition in grower pigs than in finishers. Cross-breed finisher or grower–finisher pigs fed CLA at levels of up to 1% of the diet had an increased weight gain to feed ratio and decreased 10th rib back fat, intramuscular fat, and subcutaneous fat (Thiel-Cooper *et al.*, 2001; Wiegand *et al.*, 2001). The animals had firmer bellies (which shows potential for improved sliceability and bacon yield) and the finished meat cuts were either the same weight or heavier than those from control animals. Sensory characteristics such as tenderness, juiciness, and flavour intensity were not affected. By feeding CLA, therefore, functionality is increased by reducing the fat content of the meat and by incorporating CLA into the tissues.

5. *Aquaculture*

There is a case to be made for defining fish (especially oily fish) as a functional food in its own right because of the beneficial effects of the *n*-3 fatty acids, which comprise a significant proportion of fish tissue lipids, on factors related to cardiovascular disease. Despite significant reductions in catch quotas of sea fish, particularly in the West in recent years, better health education and increasing concern regarding beef products (in particular due to bovine spongiform encephalopathy (BSE) and foot-and-mouth disease (FMD) outbreaks in Europe) have led to increased fish consumption. Some of this demand is being met through research into new fish hybrids for aquaculture. The incorporation of CLA into fish tissue is seen as a way to increase the nutritional value of the product further.

CLA has been identified in low amounts in fish and other seafood but not all of the data are reliable as the identification of CLA isomers by GC is complicated by the co-elution of stearidonic acid (18:4*n*-3) and C_{20} isomers under certain conditions (Ackman, 1999). Recent studies with juvenile hybrid striped bass and yellow perch (Twibell *et al.*, 2000, 2001) demonstrated that it was possible to accumulate CLA in fish muscle tissues and liver by modifying up to 1% of the dry diet (substituted on an energy basis for other dietary lipid sources). The results were highly dependent on the nature of the dietary lipid source and varied from species to species. CLA was incorporated into fish lipids at higher levels than into the lipids of land animals. At 1% feeding, the CLA percentage of total fatty acids was between 4.1% and 13% for a variety of fish (Choi *et al.*, 1999), with accompanying alterations in the fatty acid composition of fish muscle and liver lipids. In particular there were partitioning effects on *n*-3 fatty acids. In hybrid striped bass, *n*-3 fatty acids such as EPA and DHA accumulated in the liver, and decreased in the muscle tissue, on feeding 1% CLA. This contrasts with juvenile yellow perch, where feeding 1% CLA led to a significant decrease in both *n*-3 and *n*-6 PUFA in the liver. Fish have the potential to be a source of CLA, but care must be taken to ensure that other valuable nutritional properties are not compromised.

E. Commercial synthesis and isomer enrichment

Oils containing conjugated fatty acids have been available for many years for use as a drying agent in the paint and varnish industry. These are produced by two main methods: isomerization of linoleate-rich oil with strong alkali, and dehydration of castor oil. The latter method involves the acid-catalysed dehydration of ricinoleic acid (12-hydroxy-9-*cis*-18:1) or its ester to give mainly the 9-*cis,* 11-*trans* CLA isomer, significant amounts of the 8-*trans,* 10-*cis* CLA isomer, and non-conjugated 18:2 acids. The resultant material is not suitable for food incorporation because of its complexity and variability. It is only since the mid-1990s that the focus of CLA production has altered to yield materials suitable for human consumption.

In 1997 Christie *et al.* reported that commercial CLA samples, produced by alkali-isomerization of linoleic-rich vegetable oils and sold as dietary supplements, contained probably at least four isomers. Yurawecz *et al.* (1999b) surveyed a number of CLA supplements and also found a wide range of isomers and CLA contents. One sample contained no CLA. Thus, early commercial CLA was extremely variable. In fact, a recent report by Adlof *et al.* (2001) suggests that this problem has not been completely overcome even today. Variation in composition, coupled with limitations in analytical techniques (which were not widely known at the time), led to confusion as to which isomers were biologically active, and therefore worth producing commercially. In recent years, the focus of commercial synthesis has been to produce high-quality material containing the minimum amount of isomeric variation. Targets have been the production of the 9-*cis,* 11-*trans* isomer (as this occurs most abundantly in natural products, and so is thought not to be detrimental to human health), and more recently the 10-*trans,* 12-*cis* isomer due to its potential role in modulating body composition. A wide variety of methods have been used to produce laboratory-scale amounts of individual CLA isomers and these have been reviewed by Adlof (1999). However, the majority are not suitable for large-scale production and will not be discussed further.

1. Production of CLA

Commercial production methods are proprietary to each manufacturer, so complete methods cannot be described fully. However, the available literature indicates a number of common factors. The raw material generally used for CLA production is oil rich in linoleic acid such as safflower or sunflower oil, although the free acids or alkyl esters of such materials can also be used. Commercial free acid preparations may, however, contain undesirable amounts of monoacylglycerols and diacylglycerols. Esterified material can be converted to free fatty acids by a high-pressure splitting process at about 250°C. The free acids are then reacted (in an alcoholic solvent – discussed below) with a strong alkaline catalyst such as sodium hydroxide, which isomerizes the linoleic acid to CLA, as soap. On cooling, free CLA is recovered after addition of dilute acid and is washed, dried, deodorized and distilled to give the final product (Sæbø, 2001).

The CLA content of the final product is directly related to the linoleic acid content of the starting material. Linoleate enrichment before alkali-isomerization can be achieved by crystallization in solvents such as acetone. Urea adduct formation has also been used for enrichment on a small scale.

CLA produced by alkaline isomerization always contains a mixture of positional and geometric isomers, but the number and position of these can be limited by optimizing the processing conditions.

Several solvent systems have been used to produce CLA by alkaline catalysis. These include dimethyl sulfoxide, ethylene glycol, glycerol, and water. Economic considerations, such as the increased temperatures and pressures required, limit the use of water and ethanol on a commercial scale. However, ethylene glycol and dimethyl sulfoxide are more difficult to remove from the reaction mixture, and residual amounts would not be acceptable in a food-grade CLA product. Thus, although ethylene glycol was widely utilized to produce early CLA supplements, the less toxic propylene glycol is increasingly used nowadays. This also permits the use of potassium hydroxide as a catalyst. This is more soluble than sodium hydroxide in this solvent and is a more efficient catalyst on a molar basis (Reaney *et al.*, 1999).

Another advantage of this combination of solvent and catalyst is that it allows the production of a product containing only two main CLA isomers by manipulation of temperature and reaction time (typically 150°C for a few hours.) Use of sodium hydroxide as a catalyst in ethylene glycol leads to products containing four main CLA isomers: 8-*trans*, 12-*cis*; 9-*cis*, 11-*trans*; 10-*trans*, 12-*cis* and 11-*cis*, 13-*trans*. Potassium hydroxide, under the milder conditions described, gives a product containing near-equivalent amounts of the 9-*cis*, 11-*trans* and 10-*trans*, 12-*cis* isomers, together comprising about 90–95% of the total CLA, with only minor amounts of other isomers. A major factor in producing this simple isomer mixture is temperature. It has been shown that heating pure 10-*trans*, 12-*cis* isomer gives the 11-*cis* 13-*trans* isomer, and the 9-*cis*, 11-*trans* isomer gives the 8-*trans*, 10-*cis* isomer under similar conditions (Sæbø, 2001). Increased temperature and reaction time also lead to increased formation of the undesirable *trans*, *trans* isomers. The optimum reaction time is the minimum required for complete linoleic acid isomerization. This can be ascertained by careful monitoring of the reaction mixture, as the heating and cooling properties of the reactor and the volumes used will have significant effects on the extent of reaction.

After acidification, the lipid fraction is extracted into hexane and the solvent subsequently removed by evaporation. This gives a material consisting of 60–75% (w/w) CLA (depending on the starting materials), with the remainder consisting of unchanged saturated and monounsaturated fatty acids from the starting material. Deodorization of CLA removes volatile compounds such as hexane that may be residual solvent or secondary autoxidation products from CLA. Fractional distillation and crystallization remove metal catalysts (which may stimulate oxidation), as well as undesirable components such as dimers and polymers. The reactor type will affect the amounts of metals found.

The alkali-isomerization process leads to CLA exclusively in the free fatty acid form and the majority of commercial products currently available are therefore in this form. But there is an increased interest in using CLA incorporated into triacylglycerols since this is the form of a typical lipid in foodstuffs.

2. *Triacylglycerol formation*

CLA can be incorporated into a triacylglycerol by chemical means but enzymatic incorporation provides a cleaner and more efficient process with minimal use of solvents (see also Chapter 6). A number of enzymes, mainly from fungal sources, and in both soluble and immobilized forms, have been used to incorporate CLA into a variety of triacylglycerol products. Some of these are detailed in Table 2.

A number of points are worth mentioning. All of the reactions described took place at 50–60°C, so isomerization of the CLA was unlikely. Two of the immobilized lipases (those from *Mucor miehei* and *Candida antarctica* sp.)

Table 2. Enzymes used to incorporate CLA into triacylglycerols

	Starting material				
Source of enzyme	**Palm oil**[a]	**Butter oil**[b]	**Menhaden oil**[c]	**Tricaprylin**[d]	**Glycerol**[e]
Mucor miehei	Yes	Yes	Yes	Yes	Yes
Candida rugosa (*cylindracea*)[f]	–	Yes	Yes	–	–
Pseudomonas sp.	–	Yes	Yes	Yes	–
Candida antarctica sp.	–	Yes	–	Yes	–
Candida antarctica (lipase A)	–	Yes	Yes	–	–
Candida antarctica (lipase B)	–	Yes	Yes	–	–
Rhizopus oryzae	–	Yes	–	–	–

[a]Data taken from McNeill *et al.* (1999). [b]Data taken from Garcia *et al.* (1998, 2001). [c]Data taken from Garcia *et al.* (2000). [d]Data taken from Kim *et al.* (2001). [e]Data taken from Arcos *et al.* (1998). [f]*Candida cylindracea* has recently been renamed *Candida rugosa.*

performed well under most conditions, but while the *Mucor miehei* lipase was 1,3-specific, the *Candida antarctica* sp. lipase showed little positional specificity, and so gave the best overall incorporation (particularly into the *sn*-2 position) onto the triacylglycerol in the shortest time. *Rhizopus oryzae* gave the poorest overall performance with little reaction after 24 h. *Candida rugosa* gave incorporation into menhaden oil without loss of the *n*-3 fatty acids (so preserving any potential health benefits). Incorporation of CLA into butter oil was at the expense of short- and medium-chain fatty acids and this may have a detrimental effect on the sensory qualities of a dairy product containing this material. Garcia *et al.* (2001) reported favourable results in trials using a packed-bed reactor. For all of these reactions, increased temperature and water content proved detrimental. The molar ratio of CLA to triacylglycerol substrate required for maximum incorporation of CLA into tricaprylin was 3:1 for the 1,3-specific enzyme and 5:1 for the non-specific enzyme used. At lower proportions of CLA or at higher water content the amounts of monoacylglycerols and diacylglycerols increased in the product. However Arcos *et al.* (1998) produced a material consisting of mainly monoacylglycerols and diacylglycerols using a CLA:glycerol ratio of 1.5:1. This meets current US Food and Drug Administration (FDA) specifications for a food-grade emulsifier and could be a way to incorporate "hidden" CLA into the diet.

3. *Isomer enrichment*

While it is not possible, currently, to change the proportions of the 9-*cis*, 11-*trans* and 10-*trans*, 12-*cis* isomers by altering the conditions of alkaline hydrolysis, various methods have been described to obtain pure or enriched preparations of each individual isomer, mainly for use in clinical studies. Many of these methods are based on the use of enzymes. The most commonly used material comes from the fungus *Geotrichum candidum*. Three different preparations have been considered: *Geotrichum candidum* lipase A is non-selective, *Geotrichum candidum* lipase B (from a cloned gene) is highly selective for acids with a double bond in the 9-*cis* configuration, and a mixture of the two (the commercial product 'Amano GC-4') retains most of the selectivity of the B form. Haas *et al.* (1999) used these preparations to selectively hydrolyse the 9-*cis*, 11-*trans* isomer from a mixture of CLA methyl esters, enriching this isomer in the free acid fraction to a level of 77–94%. *Geotrichum candidum* lipase was also used by McNeill *et al.* (1999) to selectively esterify the 9-*cis*, 11-*trans* isomer with lauryl alcohol. Other enzymes that have been considered include those extracted from *Mucor miehei*, *Aspergillus niger* (Chen *et al.*, 1999), and a number of *Candida* species. *Mucor miehei* and *Candida antarctica* sp. lipases showed relatively little discrimination between isomers (Warwel and Borgdorf, 2000). However, although *Candida cylindracea* (recently renamed *Candida rugosa*) lipase, like *Geotrichum candidum* lipase B, was specific to the 9-*cis*, 11-*trans* isomer, *Candida antarctica* lipase A ('Chirazyme L-5') was found to prefer 9-*trans*-containing isomers.

Physical methods have also given good results in separating specific isomers. Berdeaux *et al.* (1998) separated the 9-*cis*, 11-*trans* and 10-*trans*, 12-*cis* isomers as fatty acid methyl esters from a CLA mixture by a series of low-temperature crystallizations from acetone with a fractional purity of >90% for each. The 10-*trans*, 12-*cis* isomer separated from the mother liquor at –58°C, followed by the 9-*cis*, 11-*trans* isomer at –62°C. The remaining mother liquor contained a mixture of 11,13 isomers. Alternatively, urea crystallization was used by Ma *et al.* (1999) to enrich the content of 10-*trans*, 12-*cis* isomer in CLA mixtures to 60% of the total CLA. O'Shea *et al.* (2000) achieved enrichment of the CLA level in anhydrous milk fat by a process of dry fractionation. A soft fat fraction was produced with a 63.2% higher CLA level than the base milk. This fraction was also enriched in 11-*trans* 18:1 and PUFA.

4. *Novel synthetic methods*

Other potential areas of interest for CLA production include the isomerization of vegetable oils using homogeneous transition metal catalysts. Tests have been conducted with a variety of rhodium and ruthenium catalysts and soybean oil (which is cheaper than sunflower or safflower oil) (Larock *et al.*, 2001). High yields of CLA were obtained in ethanol at 60°C, with >90% conjugation achieved. Concurrent experiments with ethyl linoleate gave >90% of the *cis*, *trans* and *trans*, *cis* isomers, although the full isomer profile obtained was not described.

The role of *Butyrivibrio fibrisolvens* in CLA production in ruminants was recognized by Kepler *et al.* in 1966. In recent times, other bacterial cultures have been considered for CLA production. *In vitro* cultures of various bacteria can produce CLA from linoleic acid, but increasing the linoleic acid level proves toxic to the bacteria (Jiang *et al.*, 1998), leading to the hypothesis that CLA production might be a detoxification process. Rainio *et al.* (2001) avoided toxicity by dispersing the linoleate in polyoxyethylene sorbitan monooleate detergent. Cultures of *Propionibacterium freudenreichii* subsp. *Shermanii* JS were then able to tolerate at least 1000 μg/ml of linoleic acid, leading to conversion to CLA with 57–87% efficiency. The CLA produced consisted of 85–90% of the 9-*cis,* 11-*trans* and 9-*trans,* 11-*cis* isomers. Washed cells of *Lactobacillus acidophilus*, under microaerobic conditions, were shown by Ogawa *et al.* (2001) to produce both these isomers and the 9-*trans,* 11-*trans* isomer when supplied with linoleate. A maximum level of >80% (w/w) of the total fatty acids were found to be CLA and the majority was found as free acids in or associated with the cells. This should allow for simple separation by centrifugation. Alternatively the cells themselves, which had an intracellular concentration of 5 mg/ml CLA, could be used as the CLA source.

At a recent meeting, Deng *et al.* (2001) described their work to find linoleic acid isomerases from Gram-positive bacteria that were capable of producing CLA. Although most of the strains they have isolated so far can produce the 9-*cis,* 11-*trans* isomer, only one, a soluble isomerase from *Propionibacterium acnes*, was capable of producing the 10-*trans,* 12-*cis* isomer. Lin (2000) investigated the tolerance of a number of lactic acid bacteria to sweeteners or salt in a skimmed milk medium, and their subsequent ability to produce CLA. Sucrose and sodium chloride depressed CLA production. Fructose and lactose showed similar effects, but to a lesser extent. Regardless of the conditions, however, *Lactobacillus acidophilus* gave the highest levels of CLA production and was recommended as a suitable culture for inclusion in fermented milk products such as yoghurts.

5. *Oxidative stability*

Published reports suggest a degree of confusion about the oxidative stability of CLA. Although Zhang and Chen (1997) found that the oxidation rate of free CLA in air was similar to that of DHA, Holman and Elmer (1947) reported that the oxidative stability of CLA was similar to that of linoleic acid. These differences are probably due to the measurement method used. Regardless of this, as an unsaturated fatty acid, CLA incorporated into foodstuffs should be protected from oxidation. Yang *et al.* (2000) found that, although different isomers varied in their degree of oxidative stability, CLA could be protected by the addition of 200 ppm jasmine green tea catechins, which were found to be more effective than the equivalent amount of butylated hydroxytoluene (BHT).

In an alternative approach Kim *et al.* (2000) trapped the CLA in a variety of cyclodextrin microcapsules, which dramatically reduced the rate of oxygen depletion. A patent also exists for encapsulating CLA in denatured starch for incorporation into foods (Li *et al.*, 2001). On the basis of different studies CLA has been proposed as both a pro-oxidant and an antioxidant. These claims remain controversial and it would be prudent to ensure the oxidative stability of CLA when used in functional foods.

F. Commercial availability and new product development

At the time of writing, there are four major suppliers of bulk CLA containing the 9-*cis,* 11-*trans* and 10-*trans,* 12-*cis* acids as the only major components (personal communication). These are Natural ASA of Norway, Loders Croklaan (part of the Unilever group) in the Netherlands, Bioriginal in Canada, and Rinoru Oil (part of the Mitsubishi group) in Japan. Other chemical companies also supply large quantities of CLA, but usually as a four-isomer mixture. Two of the proprietary brands names are 'Tonalin' (Natural ASA) and 'Clarinol' (Loders Croklaan). Both materials contain 80% (w/w) CLA. These products are marketed as aids to modulating body composition and the bulk material is supplied as free acid. 'Clarinol' is now also being sold as a vegetable powder for incorporation into food products. Recent collaboration between Natural ASA and BASF in Germany has led to the development of a CLA-enriched animal feed containing 'Tonalin', which should be on the market in the near future.

As stated above, the majority of CLA materials available for incorporation into foods are in the free acid form. However, a recent clinical trial, using CLA as free acid incorporated into a yoghurt-type base, found that the taste was described as very poor, compared with the equivalent triacylglycerol or ethyl ester (personal communication).

The patent literature indicates that although few products are yet on the market, a number of companies have an interest in producing functional foods containing CLA. A patent has been granted (Liu, 2001) for the esterification (or transesterification) of CLA to give sterol or stanol esters. Similar materials, containing non-conjugated fatty acids, are already available in functional food products in Europe and the USA. Unilever has been granted a patent for the incorporation of CLA mixtures rich in *trans, trans* isomers into foods or food supplements (Barclay *et al.*, 2001). Although one use for these mixtures is as a thickening agent in ice cream, the patent cites the material as having excellent anti-inflammatory properties. The potentially beneficial effects of many of the CLA isomers have not yet been thoroughly investigated, and this may become more relevant in future. ConlinCo Inc (USA) holds two patents for producing foods or other dietary materials with an increased ratio of 10-*trans,* 12-*cis* to other CLA isomers (Jerome and Skarie, 2001a, 2001b). A number of other patents have been issued for a variety of purposes, mainly regarding the use of CLA in animal husbandry and health. The full development of CLA in functional foods is still delayed, however, due to the lack of safety data.

G. Safety issues

There is little information on the toxicological effects of CLA, particularly in humans. Rodent models are contradictory, with few effects seen in a number of rat and hamster studies. However, studies in mice have demonstrated increases in liver weight, transaminase levels, and steatosis. (Tsuboyama-Kasaoka *et al.*, 2000). Conversely, CLA appears to have positive effects on immune function in chicks. A number of the more recent trials in humans have obtained some preliminary data. Kelley *et al.* (2000, 2001) fed 2.5 g of CLA isomers to 10 healthy female subjects and examined both their immune responses and their eicosanoid and cytokine production. No variations were seen with respect to control subjects. The CLA used was a mixture of four main isomers. Mixtures of two isomers, fed at levels of 4.2 g/day or up to 6.8 g/day respectively to healthy (Smedman and Vessby, 2001), overweight or obese subjects (Blankson *et al.*, 2000), showed no effects on blood lipid parameters such as LDL cholesterol level, transaminase level, and white cell count. In the latter study, the only adverse effects reported were fatigue and mild gastrointestinal disturbances, and these may have been related to the high level of encapsulated oil consumed rather than to the CLA itself. No information has yet been published about the effects of individual CLA isomers. This is an area requiring urgent clarification if CLA incorporation into functional foods is to become a reality. However, the testing of parameters such as liver steatosis in humans is not trivial, and the potentially invasive procedures required to demonstrate the safety of CLA conclusively could limit the recruitment of suitable human subjects.

H. Conclusions

There is no doubt that CLA has great potential for inclusion in functional foods; indeed, in 1999 the American Dietetic Association declared beef itself to be a functional food due to its CLA content. A recent estimate of the market potential worldwide suggested that it could be worth between US$300 million and US$2000 million per year. In animals, the beneficial effects are significant, both for food production and animal health, and should lead to animal products containing lower levels of fat for human consumption. However, the increased consumption of CLA by humans in a functional food is still some way off. There is a requirement for consistent and accurate analytical data to substantiate claims of effects in humans, and the individual effects of different isomers still have to be elucidated. Until all of these can be substantiated fully, regulatory bodies such as the FDA will not allow any health claims to be made for a CLA-containing food. Although consumers currently accept that lower-fat foods are healthier there may be another hurdle to overcome with respect to consumer acceptance. CLA are *trans* fatty acids, and should be included as such in nutritional labelling. The current public perception is that all *trans* fats are detrimental to health and should be kept at the lowest levels possible in foods. Indeed, countries such as Denmark regulate the maximum amount of *trans* fat allowable in food items. This negative perception will have to be overcome before functional foods containing CLA attain wide acceptance by the public.

I. Web sites

Further information on CLA-related research and commercial sources of CLA can be found on the web sites listed below.

1. Analysis and research

http://www.wisc.edu/fri/clarefs.htm
Comprehensive and regularly updated list of CLA publications in the scientific literature (maintained by the Food Research Institute at the University of Wisconsin–Madison).

http://www.cerin.org/periodiques/CLA/menu_CLA.asp
Details regarding CLA research conducted through the European Union scientific programmes (maintained by an independent Nutrition Research and Information Centre).

http://www.lipid.co.uk/infores/others/claintro.html
A summary of techniques for the detailed analysis of CLA isomers, with details of method development trials and links to other relevant sites (maintained by W.W. Christie through the Mylnefield Research Services site).

http://ep.espacenet.com
Access to European, World and Japanese patent information (through the European Commission).

http://www.uspto.gov/
US Patent and Trademark Office home page.

2. Commercial companies

http://www.natural.no/
Home page for Natural ASA (Norway).

http://www.lipidnutrition.com/
Home page for the nutritional lipid products of Loders Croklaan (the Netherlands).

http://www.bioriginal.com/
Home page for Bioriginal (Canada).

http://www.rinoru.co.jp/
Home page for Rinoru Oils (Japan – web site in Japanese).

References

Ackman, RG (1999) Conjugated linoleic acid (CLA) in lipids of fish tissues. In: *Advances in Conjugated Linoleic Acid Research, Volume 1* (MP Yurawecz, MM Mossoba, JKG Kramer, MW Pariza and GJ Nelson, eds), AOCS Press, Champaign, pp.283–295.

Adlof, RO (1999) Preparation of unlabeled and isotope-labeled conjugated linoleic and related fatty acid isomers. In: *Advances in Conjugated Linoleic Acid Research, Volume 1* (MP Yurawecz, MM Mossoba, JKG Kramer, MW Pariza and GJ Nelson, eds), AOCS Press, Champaign, pp.21–38.

Adlof, RO, Copes, LC and Walter, EL (2001) Changes in conjugated linoleic acid composition within samples obtained from a single source. *Lipids*, 36, 315–317.

Ahn, DU, Sell, JL, Jo, C, Chamruspollert, M and Jeffrey, M (1999) Effect of dietary conjugated linoleic acid on the quality characteristics of chicken eggs during refrigerated storage. *Poultry Sci.*, 78, 922–928.

Arcos, JA, Ottero, C and Hill, CG, Jr (1998) Rapid enzymatic production of acylglycerols from conjugated linoleic acid and glycerol in a solvent-free system. *Biotechnol. Lett.*, 20, 617–621.

Aydin, R, Pariza, MW and Cook, ME (1999) Role of dietary oils in prevention of CLA-induced chick embryonic mortality and egg properties. *FASEB J.*, 13, A541.

Aydin, R, Pariza, MW and Cook, ME (2001) Olive oil prevents the adverse effects of dietary conjugated linoleic acid on chick hatchability and egg quality. *J. Nutr.*, 131, 800–806.

Baer, RJ, Rylai, J and Schingoethe, DJ (2001) Composition and properties of milk and butter from cows fed fish oil. *J. Dairy Sci.*, 84, 345–353.

Banni, S and Martin, J-C (1998) Conjugated linoleic acid and metabolites. In: *Trans Fatty Acids in Human Nutrition* (JL Sébédio and WW Christie, eds), The Oily Press, Dundee, pp. 261–303.

Banni, S, Carta, G, Contini, MS, Angiono, E, Deiana, M, Dessi, MA, Melis, MP and Corongiu, FP (1996) Characterization of conjugated diene fatty acids in milk, dairy products, and lamb tissues. *J. Nutr. Biochem.*, 7, 150–155.

Barclay, SS, Parmar, P, Rogers, JS, Taran, V and Cain, FW (Unilever PLC and Unilever NV) (2001) Use of *trans-trans* isomers of conjugated linoleic acid. European Patent EP1097708.

Bauman, DE and Griinari, JM (2001) Regulation and nutritional manipulation of milk fat: low-fat milk syndrome. *Livestock Prod. Sci.*, 70, 15–29.

Baumgard, LH, Sangster, JK and Bauman, DE (2001) Milk fat synthesis in dairy cows is progressively reduced by increasing supplemental amounts of *trans*-10, *cis*-12 conjugated linoleic acid (CLA). *J. Nutr.*, 131, 1764–1769.

Belury, MA, Nickel, AP, Bird, CE and Wu, Y (1996) Dietary conjugated linoleic acid modulation of phorbol ester skin tumor promotion. *Nutr. Cancer*, 26, 149–157.

Berdeaux, O, Voinot, L, Angioni, E, Juanéda, P and Sébédio, JL (1998) A simple method of preparation of methyl *trans*-10, *cis*-12- and *cis*-9, *trans*-11- octadecadienoates from methyl linoleate. *J. Am. Oil Chem. Soc.*, 75, 1749–1755.

Blankson, H, Stakkestad, JA, Fagertun, H, Thom, E, Wadstein, J and Gudmundsen, O (2000) Conjugated linoleic acid reduces body fat mass in overweight and obese humans. *J. Nutr.*, 130, 2943–2948.

Chamruspollert, M and Sell, JL (1999) Transfer of dietary conjugated linoleic acid to egg yolks of chickens. *Poultry Sci.*, 78, 1138–1150.

Chen, C-A, Lu, W and Sih, CJ (1999) Synthesis of 9*Z*, 11*E*-octadecadienoic and 10*E*, 12*Z*-octadecadienoic acids, the major components of conjugated linoleic acid. *Lipids*, 34, 879–884.

Chilliard, Y, Ferlay, A, Mansbridge, RM and Doreau, M (2000) Ruminant milk fat plasticity: nutritional control of saturated, polyunsaturated, *trans* and conjugated fatty acids. *Ann. Zootech.*, 49, 181–205.

Chilliard, Y, Ferlay, A and Doreau, M (2001) Effect of different types of forages, animal fat or marine oils in cow's diet on milk fat secretion and composition, especially conjugated linoleic acid (CLA) and polyunsaturated fatty acids. *Livestock Prod. Sci.*, 70, 31–48.

Choi, B-D, Kang, S-J, Ha, Y-L and Ackerman, RG (1999) Accumulation of conjugated linoleic acid (CLA) in tissues of fish fed diets containing various levels of CLA. In: *Quality Attributes of Muscle Foods*, (YL Xiong, C-T Ho and F Shahidi, eds), Kluwer Academic/ Plenum Publishers, New York, pp.61–71.

Christie, WW, Sébédio, J-L and Juaneda, P (2001) A practical guide to the analysis of conjugated linoleic acid. *INFORM*, 12, 147–152.

Christie, WW, Dobson, G and Gunstone, FD (1997) Isomers in commercial samples of conjugated linoleic acid. *Lipids*, 32, 1231.

Cook, ME, Lee, KN, Pariza, MW and Wentworth, BC (Wisconsin Alumni Research Foundation) (1996) Method for controlling bird populations. US Patent 5,504,114.

Cook, ME, DeVoney, D, Drake, B, Pariza, MW, Whigham, L and Yang, M (1999) Dietary control of immune-induced cachexia: conjugated linoleic acid and immunity. In: *Advances in Conjugated Linoleic Acid Research, Volume 1* (MP Yurawecz, MM Mossoba, JKG Kramer, MW Pariza and GJ Nelson, eds), AOCS Press, Champaign, pp.226–237.

Deng, MD, Grund, AD, Peng, S, Langley, KM, Wassink, S, Logan, BB and Rosson, RA (2001) Purification, characterization, cloning and expression of linoleic acid isomerases. In: *1st International Conference on Conjugated Linoleic Acid (CLA), Ålesund, Norway*, Natural ASA, Norway [abstract].

Dewhurst, RJ, Scollan, ND, Youell, SJ, Tweed, SK and Humphreys, MO (2001) Influence of species cutting date and cutting interval on the fatty acid composition of grasses. *Grass Forage Sci.*, 56, 68–74.

Dhiman, TR, Satter, LD, Pariza, MW, Galli, MP, Albright, K and Tolosa, MX (2000) Conjugated linoleic acid (CLA) content of milk from cows offered diets rich in linoleic and linolenic acid. *J. Dairy Sci.*, 83, 1016–1027.

Donovan, DC, Schingoethe, DJ, Baer, RJ, Rylai, J, Hippen, AR and Franklin, ST (2000) Influence of dietary fish oil on conjugated linoleic acid and other fatty acids in milk fat from lactating dairy cows. *J. Dairy Sci.*, 83, 2620–2628.

Du, M, Ahn, DU and Sell, JL (1999) Effect of dietary conjugated linoleic acid on the composition of egg yolk lipids. *Poultry Sci.*, 78, 1639–1645.

Du, M, Ahn, DU and Sell, JL (2000) Effects of dietary conjugated linoleic acid and linoleic : linolenic acid ratio on polyunsaturated fatty acid status in laying hens. *Poultry Sci.*, 79, 1749–1756.

Du, M, Ahn, DU, Nam, KC and Sell, JL (2001) Volatile profiles and lipid oxidation of irradiated cooked chicken meat from laying hens fed diets containing conjugated linoleic acid. *Poultry Sci.*, 80, 235–241.

Dugan, MER, Aalhus, JL, Schaefer, JAL and Kramer, JKG (1997) The effects of conjugated linoleic acid on fat to lean repartitioning and feed conversion in pigs. *Can. J. Anim. Sci.*, 77, 723–725.

Fearon, AM (2001) Optimising milk fat composition and processing properties. *Austr. J. Dairy Technol.*, 76, 104–108.

Garcia, HS, Storkson, JM, Pariza, MW and Hill, CG, Jr (1998) Enrichment of butter oil with conjugated linoleic acid via enzymatic interesterification (acidolysis) reactions. *Biotechnol. Lett.*, 20, 393–395.

Garcia, HS, Arcos, JA, Ward, DJ and Hill, CG, Jr (2000) Synthesis of glycerides containing *n*-3 fatty acids and conjugated linoleic acid by solvent-free acidolysis of fish oil. *Biotechnol. Bioeng.*, 70, 587–591.

Garcia, HS, Arcos, JA, Keough, KJ and Hill, CG, Jr (2001) Immobilized lipase-mediated acidolysis of butteroil with conjugated linoleic acid: batch reactor and packed bed reactor studies. *J. Mol. Catal. B: Enzym.*, 11, 623–632.

Haas, MJ, Kramer, JKG, McNeill, G, Scott, K, Foglia, TA, Sehat, N, Fritsche, J, Mossoba, MM and Yurawecz, MP (1999) Lipase-catalysed fractionation of conjugated linoleic acid isomers. *Lipids*, 34, 979–987.

Hamilton, RJ (2001) CLA is focus for SCI seminar. *INFORM*, 12, 629–632.

Holman, RT and Elmer, OC (1947) The rates of oxidation of unsaturated fatty acids and esters. *J. Am. Oil Chem. Soc.*, 24, 127–129.

Ip, C, Singh, M, Thompson, HJ and Scimeca, JA (1994) Conjugated linoleic-acid suppresses mammary carcinogenesis and proliferative activity of the mammary-gland in the rat. *Cancer Res.*, 54, 1212–1215.

Ip, C, Briggs, SP, Saegele, AD, Thompson, HJ, Storkson, J and Scimeca, JA (1996) The efficacy of conjugated linoleic acid in mammary cancer prevention is independent of the level or type of fat in the diet. *Carcinogenesis*, 17, 1045–1050.

Ivan, M, Mir, PS, Koenig, KM, Rode, LM, Neill, L, Entz, T and Mir, Z (2001) Effects of dietary sunflower seed oil on rumen protozoa population and tissue concentration of conjugated linoleic acid in sheep. *Small Ruminant Res.*, 41, 215–227.

Jahreis, G, Kraft, J, Tischendorf, F, Schöne, F and von Loeffelholz, C (2000) Conjugated linoleic acids: physiological effects in animal and man with special regard to body composition. *Eur. J. Lipid Sci.* Technol., 102, 695–703.

Jerome, D and Skarie, C (ConlinCo) (2001a) Method of using isomer enriched conjugated linoleic acid compositions. US Patent 6,214,372.

Jerome, D and Skarie, C (ConlinCo) (2001b) Isomer enriched conjugated linoleic acid compositions. US Patent 6,242,621.

Jiang, J, Björck, L and Fondén, R (1998) Production of conjugated linoleic acid by dairy starter cultures. *J. Appl. Microbiol.*, 85, 95–102.

Kelley, DS, Taylor, PC, Rudolph, IL, Benito, P, Nelson, GJ, Mackey, BE and Erickson, KL (2000) Dietary conjugated linoleic acid did not alter immune status in young healthy women. *Lipids*, 35, 1065–1071.

Kelley, DS, Simon, VA, Taylor, PC, Rudolph, IL, Benito, P, Nelson, GJ, Mackey, BE and Erickson, KL (2001) Dietary supplementation with conjugated linoleic acid increased its concentration in human peripheral blood mononuclear cells, but did not alter their function. *Lipids*, 36, 669–674.

Kelly, MK, Kolver, ES, Baumen, DE, Van Amburgh, ME and Muller, LD (1998) Effect of intake of pasture on concentrations of conjugated linoleic acid in milk of lactating cows. *J. Dairy Sci.*, 81, 1630–1636.

Kepler, CR, Hirons, KP, McNeill, JJ and Tove, SB (1966) Intermediates and products of the biohydrogenation of linoleic acid by *Butyrivibrio fibrisolvens*. *J. Biol. Chem.*, 241, 1350–1354.

Kim, I-H, Yoon, C-S, Cho, S-H, Lee, K-W, Chung, S-H and Tae, B-S (2001) Lipase-catalysed incorporation of conjugated linoleic acid into tricaprylin. *J. Am. Oil Chem. Soc.*, 78, 542–551.

Kim, SJ, Park, GB, Kang, CB, Park, SD, Jung, MY, Kim, JO and Ha, YL (2000) Improvement of oxidative stability of conjugated linoleic acid (CLA) by microencapsulation in cyclodextrins. *J. Agric. Food Chem.*, 48, 3922–3929.

Kritchevsky, D (1999) Conjugated linoleic acid and experimental atherosclerosis in rabbits. In: *Advances in Conjugated Linoleic Acid Research, Volume 1* (MP Yurawecz, MM Mossoba, JKG Kramer, MW Pariza and GJ Nelson, eds), AOCS Press, Champaign, pp.397–403.

Larock, RC, Dong, X, Chung, S, Reddy, CK and Ehlers, LE (2001) Preparation of conjugated soybean oil and other natural oils and fatty acids by homogeneous transition metal catalysis. *J. Am. Oil Chem. Soc*, 78, 447–453.

Latour, MA, Devitt, AA, Meunier, RA, Stewart, JJ and Watkins, BA (2000) Effects of conjugated linoleic acid. 1. Fatty acid modification of yolks and neonatal fatty acid metabolism. *Poultry Sci.*, 79, 817–821.

Lavillonnière, F and Bougnoux, P (1999) Conjugated linoleic acid (CLA) and the risk of breast cancer. In: *Advances in Conjugated Linoleic Acid Research, Volume 1* (MP Yurawecz, MM Mossoba, JKG Kramer, MW Pariza and GJ Nelson, eds), AOCS Press, Champaign, pp.276–282.

Li, G, Liu, F and Shi, H (First Institute of Oceanology) (2001) Conjugated linoleic acid microcapsule. Chinese Patent CN1287114.

Lin, H, Boylston, TD, Luedecke, LO and Shultz, TD (1998) Factors affecting the conjugated linoleic acid content of cheddar cheese. *J. Agric. Food Chem.*, 46, 801–807.

Lin, TY (2000) Conjugated linoleic acid concentration as affected by lactic cultures and additives. *Food Chem.*, 69, 27–31.

Liu, L (Regents for Oklahoma State University) (2001) Sterol esters of conjugated linoleic acids and process for their production. World Patent WO 01/29060.

Ma, DWL, Wierzbicki, AA, Field, CJ and Clandinin, MT (1999) Preparation of conjugated linoleic acid from safflower oil. *J. Am. Oil Chem. Soc*, 76, 729–730.

McNeill, GP, Rawlins, C and Peilow, AC (1999) Enzymatic enrichment of conjugated linoleic acid isomers and incorporation into triglycerides. *J. Am. Oil Chem. Soc*, 76, 1265–1268.

Mir, Z, Goonewardene, LA, Okine, E, Jaegar, S and Scheer, HD (1999) Effect of feeding canola oil on constituents, conjugated linoleic acid (CLA) and long chain fatty acids in goats milk. *Small Ruminant Res.*, 33, 137–143.

Mir, Z, Rushfeldt, ML, Mir, PS, Paterson, LJ and Weselake, RJ (2000) Effect of dietary supplementation with either conjugated linoleic acid (CLA) or linoleic acid rich oil on the CLA content of lamb tissues. *Small Ruminant Res.*, 36, 25–31.

Ogawa, J, Matsumura, K, Kishino, S, Omura, Y and Shimizu, S (2001) Conjugated linoleic acid accumulation via 10-hydroxy-12-octadecaenoic acid during microaerobic transformation of linoleic acid by *Lactobacillus acidophilus*. *Appl. Environ. Microbiol.*, 67, 1246–1252.

O'Shea, M, Devery, R, Lawless, F, Keogh, K and Stanton, C (2000) Enrichment of the conjugated linoleic acid content of bovine milk by dry fractionation. *Int. Dairy J.*, 10, 289–294.

Pariza, MW, Ashoor, SH, Chu, FS and Lund, DB (1979) Effects of temperature and time on mutagen formation in pan-fried hamburger. *Cancer Lett.*, 7, 63–69.

Pariza, MW, Park, Y and Cook, ME (2001) The biologically active isomers of conjugated linoleic acid. *Prog. Lipid Res.*, 40, 283–298.

Prandini, A, Geromin, D, Conti, F, Masoero, F, Piva, A and Piva, G (2001) Survey on the level of conjugated linoleic acid in dairy products. *Ital. J. Food Sci.*, 13, 243–253.

Precht, D and Molkentin, J (2000) Recent trends in the fatty acid composition of German sunflower margarines, shortenings and cooking fats with emphasis on individual C16:1, C18:1, C18:2, C18:3 and C20:1 *trans* isomers. *Nahrung*, 44, 222–228.

Rainio, A, Vahvaselkä, M, Suomalainen, T and Laakso, S (2001) Reduction of linoleic acid inhibition in production of conjugated linoleic acid by *Propionibacterium freudenreichii* ssp. *Shermanii*. *Can. J. Microbiol.*, 47, 735–740.

Ramaswamy, N, Baer, RJ, Schingoethe, DJ, Hippen, AR, Kasperson, KM and Whitlock, LA (2001) Short communication: consumer evaluation of milk high in conjugated linoleic acid. *J. Dairy Sci.*, 84, 1604–1609.

Ramsay, TG, Evock-Clover, CM, Steele, NC and Azain, MJ (2001) Dietary conjugated linoleic acid alters fatty acid composition of pig skeletal muscle and fat. *J. Anim. Sci.*, 79, 2152–2161.

Reaney, MJT, Liu, Y-D and Westcott, ND (1999) Commercial production of conjugated linoleic acid. In: *Advances in Conjugated Linoleic Acid Research, Volume 1* (MP Yurawecz, MM Mossoba, JKG Kramer, MW Pariza and GJ Nelson, eds), AOCS Press, Champaign, pp.39–54.

Sæbø, A (2001) Commercial production of conjugated linoleic acid. *Lipid Technol. Newsl.*, 7, 9–13.

Salminen, I, Mutanen, M, Jauhiainen, M and Aro, A (1998) Dietary *trans* fatty acids increase conjugated linoleic acid levels in human serum. *J. Nutr. Biochem.*, 9, 93–98.

Simon, O, Männer, K, Schäfer, K, Sagredos, A and Eder, K (2000) Effects of conjugated linoleic acids on protein to fat proportions, fatty acids, and plasma lipids in broilers. *Eur. J. Lipid Sci. Technol.*, 102, 402–410.

Smedman, A and Vessby, B (2001) Conjugated linoleic acid supplementation in humans – metabolic effects. *Lipids*, 36, 773–781.

Sserunjogi, ML, Abrahamsen, RK and Narvhus, J (1998) A review paper: current knowledge of ghee and related products. *Int. Dairy J.*, 8, 677–688.

Szymczyk, B, Pisulewski, PM, Szczurek, W and Hanczakowski, P (2001) Effects of conjugated linoleic acid on growth performance, feed conversion efficiency, and subsequent carcass quality in broiler chickens. *Br. J. Nutr.*, 85, 465–473.

Thiel-Cooper, RL, Parrish, FC, Jr, Sparks, JC, Wiegand, BR and Ewan, RC (2001) Conjugated linoleic acid changes swine performance and carcass composition. *J. Anim. Sci.*, 79, 1821–1828.

Tsuboyama-Kasaoka, N, Takahashi, M, Tanemura, K, Kim, HJ, Tange, T, Okuyama, H, Kasai, M, Ikemoto, S and Ezaki, O (2000) Conjugated linoleic acid supplementation reduces adipose tissue by apoptosis and develops lipodystrophy in mice. *Diabetes*, 49, 1534–1542.

Twibell, RG, Watkins, BA, Rogers, L and Brown, PB (2000) Effects of dietary conjugated linoleic acids on hepatic and muscle lipids in hybrid striped bass. *Lipids*, 35, 115–161.

Twibell, RG, Watkins, BA and Brown, PB (2001) Dietary conjugated linoleic acids and lipid source alter fatty acid composition of juvenile yellow perch, *Perca flavescens*. *J. Nutr.*, 131, 2322–2328.

Voigt, J and Hagemeister, H (2001) Dietary influence on a desirable fatty acid composition in milk from dairy cattle. *J. Anim. Feed Sci.*, 10, 87–103.

Warwel, S and Borgdorf, R (2000) Substrate selectivity of lipases in the esterification of *cis/trans*-isomers and positional isomers of conjugated linoleic acid (CLA) *Biotechnol. Lett.*, 22, 1151–1155.

Watkins, BA, Li, Y and Seifert, MF (1999) Bone metabolism and dietary conjugated linoleic acid. In: *Advances in Conjugated Linoleic Acid Research, Volume 1* (MP Yurawecz, MM Mossoba, JKG Kramer, MW Pariza and GJ Nelson, eds), AOCS Press, Champaign, pp.253–275.

Wiegand, BR, Parrish, FC, Jr, Swan, JE, Larsen, ST and Baas, TJ (2001) Conjugated linoleic acid improves feed efficiency, decreases subcutaneous fat, and improves certain aspects of meat quality in stress-genotype pigs. *J. Anim. Sci.*, 79, 2187–2195.

Willett, WC, Stampfer, MJ, Manson, JE, Colditz, GA, Speizer, FE, Rosner, BA, Sampson, LA and Hennekens, CH (1993) Intake of *trans* fatty acids and risk of coronary heart disease among women. *Lancet*, 341, 581–585.

Yang, L, Leung, LK, Huang, Y and Chen, Z-Y (2000) Oxidative stability of conjugated linoleic acid isomers. *J. Agr. Food Chem.*, 48, 3072–3076.

Yurawecz, MP, Mossoba, MM, Kramer, JKG, Pariza, MW and Nelson, GJ (eds) (1999a) *Advances in Conjugated Linoleic Acid Research, Volume 1*, AOCS Press, Champaign.

Yurawecz, MP, Sehat, N, Mossoba, MM, Roach, JAG, Kramer, JKG and Ku, Y (1999b) Variations in isomer distribution in commercially available conjugated linoleic acid. *Fett/Lipid*, 101(8), 277–282.
Zhang, A and Chen, ZY (1997) Oxidative stability of conjugated linoleic acids relative to other polyunsaturated fatty acids. *J. Am. Oil Chem. Soc.*, 74, 1611–1613.

Index